Springer Optimization and Its Applications

VOLUME 47

Aims and Scope
Optimization has been expanding in all directions at an astonishing rate during the last few decades. New algorithmic and theoretical techniques have been developed, the diffusion into other disciplines has proceeded at a rapid pace, and our knowledge of all aspects of the field has grown even more profound. At the same time, one of the most striking trends in optimization is the constantly increasing emphasis on the interdisciplinary nature of the field. Optimization has been a basic tool in all areas of applied mathematics, engineering, medicine, economics, and other sciences.

The series *Springer Optimization and Its Applications* publishes undergraduate and graduate textbooks, monographs and state-of-the-art expository works that focus on algorithms for solving optimization problems and also study applications involving such problems. Some of the topics covered include nonlinear optimization (convex and nonconvex), network flow problems, stochastic optimization, optimal control, discrete optimization, multi-objective programming, description of software packages, approximation techniques and heuristic approaches.

For further volumes:
http://www.springer.com/series/7393

Boris S. Mordukhovich

VARIATIONAL ANALYSIS AND GENERALIZED DIFFERENTIATION IN OPTIMIZATION AND CONTROL

In Honor of Boris S. Mordukhovich

Edited By

REGINA S. BURACHIK
School of Mathematics and Statistics
University of South Australia, Australia

JEN-CHIH YAO
Department of Applied Mathematics, National Sun Yat-Sen University, Kaohsiung, Taiwan

Springer

Editors
Regina S. Burachik
School of Mathematics and Statistics
University of South Australia
Mawson Lakes SA 5095
Australia
regina.burachik@unisa.edu.au

Jen-Chih Yao
Department of Applied Mathematics
National Sun Yat-Sen University
Kaohsiung 80424
Taiwan, R.O.C.
yaojc@math.nsysu.edu.tw

ISSN 1931-6828
ISBN 978-1-4419-0436-2 e-ISBN 978-1-4419-0437-9
DOI 10.1007/978-1-4419-0437-9
Springer New York Dordrecht Heidelberg London

Mathematics Subject Classification (2010): 49Jxx, 49J53

Cover illustration: 'Mordukhovich subdifferentials of f if $r = 2\dots$'

Printed on acid-free paper

Springer is part of Springer Science+Business Media (www.springer.com)

Preface

This special volume is dedicated to Boris M. Mordukhovich, on the occasion of his 60th birthday, and aims to celebrate his fundamental contributions to variational analysis, generalized differentiation and their applications. A main example of these contributions is Boris' recent opus magnus "Variational Analysis and Generalized Differentiation" (vols. I and II) [2,3]. A detailed explanation and careful description of Boris' research and achievements can be found in [1].

Boris' active work and jovial attitude have constantly inspired researchers of several generations, with whom he has generously shared his knowledge and enthusiasm, along with his well-known warmth and human touch.

Variational analysis is a rapidly growing field within pure and applied mathematics, with numerous applications to optimization, control theory, economics, engineering, and other disciplines. Each of the 12 chapters of this volume is a carefully reviewed paper in the field of variational analysis and related topics.

Many chapters of this volume were presented at the **International Symposium on Variational Analysis and Optimization** (ISVAO), held in the Department of Applied Mathematics, National Sun Yat-sen University, Kaohsiung, Taiwan, from November 28 to November 30, 2008. The symposium was organized in honour of Boris' 60th birthday. It brought together Boris and other researchers to discuss state-of-the-art results in variational analysis and its applications, with emphasis on optimization and control. We thank the organizers and participants of the symposium, who made the symposium a highly beneficial and enjoyable event.

We are also grateful to all the authors of this special volume, who have taken the opportunity to celebrate Boris' birthday and his decades of contributions to the area.

References

1. R. Henrion, A. Kruger and J. Outrata, Boris Mordukhovich, The Never Tiring Traveller, Celebrates His Sixtieth Birthday, *Set-Valued Analysis*, Vol. 16, Numbers 2-3, pp. 125–127, 2008.

2. B. S. Mordukhovich, *Variational analysis and generalized differentiation, I: Basic theory*, Grundlehren Series (Fundamental Principles of Mathematical Sciences), Vol. 330, Springer, Berlin, 2006.
3. B. S. Mordukhovich, *Variational analysis and generalized differentiation, II: Applications*, Grundlehren Series (Fundamental Principles of Mathematical Sciences), Vol. 331, Springer, Berlin, 2006.

Guest Editors:
Regina S. Burachik
Jen-Chih Yao

Contents

List of Contributors

Qamrul Hasan Ansari Department of Mathematics, Aligarh Muslim University, Aligarh 202 002, India, qhansari@gmail.com

Robert Baier Chair of Applied Mathematics, University of Bayreuth, 95440 Bayreuth, Germany, robert.baier@uni-bayreuth.de

Jonathan M. Borwein CARMA, University of Newcastle, Newcastle, NSW, Australia, jonathan.borwein@newcastle.edu.au

Regina S. Burachik Mathematics and Statistics, University of South Australia, Mawson Lakes SA 5095, Australia, regina.burachik@unisa.edu.au

Adela Capătă Department of Mathematics, Technical University of Cluj-Napoca, Str. Constantin Daicoviciu 15, 400020 Cluj-Napoca, Romania, adela.capata@math.utcluj.ro

S. Dempe Department of Mathematics and Computer Science, TU Bergakademie Freiberg, Freiberg, Germany, dempe@tu-freiberg.de

N. Dinh Department of Mathematics, International University, Vietnam National University of Ho Chi Minh City, Ho Chi Minh City, Vietnam, ndinh02@yahoo.fr

J. Dutta Department of Mathematics and Statistics, Indian Institute of Technology, Kanpur, India, jdutta@iitk.ac.in

Ya Ping Fang Department of Mathematics, Sichuan University, Chengdu, Sichuan, China, fabhcn@yahoo.com.cn

Elza Farkhi School of Mathematical Sciences, Sackler Faculty of Exact Sciences, Tel Aviv University, 69978 Tel Aviv, Israel, elza@post.tau.ac.il

V. Jeyakumar Department of Applied Mathematics, University of New South Wales, Sydney 2052, Australia, v.jeyakumar@unsw.edu.au

Gábor Kassay Faculty of Mathematics and Computer Science, Babeş-Bolyai University, Cluj-Napoca, Romania, kassay@math.ubbcluj.ro

C. Yalçın Kaya Mathematics and Statistics, University of South Australia, Mawson Lakes SA 5095, Australia, yalcin.kaya@unisa.edu.au

G.M. Lee Department of Applied Mathematics, Pukyong National University, Busan 608-737, Korea, gmlee@pknu.ac.kr

G. Li Department of Applied Mathematics, University of New South Wales, Sydney 2052, Australia, g.li@unsw.edu.au Bradley University, Peoria, IL 61625, USA, mcasey@bradley.edu

Boglárka Mosoni Faculty of Mathematics and Computer Science, Babeş-Bolyai University, Cluj-Napoca, Romania, mo_bogi@yahoo.com

Libin Mou Bradley University, Peoria, IL 61625, USA, mou@bradley.edu

Vera Roshchina Centro de Investigação em Matemática e Aplicações, Universidade de Évora, Colégio Luís Verney, Rua Romão Ramalho 59, 7000-671 Évora, Portugal, vera.roshchina@gmail.com

Ilya Shvartsman Department of Mathematics and Computer Science, Penn State Harrisburg, Middletown, PA 17057, USA, ius13@psu.edu

Xiao Qi Yang Department of Applied Mathematics, The Hong Kong Polytechnic University, Hung Hum, Kowloon, Hong Kong, mayangxq@polyu.edu.hk

Jen-Chih Yao Department of Applied Mathematics, National Sun Yat-Sen University, Kaohsiung 80424, Taiwan, yaojc@math.nsysu.edu.tw

N.D. Yen Institute of Mathematics, Vietnamese Academy of Science and Technology, 18 Hoang Quoc Viet, Hanoi 10307, Vietnam, ndyen@math.ac.vn

Alexander J. Zaslavski Department of Mathematics, The Technion-Israel Institute of Technology, 32000 Haifa, Israel, ajzasl@tx.technion.ac.il

Systems of Vector Quasi-equilibrium Problems and Their Applications

Qamrul Hasan Ansari and Jen-Chih Yao

Abstract In this survey chapter, we present systems of various kinds of vector quasi-equilibrium problems and give existence theory for their solutions. Some applications to systems of vector quasi-optimization problems, quasi-saddle point problems for vector-valued functions and Debreu type equilibrium problems, also known as constrained Nash equilibrium problems, for vector-valued functions are presented. The investigations of this chapter are based on our papers: Ansari (J Math Anal Appl 341:1271–1283, 2008); Ansari et al. (J Global Optim 29:45–57, 2004); Ansari and Khan (*Mathematical Analysis and Applications*, edited by S. Nanda and G.P. Rajasekhar, Narosa, New Delhi, 2004, pp. 1–13); and Ansari et al. (J Optim Theory Appl 127:27–44, 2005).

1 Introduction

In the last two decades, vector variational inequalities (VVI) have been investigated [2,32,47,48,55,57,62,65,87,97] and used as tools to solve vector optimization problems (VOP) for differentiable and convex or nonconvex vector-valued functions. A generalized form of VVI for multivalued maps is called a generalized vector variational inequality (GVVI). GVVI has been used to study VOP for nondifferentiable and nonconvex vector-valued functions. The weak (respectively, strong) solution of Stampacchia GVVI provides a sufficient condition (respectively, necessary and sufficient conditions) for a solution of VOP; see, for example, [17, 23, 30, 59, 60] and the references therein.

Qamrul Hasan Ansari
Department of Mathematics, Aligarh Muslim University, Aligarh 202 002, India
e-mail: qhansari@gmail.com

Jen-Chih Yao
Department of Applied Mathematics, National Sun Yat-Sen University, Kaohsiung 80424, Taiwan
e-mail: yaojc@math.nsysu.edu.tw

R.S. Burachik and J.-C. Yao (eds.), *Variational Analysis and Generalized Differentiation in Optimization and Control*, Springer Optimization and Its Applications 47,
DOI 10.1007/978-1-4419-0437-9_1,

In the recent years, vector equilibrium problems (VEP) have been studied in [3, 19, 28, 44, 48–50, 94] and the references therein. It is a unified model of several problems, namely, vector variational inequality problems, vector variational-like inequality problems (also called vector pre-variational inequality problems), vector complementarity problems, vector saddle point problems and vector optimization problems. A comprehensive bibliography on VEP, vector variational inequalities, vector variational-like inequalities and their generalizations can be found in [48]. For further details on generalized vector variational inequality problems, generalized vector variational-like inequality problems and vector equilibrium problems, we refer to [2,17,23,28,37,48,50,56,58,61,62,64,83,94] and the references therein. In [24], we extended a quasi-equilibrium problem, studied in [41, 72], to the case of vector-valued functions, called a *vector quasi-equilibrium problem* (VQEP). We established some existence results for a solution of VQEP with or without a generalized pseudomonotonicity assumption. As a result, we derived the existence results for solutions of vector quasi-optimization problems, vector quasi-saddle point problems, vector quasi-variational inequality problems and vector quasi-variational-like inequality problems [55, 63, 65].

When the involved bifunction in the formulation of VEP (respectively, VQEP) is a multivalued map, then VEP (respectively, VQEP) is called a generalized vector equilibrium problem (GVEP) [respectively, generalized vector quasi-equilibrium problem (GVQEP)]. The GVEP (respectively, GVQEP) includes as special cases generalized implicit vector variational inequality problems, GVVI problems, generalized vector variational-like inequality problems and vector equilibrium problems (respectively, generalized implicit vector quasi-variational inequality problems, generalized vector quasi-variational inequality problems, generalized vector quasi-variational-like inequality problems and vector quasi-equilibrium problems). GVEP and GVQEP has been studied in [7, 8, 11, 13, 18, 21, 33, 46, 57, 74, 83–85, 93] and the references therein.

The system of vector equilibrium problems (SVEP), that is, a family of equilibrium problems for vector-valued bifunctions defined on a product set, is introduced in [14] with applications in vector optimization and the Nash equilibrium problem [80–82] for vector-valued functions. The SVEP contains a system of equilibrium problems, a system of vector variational inequalities, a system of vector variational-like inequalities, a system of optimization problems and the Nash equilibrium problem for vector-valued functions as special cases. In the recent past, systems of scalar (vector) equilibrium problems are used as tools to solve the Nash equilibrium problem for vector-valued functions; see, for example, [14, 15, 22, 39, 98, 99] and the references therein. But, by using SVEP, we cannot establish the existence of a solution of Debreu type equilibrium problem [38], also known as constrained Nash equilibrium problem, for vector-valued functions that extends the classical concept of the Nash equilibrium problem for a noncooperative game. For this purpose, in [5], we introduced a system of vector quasi-equilibrium problems (SVQEP) with or without involving Φ-condensing maps and proved the existence of its solution. Consequently, we established some existence results for a solution of a system of vector quasi-variational-like inequalities. The equivalence between a system of vector quasi-variational-like inequalities and the Debreu type equilibrium problem for vector-valued functions (Debreu VEP) is presented. As an application, we derived some existence results for a solution of the Debreu VEP.

In [15], we introduced a system of generalized vector equilibrium problems (SGVEP) which contains a system of generalized implicit vector variational inequality problems, a system of generalized vector variational inequalities, a system of generalized vector variational-like inequalities and SVEP as special cases. We established some existence results for a solution of SGVEP by using a maximal element theorem for a family of multivalued maps due to Deguire et al. [39]. We also derived some existence results for a solution of a system of generalized implicit vector variational inequality problems, a system of generalized vector variational inequalities, a system of generalized vector variational-like inequalities and SVEP. As an application, we gave some existence results for a solution of the Nash equilibrium problem for differentiable (in some sense) vector-valued functions.

In [10], we introduced a system of generalized vector quasi-equilibrium problems (SGVQEP). It is a very general and unified model of several problems, namely, a system of generalized implicit vector quasi-variational inequality problems, a system of generalized vector quasi-variational inequalities, a system of generalized vector quasi-variational-like inequalities, SVEP, SVQEP and SGVEP. We established some existence results for a solution of SGVQEP with or without involving Φ-condensing maps. As consequences, we proved the existence of solutions of several known problems mentioned above. As applications of our results, we derived the existence results for a solution of Debreu VEP for nondifferentiable (in some sense) functions.

In 1994, Husain and Tarafdar [52] introduced simultaneous variational inequalities and gave some applications to minimization problems. These are further studied by Fu [45] for the vector-valued case with applications to vector complementarity problems. Recently, Lin [67] considered and studied simultaneous vector quasi-equilibrium problems and proved existence results for their solutions. By using these results, Lin derived existence results for a solution of a vector quasi-saddle point problem. In [12], we considered systems of simultaneous generalized vector quasi-equilibrium problems (SSGVQEP) which contain simultaneous generalized vector quasi-equilibrium problems [67], generalized vector quasi-equilibrium problems [46], systems of vector quasi-equilibrium problems [5], systems of generalized vector quasi-variational-like inequalities [10] and simultaneous vector variational inequalities [45] as special cases. By using Kakutani fixed point theorem [54], we established an existence result for solutions of SSGVQEP. We derived several existence results for solutions of above-mentioned problems. These existence results either improve or extend known results in the literature. We also considered systems of vector quasi-saddle point problems (SVQSPP) and systems of quasi-minimax inequalities (SQMI). As applications of our existence results for solutions of SSGVQEP, we proved existence of solutions of SVQSPP and SQMI. We gave another application of our results to establish existence of a solution of Debreu VEP.

Because of the applications to vector optimization, game theory and economics, saddle point problems for vector-valued functions, the theory of (vector) equilibrium problems is emerged as a new direction for the researchers; see the references in this chapter.

In this survey chapter, we present systems of various kinds of vector quasi-equilibrium problems and give existence theory for their solutions and some applications to systems of (quasi-) vector optimization problems, systems of quasi-saddle point problems for vector-valued functions and Debreu VEP. The investigations of this chapter are based on our papers [4, 5, 10, 12].

2 Preliminaries

Throughout the chapter, we use the following notations. Let A be a nonempty subset of a topological vector space $\mathscr{X}$, we denote by int A, $\overline{A}$, co$A$ and $\overline{\text{co}}A$, the interior of A in $\mathscr{X}$, the closure of A in $\mathscr{X}$, the convex hull of A, and the closed convex hull of A, respectively. The family of all subsets of A is denoted by 2^A. If X and Y are topological vector spaces, then $L(X,Y)$ denotes the family of all continuous linear maps from X to Y.

Definition 1 ([26, 27]). Let $\mathscr{X}$ and $\mathscr{Y}$ be topological spaces. A multivalued map $T : \mathscr{X} \to 2^{\mathscr{Y}}$ is called *upper semicontinuous at* $x_0 \in \mathscr{X}$ if for any open set $V \subseteq \mathscr{Y}$ containing $T(x_0)$, there exists an open neighbourhood U of x_0 in $\mathscr{X}$ such that $T(x) \subseteq V$ for all $x \in U$.

T is called *lower semicontinuous at* $x \in \mathscr{X}$ if for any open set $V \subseteq \mathscr{Y}$ such that $V \cap T(x_0) \neq \emptyset$, there exists an open neighbourhood U of x_0 in $\mathscr{X}$ such that $T(x) \cap V \neq \emptyset$ for all $x \in U$.

It is said to be *upper (lower) semicontinuous on* $\mathscr{X}$ if it is upper (lower) semicontinuous at every point $x \in \mathscr{X}$.

Further, T is said to be *continuous* on X if it is upper semicontinuous as well as lower semicontinuous on X.

Lemma 1 ([26]). *A multivalued map* $T : \mathscr{X} \to 2^{\mathscr{Y}}$ *is* lower semicontinuous at $x \in \mathscr{X}$ *if and only if for any* $y \in T(x)$ *and for any* $x_n \in \mathscr{X}$ *such that* $x_n \to x$, *there exists* $y_n \in T(x_n)$ *such that* $y_n \to y$.

Definition 2. Let $\mathscr{X}$ and $\mathscr{Y}$ be two topological spaces. A multivalued map $T : \mathscr{X} \to 2^{\mathscr{Y}}$ is said to be:

(i) *Compact* if there exists a compact subset $\mathscr{K} \subseteq \mathscr{Y}$ such that $T(\mathscr{X}) \subseteq \mathscr{K}$
(ii) *Closed* if its graph $Gr(T) = \{(x,y) \mid x \in \mathscr{X},\ y \in T(x)\}$ is closed in $\mathscr{X} \times \mathscr{Y}$

Lemma 2 ([79]). *Let* $(E, \|\cdot\|)$ *be a normed vector space and* $\mathscr{H}$ *be a Hausdorff metric on the collection* $\mathscr{CB}(E)$ *of all nonempty, closed and bounded subsets of* E, *induced by a metric* d *in terms of* $d(x,y) = \|x-y\|$, *which is defined as*

$$\mathscr{H}(U,V) = \max\left\{\sup_{x\in U}\inf_{y\in V}\|x-y\|, \sup_{y\in V}\inf_{x\in U}\|x-y\|\right\},$$

for all $U, V \in \mathscr{CB}(E)$. *If* U *and* V *are compact sets in* E, *then for all* $x \in U$, *there exists* $y \in V$ *such that*

$$\|x-y\| \leq \mathscr{H}(U,V).$$

Definition 3 ([79]). Let (E,d) be a metric space and $\mathscr{H}$ be a Hausdorff metric on $\mathscr{CB}(E)$. A multivalued map $T : E \to \mathscr{CB}(E)$ is said to be *continuous* (in the sense of Nadler) on E if for every $\varepsilon > 0$, there exists a $\delta > 0$ such that for all $x, y \in E$

$$\mathscr{H}(T(x),T(y)) < \varepsilon \quad \text{whenever} \quad d(x,y) < \delta.$$

Remark 1. The notions of continuity in the sense of Definitions 1 and 3 are equivalent if T is compact valued.

Definition 4 ([101]). Let Ω be a nonempty convex subset of a normed space $(E, \|\cdot\|)$ and Υ be a normed linear space. A nonempty compact-valued multifunction $T : \Omega \to 2^{L(E,\Upsilon)}$ is said to be $\mathscr{H}$*-hemicontinuous* if for any $x, y \in \Omega$, the mapping $\alpha \mapsto \mathscr{H}(T(x+\alpha(y-x)), T(x)))$ is continuous at 0^+, where $\mathscr{H}$ is the Hausdorff metric defined on $\mathscr{CB}(E)$.

Definition 5 ([88, 89]). Let $\mathscr{E}$ be a Hausdorff topological vector space and L a lattice with least element, denoted by $\mathbf{0}$. A mapping $\Phi : 2^{\mathscr{E}} \to L$ is called a *measure of noncompactness* provided that the following conditions hold for any $M, N \in 2^{\mathscr{E}}$:

(i) $\Phi(M) = \mathbf{0}$ if and only if M is precompact (i.e., it is relatively compact).
(ii) $\Phi(\overline{\text{conv}}M) = \Phi(M)$, where $\overline{\text{conv}}M$ denotes the closed convex hull of M.
(iii) $\Phi(M \cup N) = \max\{\Phi(M), \Phi(N)\}$.

It follows from (iii) that if $M \subseteq N$, then $\Phi(M) \leq \Phi(N)$.

Definition 6 ([88, 89]). Let $\Phi : 2^{\mathscr{E}} \to L$ be a measure of noncompactness on $\mathscr{E}$ and $D \subseteq \mathscr{E}$. A multivalued map $T : D \to 2^{\mathscr{E}}$ is called Φ*-condensing* provided that if $M \subseteq D$ with $\Phi(T(M)) \geq \Phi(M)$ then M is relatively compact.

Remark 2. Note that every multivalued map defined on a compact set is necessarily Φ-condensing. If $\mathscr{E}$ is locally convex, then a compact multivalued map (i.e., $T(D)$ is precompact) is Φ-condensing for any measure of noncompactness Φ. Obviously, if $T : D \to 2^{\mathscr{E}}$ is Φ-condensing and if $S : D \to 2^{\mathscr{E}}$ satisfies $S(x) \subseteq T(x)$ for all $x \in D$, then S is also Φ-condensing.

The following maximal element theorem for a family of multivalued maps is a main tool to study systems of vector quasi-equilibrium problems and their generalizations.

Theorem 1 ([39, 69]). *For each* $i \in I$, *let* K_i *be a nonempty convex subset of a Hausdorff topological vector space* X_i. *Let* $K = \prod_{i \in I} K_i$. *For each* $i \in I$, *let* $S_i, T_i : K \to 2^{K_i}$ *be multivalued maps satisfying the following conditions:*

(i) *For each* $i \in I$ *and for all* $x \in K$, $\text{co}S_i(x) \subseteq T_i(x)$, *where* $\text{co}S_i(x)$ *denotes the convex hull of* $S_i(x)$.

(ii) *For each* $i \in I$ *and for all* $x = (x_i)_{i \in I} \in K$, $x_i \notin T_i(x)$, *where* x_i *is the ith component of* x.
(iii) *For each* $i \in I$ *and for all* $y_i \in K_i$, $S_i^{-1}(y_i) = \{x \in K : y_i \in S_i(x)\}$ *is open in* K.
(iv) *There exist a nonempty compact subset* M *of* K *and a nonempty compact convex subset* N_i *of* K_i *for each* $i \in I$ *such that for all* $x \in K \setminus M$, *there exists* $i \in I$ *such that* $S_i(x) \cap N_i \neq \emptyset$.

Then there exists $\bar{x} \in K$ *such that* $S_i(\bar{x}) = \emptyset$ *for all* $i \in I$.

Remark 3. If for each $i \in I$, K_i is a nonempty, closed and convex subset of a locally convex Hausdorff topological vector space X_i, then condition (iv) of Theorem 1 can be replaced by the following condition:

(iv)′ The multivalued map $S : K \longrightarrow 2^K$ defined as $S(x) := \prod_{i \in I} S_i(x)$ for all $x \in K$, is $\boldsymbol{\Phi}$-condensing.

(See Corollary 4 in [29]).

Let $\mathscr{Z}$ be a topological vector space and P a closed convex cone in $\mathscr{Z}$ with $\operatorname{int} P \neq \emptyset$. Then, P induces the vector ordering in $\mathscr{Z}$ by setting, $\forall\, x, y \in P$,

$$x \leq_P y \quad \Leftrightarrow \quad y - x \in P;$$

$$x \not\leq_P y \quad \Leftrightarrow \quad y - x \notin P.$$

Since $\operatorname{int} P \neq \emptyset$, we also have the weak ordering in $\mathscr{Z}$ by setting, $\forall\, x, y \in P$,

$$x <_P y \quad \Leftrightarrow \quad y - x \in \operatorname{int} P;$$

$$x \not<_P y \quad \Leftrightarrow \quad y - x \notin \operatorname{int} P.$$

The ordering $\geq_P$, $\not\geq_P$, $>_P$, $\not>_P$ are defined similarly. A cone P is called *pointed* if $P \cap (-P) = \{\mathbf{0}\}$, where $\mathbf{0}$ is the zero element of $\mathscr{Z}$.

Definition 7 ([28, 76, 94]). Let $\mathscr{M}$ be a nonempty subset of a topological vector space $\mathscr{E}$, and let $\mathscr{Z}$ be a topological vector space with a proper, closed and convex cone P with apex at the origin and $\operatorname{int} P \neq \emptyset$. A vector-valued function $\phi : \mathscr{M} \longrightarrow \mathscr{Z}$ is said to be *P-lower semicontinuous* (respectively, *P-upper semicontinuous*) *at* $x_0 \in \mathscr{M}$ if and only if for any neighbourhood V of $\phi(x_0)$ in $\mathscr{Z}$, $\exists$ a neighbourhood U of x_0 in $\mathscr{E}$ such that

$$\phi(x) \in V + P, \quad \forall\, x \in U \cap \mathscr{M}$$

$$(\text{respectively}, \quad \phi(x) \in V - P, \quad \forall\, x \in U \cap \mathscr{M}).$$

Furthermore, ϕ is *P-lower semicontinuous* (respectively, *P-upper semicontinuous*) *on* $\mathscr{M}$ if and only if it is P-lower semicontinuous (respectively, P-upper semicontinuous) at each $x \in \mathscr{M}$.

ϕ is *P-continuous on* $\mathscr{M}$ if and only if it is both P-lower semicontinuous and P-upper semicontinuous on $\mathscr{M}$.

Remark 4. In [28], it is shown that a function $\phi : \mathscr{M} \to \mathscr{Z}$ is P-lower semicontinuous if and only if $\forall\ \alpha \in \mathscr{Z}$, the set

$$L(\alpha) := \{x \in \mathscr{M} : \phi(x) - \alpha \notin \operatorname{int} P\}$$

is closed in $\mathscr{M}$.

Similarly, we can show that ϕ is P-upper semicontinuous if and only if $\forall\ \alpha \in \mathscr{Z}$, the set

$$U(\alpha) := \{x \in \mathscr{M} : \phi(x) - \alpha \notin -\operatorname{int} P\}$$

is closed in $\mathscr{M}$.

Definition 8 ([28,43,76]).[1] Let $(\mathscr{Z}, P)$ be an ordered topological vector space and $\mathscr{K}$ a nonempty convex subset of a vector space $\mathscr{X}$. A map $\phi : \mathscr{K} \to \mathscr{Z}$ is said to be:

(i) *P-convex* if $\forall\ x, y \in \mathscr{K}$ and $t \in [0,1]$, we have

$$\phi(tx + (1-t)y) \leq_P t\phi(x) + (1-t)\phi(y).$$

(ii) *Properly P-quasiconvex* if $\forall\ x, y \in \mathscr{K}$ and $t \in [0,1]$, we have either

$$\phi(tx + (1-t)y) \leq_P \phi(x)$$

or

$$\phi(tx + (1-t)y) \leq_P \phi(y).$$

(iii) *Properly P-quasiconcave* if $-\phi$ is properly quasiconvex.

(iv) *Natural P-quasiconvex* (or *natural P-quasifunction*) if $\forall\ x, y \in \mathscr{K}$ and $\forall\ t \in [0,1]$,

$$\phi(tx + (1-t)y) \in \operatorname{co}\{\phi(x), \phi(y)\} - P.$$

(v) *P-quasiconvex* (or *P-quasifunction*) if $\forall\ \alpha \in \mathscr{Z}$, the set $\{x \in \mathscr{K} : \phi(x) - \alpha \in -P\}$ is convex.

Remark 5. (a) Every P-convex function is natural P-quasiconvex and every natural P-quasiconvex function is P-quasiconvex, but converse assertions are not true; see, for example, Remark 2.1 in [95].

(b) ϕ is a natural P-quasiconvex function if and only if $\forall\ x, y \in \mathscr{K}$ and $\forall\ t \in [0,1]$, $\exists\ s \in [0,1]$ such that

$$\phi(tx + (1-t)y) \in s\phi(x) + (1-s)\phi(y) - P.$$

(c) If ϕ is a P-quasiconvex function, then the set $\{x \in \mathscr{K} : \phi(x) - \alpha \in -\operatorname{int} P\}$ is also convex for all $\alpha \in \mathscr{Z}$.

---

[1] The terms P-convex, natural P-quasiconvex and P-quasiconvex are used in [28,43,76] instead of P-function, natural P-quasifunction and P-quasifunction which are suggested by Prof. F. Giannessi.

Example 1. Let $\mathscr{K} = [0,1]$, $\mathscr{Z} = \mathbb{R}^2$, $P = \mathbb{R}^2_+ = \{(y_1,y_2) \in \mathbb{R}^2 : y_1 \geq 0,\ y_2 \geq 0\}$ and define a function $\phi : \mathscr{K} \to \mathscr{Z}$ by $\phi(x) = \left(x^2, 1-x^2\right)$. Then, the function ϕ is continuous and natural P-quasiconvex, but neither P-convex nor properly P-quasiconvex.

Example 2. Let $\mathscr{K}, \mathscr{Z}, P$ be the same as in Example 1. We define functions $\xi : \mathscr{K} \to \mathscr{Z}$ by

$$\xi(x) = \left(\cos\left(\frac{\pi x}{2}\right), \sin\left(\frac{\pi x}{2}\right)\right)$$

and the function $\tau : \mathscr{K} \to \mathscr{Z}$ by

$$\tau(x) = (\cos(2\pi x), \sin(2\pi x)).$$

Then, the function ξ is continuous and P-quasiconvex, but not natural P-quasiconvex, and the function τ is continuous, but not natural P-quasiconvex and hence, not P-convex.

Throughout the chapter, all topological spaces are assumed to be Hausdorff.

3 System of Vector Quasi-equilibrium Problems

Throughout the chapter, unless otherwise specified, we use the following notations. Let I be any index set (countable or uncountable). For each $i \in I$, let X_i be a Hausdorff topological vector space and K_i be a nonempty convex subset of X_i. We set $K = \prod_{i\in I} K_i$, $X = \prod_{i\in I} X_i$ and $K^i = \prod_{j\in I,\ j\neq i} K_j$, and we write $K = K^i \times K_i$. For $x \in K$, x^i denotes the projection of x onto K^i and hence we also write $x = (x^i, x_i)$. For each $i \in I$, let Y_i be a topological vector space and $C_i : K \to 2^{Y_i}$ be a multivalued map such that for each $x \in K$, $C_i(x)$ is a proper, closed and convex cone with apex at the origin and $\operatorname{int} C_i(x) \neq \emptyset$. For each $i \in I$, let $P_i = \bigcap_{x\in K} C_i(x)$. For each $i \in I$, we denote by $L(X_i, Y_i)$ the space of all continuous linear operators from X_i into Y_i. We denote by $\langle s_i, x_i\rangle$ the evaluation of $s_i \in L(X_i, Y_i)$ at $x_i \in X_i$. We also assume that $\forall i \in I$, $A_i : K \to 2^{K_i}$ is a multivalued map such that $\forall\, x \in K$, $A_i(x)$ is nonempty and convex, $A_i^{-1}(y_i)$ is open in K $\forall\, y_i \in K_i$ and the set $\mathscr{F}_i := \{x \in K : x_i \in A_i(x)\}$ is closed in K, where x_i is the ith component of x.

We consider the following *system of vector quasi-equilibrium problems* (SVQEP) [5], that is, to find $\bar{x} \in K$ such that for each $i \in I$,

$$\bar{x}_i \in A_i(\bar{x}) \quad : \quad f_i(\bar{x}, y_i) \notin -\operatorname{int} C_i(\bar{x}), \quad \forall\, y_i \in A_i(\bar{x}).$$

If for each $i \in I$, $Y_i = \mathbb{R}$ and $C_i(x) = \mathbb{R}_+$ $\forall\, x \in K$, then SVQEP is known as a *system of quasi-equilibrium problems*; see [9, 98] and the references therein.

If for each $i \in I$ and $\forall x \in K$, $A_i(x) = K_i$ and $C_i(x) = P_i$ a fixed proper closed convex cone with nonempty interior, then SVQEP reduces to a *system of vector equilibrium problems* (SVEP) [14] of finding $\bar{x} \in K$ such that for each $i \in I$,

$$f_i(\bar{x}, y_i) \notin -\operatorname{int} P_i, \quad \forall\, y_i \in K_i.$$

If the index set I is singleton, then SVQEP becomes a vector quasi-equilibrium problem [24] which contains vector quasi-optimization problems, vector quasi-variational inequality problems, vector quasi-variational-like inequality problems and vector quasi-saddle point problems as special cases.

Examples of SVQEP

(1) For each $i \in I$, let $T_i : K \to L(X_i, Y_i)$ and $\eta_i : K_i \times K_i \to X_i$ be two maps. If for each $i \in I$,

$$f_i(x, y_i) = \langle T_i(x), \eta_i(y_i, x_i) \rangle,$$

then SVQEP is equivalent to the following problem of finding $\bar{x} \in K$ such that $\forall\, i \in I$,

$$\bar{x}_i \in A_i(\bar{x}) \ : \ \langle T_i(\bar{x}), \eta_i(y_i, \bar{x}_i) \rangle \notin -\text{int}\, C_i(\bar{x}), \quad \forall\, y_i \in A_i(\bar{x}).$$

It is known as a *system of vector quasi-variational-like inequalities* (SVQVLI). When $\eta_i(y_i, x_i) = y_i - x_i$, then SVQVLI is called a *system of vector quasi-variational inequalities* (SVQVI). If for each $i \in I$, $Y_i = \mathbb{R}$ and $C_i(x) = \mathbb{R}_+ \ \forall\, x \in K$, SVQVI is studied in [9, 98].

If for each $i \in I$, $A_i(x) = K_i \ \forall\, x \in K$, SVQVLI and SVQVI reduce to the following system of vector variational-like inequalities and the system of vector variational inequalities, respectively, studied in [14].

The *system of vector variational-like inequalities* (SVVLI): find $\bar{x} \in K$ such that for each $i \in I$,

$$\langle T_i(\bar{x}), \eta(y_i, \bar{x}_i) \rangle \notin -\text{int}\, C_i(\bar{x}), \quad \text{for all } \ y_i \in K_i.$$

The *system of vector variational inequalities* (SVVI): find $\bar{x} \in K$ such that for each $i \in I$,

$$\langle T_i(\bar{x}), y_i - \bar{x}_i \rangle \notin -\text{int}\, C_i(\bar{x}), \quad \text{for all } \ y_i \in K_i.$$

If for each $i \in I$, $Y_i = \mathbb{R}$ and int $C_i(x) = \mathbb{R}_+$, then SVVI becomes the systems of variational inequalities studied in [20, 35, 86].

In case the index set I is a singleton, SVVI reduces to a vector variational inequality first considered in [47]; see also [48] and the references therein.

(2) For each $i \in I$, let $\varphi_i : K \to Y$ be a given function. The *system of vector quasi-optimization problems* (SVQOP) is to find $\bar{x} \in K$ such that for each $i \in I$,

$$\varphi_i(y) - \varphi_i(\bar{x}) \notin -\text{int}\, C_i(\bar{x}), \quad \text{for all } \ y \in A(\bar{x}).$$

We can choose $y \in K$ in such a way that $y^i = \bar{x}^i$. Then we have *Debreu VEP* also known as *constrained Nash equilibrium problem* for vector-valued functions which is to find $\bar{x} \in K$ such that for each $i \in I$,

$$\varphi_i(\bar{x}^i, y_i) - \varphi_i(\bar{x}) \notin -\text{int}\, C_i(\bar{x}), \quad \text{for all } \ y_i \in A_i(\bar{x}).$$

For each $i \in I$ and for all $x \in K$, if $A_i(x) = K_i$, then Debreu VEP reduces to the following *Nash equilibrium problem* for vector-valued functions: Find $\bar{x} \in K$ such that $\forall\, i \in I$,

$$\varphi_i(\bar{x}^i, y_i) - \varphi_i(\bar{x}) \notin -\text{int}\, C_i(\bar{x}), \quad \forall\, y_i \in K_i.$$

It is clear that every solution of SVQOP is also a solution of Debreu VEP but the converse is not true.

Of course, if for each $i \in I$, φ_i is a scalar-valued function, then Debreu VEP is the same as one introduced and studied by Debreu in [38], see also [80–82]. In this case, a large number of papers have already been appeared in the literature; see [9,98] and the references therein.

Section 3.1 deals with the existence theory of solutions of SVEP and SVQEP with or without involving Φ-condensing maps. Consequently, we get some existence results for a solution of SVQVLI. In Sect. 3.2, we first establish an equivalence between SVQVLI and Debreu VEP and then we derive some existence results for a solution of the Debreu VEP for convex or nonconvex functions.

3.1 Existence Results for Solutions of SVEP and SVQEP

We present the following existence results for solutions of SVEP which are established in [14] by utilizing scalarization technique and by using collectively fixed point theorem for a family of multivalued maps [22].

Theorem 2 ([14]). *Let Y be a topological vector space and $C \subset Y$ be a proper, closed convex cone with apex at the origin $\mathbf{0}$ and $\text{int}\, C \neq \emptyset$. For each $i \in I$, let K_i be a nonempty compact convex subset of X_i and let $f_i : K \times K_i \to Y$ be a bifunction such that $f_i(x, x_i) = \mathbf{0}$ for all $x = (x^i, x_i) \in K$. Assume that the following conditions are satisfied:*

(i) *For each $i \in I$ and $\forall x \in K$, the function $y_i \mapsto f_i(x, y_i)$ is C-quasiconvex.*
(ii) *For each $i \in I$, f_i is continuous on $K \times K_i$.*

Then the solution set of SVEP is nonempty and compact.

In case K_i is not necessarily compact, we have the following result.

Theorem 3 ([14]). *Let Y be a topological vector space and $C \subset Y$ be a proper, closed convex cone with apex at the origin $\mathbf{0}$ and $\text{int}\, C \neq \emptyset$. For each $i \in I$, let K_i be a nonempty convex subset of X_i and let $f_i : K \times K_i \to Y$ be a bifunction such that $f_i(x, x_i) = \mathbf{0}$ for all $x = (x^i, x_i) \in K$. Assume that the following conditions are satisfied:*

(i) *For each $i \in I$ and $\forall x \in K$, the function $y_i \mapsto f_i(x, y_i)$ is C-quasiconvex.*
(ii) *For each $i \in I$, f_i is continuous on each compact convex subset of $K \times K_i$.*

(iii) *For each $i \in I$, there exists a nonempty compact convex subset B_i of K_i, and let $B = \prod_{i\in I} B_i \subset K$ such that for each $x \in K \setminus B$, there exists $\tilde{y}_i \in B_i$ such that*

$$f_i(x, \tilde{y}_i) \in -\operatorname{int} C.$$

Then there exists a solution $\bar{x} \in B$ of SVEP.

Remark 6. Let I be a finite index set and for each $i \in I$, let X_i be a reflexive Banach space with norm $||\cdot||_i$ equipped with the weak topology. Consider a Banach space Y equipped with the norm topology. The norm on $X = \prod_{i\in I} X_i$ will be denoted by $||\cdot||$. Then assumption (iii) in Theorem 3 can be replaced by the following condition:

(iii)$'$ There exists $r > 0$ such that for all $x \in K$, $||x|| > r$, there exists $\tilde{y}_i \in K_i$, $||\tilde{y}_i||_i \leq r$ such that

$$f_i(x, \tilde{y}_i) \in -\operatorname{int} C.$$

We present the following existence result for solutions of SVQEP without involving Φ-condensing maps. In [5], we proved this result by using maximal element Theorem 1.

Theorem 4 ([5]). *For each $i \in I$, let K_i be a nonempty and convex subset of a Hausdorff topological vector space X_i and $f_i : K \times K_i \longrightarrow Y_i$ be a bifunction. Assume that the following conditions hold:*

(i) *For each $i \in I$ and $\forall\, x \in K$, $f_i(x, x_i) \notin -\operatorname{int} C_i(x)$, where x_i is the ith component of x.*
(ii) *For each $i \in I$ and $\forall\, x \in K$, the vector-valued function $y_i \mapsto f_i(x, y_i)$ is a natural P_i-quasiconvex function.*
(iii) *For each $i \in I$ and $\forall\, y_i \in K_i$, the set $\{x \in K : f_i(x, y_i) \notin -\operatorname{int} C_i(x)\}$ is closed in K.*
(iv) *There exist a nonempty and compact subset N of K and a nonempty, compact and convex subset B_i of K_i $\forall\, i \in I$, such that $\forall\, x \in K \setminus N$ $\exists\, i \in I$ and $\exists\, \tilde{y}_i \in B_i$, such that $\tilde{y}_i \in A_i(x)$ and $f_i(x, \tilde{y}_i) \in -\operatorname{int} C_i(x)$.*

Then SVQEP has a solution.

Remark 7. (1) The condition (iii) of Theorem 4 is satisfied if the following conditions hold $\forall\, i \in I$:

(a) The multivalued map $W_i : K \longrightarrow 2^{Y_i}$ defined by $W_i(x) = Y_i \setminus \{-\operatorname{int} C_i(x)\}$ $\forall\, x \in K$, is closed in $K \times K_i$.
(b) For all $y_i \in K_i$, $f_i(\cdot, y_i) : K \longrightarrow Y_i$ is continuous (in the usual sense) on K.

(2) If for each $i \in I$ and $\forall\, x \in K$, $C_i(x) = C_i$, a (fixed) proper, closed and convex cone in Y_i, then conditions (ii) and (iii) of Theorem 4 can be replaced, respectively, by the following conditions:

(c) For each $i \in I$ and $\forall\, x \in K$, the vector-valued function $y_i \mapsto f_i(x, y_i)$ is a C_i-quasiconvex function.

(d) For each $i \in I$ and $\forall\, y_i \in K_i$ the vector-valued function $x \mapsto f_i(x, y_i)$ is C_i-upper semicontinuous on K.

(3) Theorem 4 extends and generalizes Theorem 6 in [9], Theorem 2.1 in [14] and Corollary 3.1 in [24] in several ways.

(4) If for each $i \in I$, K_i is a nonempty, compact and convex subset of a Hausdorff topological vector space X_i, then the conclusion of Theorem 4 holds without condition (iv).

We mention the following existence result for a solution of SVQEP involving Φ-condensing maps. We proved this result by using maximal element Theorem 1 with Remark 3.

Theorem 5 ([5]). *For each $i \in I$, let K_i be a nonempty, closed and convex subset of a locally convex Hausdorff topological vector space X_i, $f_i : K \times K_i \to Y_i$ be a bifunction and let the multivalued map $A = \prod_{i\in I} A_i : K \to 2^K$ defined as $A(x) = \prod_{i\in I} A_i(x)\ \ \forall\, x \in K$, be Φ-condensing. Assume that the conditions (i), (ii) and (iii) of Theorem 4 hold. Then SVQEP has a solution.*

In order to derive the existence results for solutions of systems of vector quasi-variational (-like) inequalities, we define a topology on the space $L(\mathscr{E}, \mathscr{Z})$ by the following way:

Let $\mathscr{E}$ and $\mathscr{Z}$ be Hausdorff topological vector spaces. Let σ be the family of bounded subsets of $\mathscr{E}$ whose union is total in $\mathscr{E}$, that is, the linear hull of $\bigcup\{U : U \in \sigma\}$ is dense in $\mathscr{E}$. Let $\mathscr{B}$ be a neighbourhood base of 0 in $\mathscr{Z}$. When U runs through σ, V through $\mathscr{B}$, thc family

$$M(U,V) = \{\xi \in L(\mathscr{E}, \mathscr{Z}) : \cup_{x\in U} \langle \xi, x\rangle \subseteq V\}$$

is a neighbourhood base of 0 in $L(\mathscr{E}, \mathscr{Z})$ for a unique translation-invariant topology, called the *topology of uniform convergence* on the sets $U \in \sigma$, or, briefly, the σ-topology (see [42, pp. 79–80] and also [91]).

Lemma 3 ([42]). *Let $\mathscr{E}$ and $\mathscr{Z}$ be Hausdorff topological vector spaces and $L(\mathscr{E}, \mathscr{Z})$ be the topological vector space under the σ-topology. Then, the bilinear mapping $\langle ., . \rangle : L(\mathscr{E}, \mathscr{Z}) \times \mathscr{E} \to \mathscr{Z}$ is continuous on $L(\mathscr{E}, \mathscr{Z}) \times \mathscr{E}$.*

Throughout the chapter, we assume that $L(\mathscr{E}, \mathscr{Z})$ is equipped with σ-topology.

In addition to the assumptions on $C_i : K \to 2^{Y_i}$, in the following corollaries, we further assume that $C_i(x)$ is pointed, $\forall\, i \in I$ and $\forall\, x \in K$. Then the following results can be easily derived, respectively, from Theorems 4 and 5 by setting

$$f_i(x, y_i) = \langle T_i(x), \eta_i(y_i, x_i)\rangle.$$

Corollary 1 ([5]). *For each $i \in I$, let K_i, X_i and W_i be the same as in Theorem 4 and Remark 7, respectively. For each $i \in I$, let $\eta_i : K_i \times K_i \to X_i$ be continuous in the second variable such that $\eta_i(x_i, x_i) = 0\ \ \forall\, x_i \in K_i$, and let $T_i : K \to L(X_i, Y_i)$ be* *continuous on K such that the map $y_i \mapsto \langle T_i(x), \eta_i(y_i,x_i)\rangle$ is a natural P_i-quasiconvex function, $\forall\, x \in K$. Assume that there exist a nonempty and compact subset N of K and a nonempty, compact and convex subset B_i of K_i $\forall\, i \in I$, such that $\forall\, x \in K \setminus N$ $\exists\, i \in I$ and $\exists\, \tilde{y}_i \in B_i$ such that $\tilde{y}_i \in A_i(x)$ and $\langle T_i(x), \eta_i(\tilde{y}_i,x_i)\rangle \in -\text{int}\, C_i(x)$. Then SVQVLI has a solution.*

Corollary 2 ([5]). *For each $i \in I$, let K_i, X_i, A_i, A and η_i, T_i, W_i, $L(X_i,Y_i)$ be the same as in Theorem 5 and Corollary 1, respectively. Then SVQVLI has a solution.*

Remark 8. To the best of our knowledge, there is only one paper [40] appeared in the literature on the scalar quasi-variational-like inequality problems involving Φ-condensing maps. Since the approach in this chapter is different from the one adopted in [40], Corollary 2 is a new result in the literature, not only for the vector case but also for the scalar one.

3.2 Applications of SVQEP

Let $I = \{1,2,\ldots,n\}$ be a finite index set and for each $i \in I$, let X_i be a normed space and $X = \prod_{i\in I} X_i$. Let Z be a normed space. We recall the following definition.

Definition 9 ([100]). The function $\phi : X \to Z$ is said to be *partial Gâteaux differentiable at* $x = (x_1,\ldots,x_{j-1},x_j,x_{j+1},\ldots,x_n) \in X$ w. r. t. the jth variable x_j if

$$\langle D_{x_j}\phi(x), h_j\rangle = \lim_{t\to 0} \frac{\phi(x_1,\ldots,x_{j-1},x_j+th_j,x_{j+1},\ldots,x_n) - \phi(x)}{t} \quad \text{exists,}$$

for all $h_j \in X_j$. $D_{x_j}\phi(x) \in L(X_j,Z)$ is called the *partial Gâteaux derivative of ϕ at* $x \in X$ w.r.t. the jth variable x_j.

ϕ is called *partial Gâteaux differentiable on X* if it is partial Gâteaux differentiable at each point of K w.r.t. each variable.

Definition 10 ([96]). Let E be a normed space, Z a normed space with a closed and convex cone P with apex at the origin, M a nonempty subset of E, $\eta : M \times M \to E$ a function. A Gâteaux differentiable function $\phi : M \to Z$ is said to be *P-invex* w.r.t. η if $\forall\, x,y \in M$,

$$\phi(y) - \phi(x) - \langle D_x\phi(x), \eta(y,x)\rangle \in P,$$

where $D_x\phi(x)$ denotes the Gâteaux derivative of ϕ at x.

Definition 11 ([78]). A subset M of a vector space E is said to be *invex w.r.t.* $\eta : M \times M \to E$ if $\forall\, x,y \in M$ and $\forall\, t \in [0,1]$, $x + t\eta(y,x) \in M$.

Definition 12 ([96]). Let M be an invex set in a normed space E w.r.t. $\eta : M \times M \to E$. A vector-valued function $\phi : M \to Z$ is said to be *P-preinvex* if $\forall\, x,y \in M$ and $\forall\, t \in [0,1]$,

$$t\phi(y) + (1-t)\phi(x) - \phi(x + t\eta(y,x)) \in P.$$

Remark 9. It can be easily seen that if M is an invex subset of E w.r.t. $\eta : M \times M \to E$ and $\phi : M \to Z$ is Gâteaux differentiable on M and P-preinvex, then ϕ is P-invex w.r.t. η. But the converse assertion may not be true.

We have the following result which provides a sufficient condition for a solution of Debreu VEP.

Proposition 1 ([5]). *Let I be a finite index set. For each $i \in I$, let X_i and Y_i be normed spaces, K_i a nonempty and convex subset of X_i, $K = \prod_{i\in I} K_i$, $A_i : K \to 2^{K_i}$ nonempty convex-valued multivalued map, $\eta_i : K_i \times K_i \to X_i$, and $\varphi_i : K \to Y_i$ partial Gâteaux differentiable on each open subset of K and P_i-invex w.r.t. η_i in each argument. Then every solution of SVQVLI with $T_i(x) = D_{x_i}\varphi_i(x)$ is also a solution of Debreu VEP.*

Proof. Assume that $\bar{x} \in K$ is a solution of SVQVLI with $T_i(x) = D_{x_i}\varphi_i(x)$. Then for each $i \in I$,

$$\bar{x}_i \in A_i(\bar{x}) \ : \ \langle D_{x_i}\varphi_i(\bar{x}), \eta_i(y_i,\bar{x}_i)\rangle \notin -\text{int}\, C_i(\bar{x}), \quad \forall\, y_i \in A_i(\bar{x}). \tag{1}$$

Since for each $i \in I$, φ_i is P_i-invex w.r.t. η_i in each argument, we have

$$\varphi_i(\bar{x}^i, y_i) - \varphi_i(\bar{x}) - \langle D_{x_i}\varphi_i(\bar{x}), \eta_i(y_i,\bar{x}_i)\rangle \in P_i \subseteq C_i(\bar{x}). \tag{2}$$

Since $a - b \in P$ and $b \notin -\text{int}\, P \Rightarrow a \notin -\text{int}\, P$, it follows from (1) and (2) that

$$\bar{x}_i \in A_i(\bar{x}) \ : \ \varphi_i(\bar{x}^i, y_i) - \varphi_i(\bar{x}) \notin -\text{int}\, C_i(\bar{x}), \quad \forall\, y_i \in A_i(\bar{x}).$$

Hence $\bar{x} \in K$ is a solution of Debreu VEP.

The next result provides the equivalence between SVQVLI and Debreu VEP.

Proposition 2 ([5]). *Let I be a finite index set. For each $i \in I$, let X_i and Y_i be normed spaces, $K_i \subseteq X_i$ nonempty invex w.r.t. $\eta_i : K_i \times K_i \to X_i$, $K = \prod_{i\in I} K_i$, $A_i : K \to 2^{K_i}$ nonempty invex valued multivalued map and $\varphi_i : K \to Y_i$ partial Gâteaux differentiable on each open subset of K and P_i-preinvex in each argument. Then $\bar{x} \in K$ is a solution of SVQVLI with $T_i(x) = D_{x_i}\varphi_i(x)$ if and only if it is a solution of Debreu VEP.*

Proof. Assume that $\bar{x} \in K$ is a solution of SVQVLI. Then by Proposition 1, $\bar{x} \in K$ is a solution of Debreu VEP.

Conversely, let $\bar{x} \in K$ be a solution of Debreu VEP. Then for each $i \in I$,

$$\bar{x}_i \in A_i(\bar{x}) \ : \ \varphi_i(\bar{x}^i, y_i) - \varphi_i(\bar{x}) \notin -\text{int}\, C_i(\bar{x}), \quad \forall\, y_i \in A_i(\bar{x}). \tag{3}$$

Since $\bar{x}_i, y_i \in A_i(\bar{x})$ and each $A_i(\bar{x})$ is invex, we have $\bar{x}_i + t\eta_i(y_i,\bar{x}_i) \in A_i(\bar{x})\ \forall\, t \in [0,1]$. Therefore, from (3), we get

$$\varphi_i(\bar{x}^i, \bar{x}_i + t\eta_i(y_i,\bar{x}_i)) - \varphi_i(\bar{x}) \in W_i(\bar{x}) = Y_i \setminus \{-\text{int}\, C_i(\bar{x})\}.$$

Since for each $i \in I$, $W_i(\bar{x})$ is a closed cone, we have

$$\lim_{t\to 0} \frac{\varphi_i(\bar{x}^i, \bar{x}_i + t\eta_i(y_i, \bar{x}_i)) - \varphi_i(\bar{x})}{t} \in W_i(\bar{x}).$$

From the partial Gâteaux differentiability of each φ_i, we get, $\forall\, i \in I$

$$\bar{x}_i \in A_i(\bar{x}) \;:\; \langle D_{x_i}\varphi_i(\bar{x}), \eta_i(y_i, \bar{x}_i)\rangle \notin -\mathrm{int}\, C_i(\bar{x}), \quad \forall\, y_i \in A_i(\bar{x}).$$

Hence $\bar{x} \in K$ is a solution of SVQVLI with $T_i(\bar{x}) = D_{x_i}\varphi_i(\bar{x})\ \forall\, i \in I$.

Remark 10. If for each $i \in I$ and $\forall\, x \in K$, $\eta_i(y_i, x_i) = y_i - x_i$, $A_i(x) = K_i$, $C_i(x) = \mathbb{R}_+$ and $Y_i = \mathbb{R}$, then Proposition 2 reduces to Proposition 4 in [25, p. 269]. Hence Proposition 2 extends Proposition 4 in [25] in several ways.

By using Proposition 1 and Corollary 1, we can easily derive the following existence result for a solution of Debreu VEP.

Theorem 6 ([5]). *Let I be a finite index set. For each $i \in I$, let X_i and Y_i be normed spaces, K_i be a nonempty convex subset of X_i, $K = \prod_{i\in I} K_i$ and W_i be the same as in Remark 7. For each $i \in I$, let $\eta_i : K_i \times K_i \to X_i$ be continuous in the second argument such that $\eta_i(x_i, x_i) = 0\ \ \forall\, x_i \in K_i$, and $\varphi_i : K \to Y_i$ partial Gâteaux differentiable on K and P_i-invex in each variable such that the function $y_i \mapsto \langle D_{x_i}\varphi_i(x), \eta_i(y_i, x_i)\rangle$ is a natural P_i-quasiconvex function, $\forall\, x \in K$. Assume that there exist a nonempty and compact subset N of K and a nonempty, compact and convex subset B_i of $K_i\ \ \forall\, i \in I$, such that $\forall\, x \in K\setminus N\ \exists\, i \in I$ and $\exists\, \tilde{y}_i \in B_i$, such that $\tilde{y}_i \in A_i(x)$ and $\langle D_{x_i}\varphi_i(x), \eta_i(\tilde{y}_i, x_i)\rangle \in -\mathrm{int}\, C_i(x)$. Then the Debreu VEP has a solution.*

If the index set I need not be finite and for each $i \in I$, φ_i need not be partial Gâteaux differentiable, then we can also easily derive the following existence results for a solution of Debreu VEP from Theorems 4 and 5 by setting, $\forall\, i \in I$,

$$f_i(x, y_i) = \varphi_i(x^i, y_i) - \varphi_i(x).$$

Theorem 7 ([5]). *For each $i \in I$, let K_i, K, X_i and W_i be the same as in Theorem 4 and Remark 7, respectively, and let $\varphi_i : K \to Y_i$ be a vector-valued function. Assume that the following conditions hold:*

(i) *For each $i \in I$, φ_i is a natural P_i-quasiconvex function in the ith argument.*
(ii) *For each $i \in I$, φ_i is continuous on K.*
(iii) *There exist a nonempty and compact subset N of K and a nonempty, compact and convex subset B_i of $K_i\ \forall\, i \in I$, such that $\forall\, x \in K\setminus N\ \exists\, i \in I$ and $\exists\, \tilde{y}_i \in B_i$, such that $\tilde{y}_i \in A_i(x)$ and $\varphi_i(x^i, \tilde{y}_i) - \varphi_i(x) \in -\mathrm{int}\, C_i(x)$.*

Then Debreu VEP has a solution.

Theorem 8 ([5]). *For each $i \in I$, let K_i, K, X_i and W_i be the same as in Theorems 5 and 7, respectively. Assume that the conditions (i) and (ii) of Theorem 5 hold. Then Debreu VEP has a solution.*

Remark 11. (1) If for each $i \in I$ and $\forall\, x \in K$, $C_i(x) = C_i$, a (fixed) proper, closed and convex cone in Y_i, then conditions (i) and (ii) in Theorem 5, and subsequently, in Theorem 6 can be replaced, respectively, by the following conditions:

(i)′ For each $i \in I$ and $\forall\, x \in K$, φ_i is a C_i-quasiconvex function in the ith argument.
(ii)′ For each $i \in I$, φ_i is C_i-upper semicontinuous on K.

(2) Theorem 6 provides the existence of a solution of Debreu VEP involving Φ-condensing map and, consequently, for scalar-valued functions. Therefore, Theorem 6 is a new result in the literature.

4 System of Generalized Vector Quasi-equilibrium Problems

For each $i \in I$, let $F_i : K \times K_i \to 2^{Y_i}$ and $A_i : K \to 2^{K_i}$ be multivalued maps with nonempty values. We consider the following *system of generalized vector quasi-equilibrium problems [10]:*

$$\text{(SGVQEP)} \quad \begin{cases} \text{Find } \bar{x} \in K \text{ such that for each } i \in I,\ \bar{x}_i \in A_i(\bar{x}) \quad \text{and} \\ F_i(\bar{x}, y_i) \not\subseteq -\operatorname{int} C_i(\bar{x}), \quad \forall y_i \in A_i(\bar{x}). \end{cases}$$

If for each $i \in I$ and $\forall\, x \in K$, $A_i(x) = K_i$, then SGVQEP reduces to the following *system of generalized vector equilibrium problems* (SGVEP) [15]:

$$\text{(SGVEP)} \quad \begin{cases} \text{Find } \bar{x} \in K \text{ such that for each } i \in I, \\ F_i(\bar{x}, y_i) \not\subseteq -\operatorname{int} C_i(x), \quad \forall y_i \in K_i. \end{cases}$$

It is introduced and studied in [15] with applications to the Nash equilibrium problem for vector-valued functions.

If I is a singleton set, then SGV(Q)EP reduces to a generalized vector (quasi-) equilibrium problem which contains generalized implicit vector (quasi-) variational inequality problems, generalized vector (quasi-) variational inequality problems, generalized vector (quasi-) variational-like inequality problems and vector (quasi-) equilibrium problems as special cases. For further detail on generalized vector (quasi-) equilibrium problems and their applications, we refer [7, 8, 11, 13, 18, 21, 46, 84, 85, 93, 101] and the references therein.

Examples of SGVQEP

For each $i \in I$, let D_i be a nonempty subset of $L(X_i, Y_i)$. For each $i \in I$, let $T_i : K \to 2^{D_i}$ be a multivalued map with nonempty values. For each $i \in I$, let $\psi_i : D_i \times K_i \times K_i \to Y_i$ be a vector-valued map. The problem of *system of generalized implicit vector quasi-variational inequalities* (SGIVQVIP) is to find $\bar{x} \in K$ such that for each $i \in I$, $\bar{x}_i \in A_i(\bar{x})$ and

$$\forall y_i \in A_i(\bar{x}),\ \exists \bar{u}_i \in T_i(\bar{x}) \ :\ \psi_i(\bar{u}_i, \bar{x}_i, y_i) \notin -\operatorname{int} C_i(\bar{x}).$$

Setting for each $i \in I$,

$$F_i(x, y_i) = \psi_i(T_i(x), x_i, y_i) = \{\psi_i(u_i, x_i, y_i) : u_i \in T_i(x)\}.$$

Then SGVQEP coincides with SGIVQVIP.

For $Y_i = \mathbb{R}$ and $C_i(x) = \mathbb{R}_-$ for all $x \in K$ and for each $i \in I$, SGIVQVIP is called the *problem of system of generalized implicit quasi-variational inequalities*. Further, for all $x \in K$ and for each $i \in I$, $A_i(x) = K_i$, it is called the *problem of system of generalized implicit variational inequalities*. Such problem is studied in [22] with application to Nash equilibrium problem [81].

If I is a singleton set, SGIVQVIP reduces to *generalized implicit vector quasi-variational inequality problem.*

The SGIVQVIP contains the following problems as special cases:

(i) For each $i \in I$, let $\theta_i : K \times D_i \to D_i$ and $\eta_i : K_i \times K_i \to X_i$ be bifunctions. If for each $i \in I$,

$$\psi_i(T_i(x), x_i, y_i) = \langle \theta_i(x, T_i(x)), \eta_i(y_i, x_i) \rangle = \{\langle \theta_i(x, u_i), \eta_i(y_i, x_i) \rangle : u_i \in T_i(x)\},$$

then SGIVQVIP reduces to the *problem of system of generalized vector quasi-variational-like inequalities* (SGVQVLIP) (I) which is to find $\bar{x} \in K$ such that for each $i \in I$, $\bar{x}_i \in A_i(\bar{x})$ and

$$\forall y_i \in A_i(\bar{x}), \ \exists \bar{u}_i \in T_i(\bar{x}) \ : \ \langle \theta_i(\bar{x}, \bar{u}_i), \eta_i(y_i, \bar{x}_i) \rangle \notin -\mathrm{int}\, C_i(\bar{x}).$$

If I is a singleton set, then SGVQVLIP(I) becomes the *generalized vector quasi-variational-like inequality problem*. The strong solution (i.e., $\bar{u}_i$ does not depend on y_i) of SGVQVLIP(I) is studied by Chen et al. [31] and Lee et al. [65], see also the references therein.

If for each $i \in I$, $\theta_i(x, u_i) = u_i$ for all $x \in K$, then SGVQVLIP(I) becomes the following problem denoted by SGVQVLIP(II): Find $\bar{x} \in K$ such that for each $i \in I$, $\bar{x}_i \in A_i(\bar{x})$ and

$$\forall y_i \in A_i(\bar{x}), \ \exists \bar{u}_i \in T_i(\bar{x}) \ : \ \langle \bar{u}_i, \eta_i(y_i, \bar{x}_i) \rangle \notin -\mathrm{int}\, C_i(\bar{x}).$$

For $Y_i = \mathbb{R}$, $C_i(x) = \mathbb{R}_-$ and $A_i(x) = K_i$ for all $x \in K$ and for each $i \in I$, this problem is studied in [22] with application to the Nash equilibrium problem [81].

(ii) If for each $i \in I$,

$$\psi_i(T_i(x), x_i, y_i) = \langle T_i(x), y_i - x_i \rangle = \{\langle u_i, y_i - x_i \rangle : u_i \in T_i(x)\},$$

then SGIVQVIP reduces to the *problem of system of generalized vector quasi-variational inequalities* (SGVQVIP) which is to find $\bar{x} \in K$ such that for each $i \in I$, $\bar{x}_i \in A_i(\bar{x})$ and

$$\forall y_i \in A_i(\bar{x}), \ \exists \bar{u}_i \in T_i(\bar{x}) \ : \ \langle \bar{u}_i, y_i - \bar{x}_i \rangle \notin -\mathrm{int}\, C_i(\bar{x}).$$

For each $i \in I$, if F_i is a single-valued map, then SGVQEP reduces to SVQEP (Sect. 3).

4.1 Existence Results for Solutions of SGVQEP

The following results provide the existence of a solution of SGVQEP with or without Φ-condensing maps.

Theorem 9 ([10]). *For each $i \in I$, let K_i be a nonempty convex subset of a Hausdorff topological vector space X_i and let $F_i : K \times K_i \to 2^{Y_i}$ be a multivalued map with nonempty values. For each $i \in I$, assume that the following conditions hold:*

(i) *For all $x \in K$, $F_i(x, x_i) \not\subseteq -\text{int}\, C_i(x)$, where x_i is the ith component of x.*
(ii) *For all $x \in K$, the set $\{y_i \in K_i : F_i(x, y_i) \subseteq -\text{int}\, C_i(x)\}$ is convex.*
(iii) *For all $y_i \in K_i$, the set $\{x \in K : F_i(x, y_i) \not\subseteq -\text{int}\, C_i(x)\}$ is closed in K.*
(iv) *There exist a nonempty compact subset N of K and a nonempty compact convex subset B_i of K_i for each $i \in I$ such that for each $x \in K \setminus N$ there exist $i \in I$ and $\tilde{y}_i \in B_i$ satisfying $\tilde{y}_i \in A_i(x)$ and $F_i(x, \tilde{y}_i) \subseteq -\text{int}\, C_i(x)$.*

Then the SGVQEP has a solution.

Theorem 10 ([10]). *For each $i \in I$, let K_i be a nonempty, closed and convex subset of a locally convex Hausdorff topological vector space X_i, $F_i : K \times K_i \to 2^{Y_i}$ a multivalued map with nonempty values and let the multivalued map $A = \prod_{i \in I} A_i : K \to 2^K$ defined as $A(x) = \prod_{i \in I} A_i(x)$ for all $x \in K$, be Φ-condensing. Assume that the conditions (i)–(iii) of Theorem 9 hold. Then the SGVQEP has a solution.*

In order to verify condition (ii) in Theorems 9 and 10, we introduce the following concept.

Definition 13 ([21]). Let W and Z be topological vector spaces and M be a nonempty convex subset of W and let $P : M \to 2^Z$ be a multivalued map such that for each $x \in M$, $P(x)$ is a closed, convex cone with nonempty interior. For each fixed $x \in M$, a multivalued map $F : M \times M \to 2^Z \setminus \{\emptyset\}$ is called *$P(x)$-quasiconvex-like* if for all $y_1, y_2 \in M$ and $t \in [0,1]$, we have either

$$F(x, ty_1 + (1-t)y_2) \subseteq F(x, y_1) - P(x),$$

or

$$F(x, ty_1 + (1-t)y_2) \subseteq F(x, y_2) - P(x).$$

To show the class of $P(x)$-quasiconvex-like multivalued is nonempty, we give the following example.

Example 3. Let $M = [0,1]$, $P(x) = [0, +\infty)$ for all $x \in M$. We define $F : M \times M \to 2^{\mathbb{R}}$ by

$$F(x, y) = [x, y+1] \quad \text{for all } x, y \in M.$$

For all $x, y_1, y_2 \in M$ and $0 \leq t \leq 1$, we note that

$$\text{if } y_1 \leq y_2 \quad \text{then } ty_1 + (1-t)y_2 \leq y_2$$

and

$$\text{if } y_1 > y_2 \quad \text{then } ty_1 + (1-t)y_2 \leq y_1.$$

Therefore, we have for each $\alpha \in F(x, ty_1 + (1-t)y_2)$,

$$\alpha = \begin{cases} (y_2+1) - [(y_2+1) - \alpha], & y_1 \leq y_2, \\ (y_1+1) - [(y_1+1) - \alpha], & y_1 > y_2. \end{cases}$$

Hence, we have either $F(x, ty_1 + (1-t)y_2) \subseteq F(x, y_1) - P(x)$ or $F(x, ty_1 + (1-t)y_2) \subseteq F(x, y_2) - P(x)$. Thus, F is $P(x)$-quasiconvex-like.

Remark 12. (a) If for each $i \in I$, F_i is $C_i(x)$-quasiconvex-like, then the set $\{y_i \in K_i : F_i(x, y_i) \subseteq -\text{int } C_i(x)\}$ is convex, for all $x \in K$ (see, e.g., the proof of Theorem 2.1 in [21]).

(b) If for each $i \in I$, X_i is locally convex Hausdorff topological vector space, the multivalued map $W_i : K \to 2^{Y_i}$ defined by $W_i(x) = Y_i \setminus \{-\text{int } C_i(x)\}$ for all $x \in K$, is closed on K and for all $y_i \in K_i$, $F_i(., y_i)$ is upper semicontinuous on K, then condition (iii) of Theorem 9 is satisfied; see, for example, the proof of Theorem 2.1 in [21].

In order to establish existence results for a solution of SGIVQVIP, we modify the definition of $P(x)$-quasiconvex-like multivalued bifunction to a single-valued trifunction.

Definition 14 ([11]). Let W and Z be topological vector spaces, M a nonempty convex subset of W and D a nonempty subset of $L(W, Z)$. Let $T : M \to 2^D \setminus \{\emptyset\}$ and $P : M \to 2^Z$ be multivalued maps such that for each $x \in M$, $P(x)$ is a closed, convex cone with nonempty interior. For each fixed $x \in M$, a function $\psi : D \times M \times M \to Z$ is called *$P(x)$-quasiconvex-like* if for all $y_1, y_2 \in M$ and $t \in [0,1]$, we have either for all $u \in T(x)$,

$$\psi(u, x, ty_1 + (1-t)y_2) \in \psi(u, x, y_1) - P(x),$$

or

$$\psi(u, x, ty_1 + (1-t)y_2) \in \psi(u, x, y_2) - P(x).$$

From Theorems 9 and 10, we derive the following existence result for a solution of SGIVQVIP.

Corollary 3 ([10]). *For each $i \in I$, let K_i be a nonempty convex subset of a locally convex topological vector space X_i and let D_i be a nonempty subset of $L(X_i, Y_i)$. For each $i \in I$, $T_i : K \to 2^{D_i}$ be an upper semicontinuous multivalued map with nonempty values and $\psi_i : D_i \times K_i \times K_i \to Y_i$ be a vector-valued map. For each $i \in I$, assume that:*

(i) *The multivalued map $W_i : K \to 2^{Y_i}$ defined by $W_i(x) = Y_i \setminus \{-\text{int } C_i(x)\}$ for all $x \in K$, is closed on K.*
(ii) *For all $x \in K$ and $u_i \in T_i(x)$, $\psi_i(u_i, x_i, x_i) \notin -\text{int } C_i(x)$, where x_i is the ith component of x.*
(iii) *ψ_i is $C_i(x)$-quasiconvex-like.*

(iv) *For all* $y_i \in K_i$, *the map* $(u_i, x_i) \mapsto \psi_i(u_i, x_i, y_i)$ *is upper semicontinuous on* $D_i \times K_i$.

(v) *There exist a nonempty compact subset* N *of* K *and a nonempty compact convex subset* B_i *of* K_i *for each* $i \in I$ *such that for each* $x \in K \setminus N$ *there exist* $i \in I$ *and* $\tilde{y}_i \in B_i$ *satisfying* $\tilde{y}_i \in A_i(x)$ *and* $\psi_i(u_i, x_i, \tilde{y}_i) \in -\text{int}\, C_i(x)$ *for all* $u_i \in T_i(x)$.

Then the SGIVQVIP has a solution.

Corollary 4 ([10]). *For each* $i \in I$, *let* $K_i, X_i, D_i, \psi_i, T_i$ *and* W_i *be the same as in Corollary 3 and let the multivalued map* $A = \prod_{i\in I} A_i : K \to 2^K$ *defined as* $A(x) = \prod_{i\in I} A_i(x)$ *for all* $x \in K$, *be* Φ*-condensing. Assume that the conditions (i)–(iv) of Corollary 3 hold. Then the SGIVQVIP has a solution.*

We derive the existence results for a solution of SGVQVLIP by using Corollaries 3 and 4.

Corollary 5 ([10]). *For each* $i \in I$, *let* Y_i *be a Hausdorff topological vector space and let* K_i, X_i, D_i, T_i *and* W_i *be the same as in Corollary 3. For each* $i \in I$, *let* $\eta_i : K_i \times K_i \to X_i$ *be affine in the first argument and continuous in the second argument such that* $\eta_i(x_i, x_i) = 0$ *for all* $x_i \in K_i$. *Assume that there exist a nonempty compact subset* N *of* K *and a nonempty compact convex subset* B_i *of* K_i *for each* $i \in I$ *such that for each* $x \in K \setminus N$ *there exist* $i \in I$ *and* $\tilde{y}_i \in B_i$ *satisfying* $\tilde{y}_i \in A_i(x)$ *and* $\langle u_i, \eta_i(\tilde{y}_i, x_i)\rangle \in -\text{int}\, C_i(x)$ *for all* $u_i \in T_i(x)$. *Then the SGVQVLIP has a solution.*

Corollary 6 ([10]). *For each* $i \in I$, *let* $K_i, X_i, Y_i, D_i, \eta_i, T_i$ *and* W_i *be the same as in Corollary 4. For each* $i \in I$, *let* $\eta_i : K_i \times K_i \to X_i$ *be affine in the first argument and continuous in the second argument such that* $\eta_i(x_i, x_i) = 0$ *for all* $x_i \in K_i$. *Let the multivalued map* $A = \prod_{i\in I} A_i : K \to 2^K$ *defined as* $A(x) = \prod_{i\in I} A_i(x)$ *for all* $x \in K$, *be* Φ*-condensing. Then SGVQVIP has a solution.*

The following results provide the existence of a solution of SGVQVIP with or without Φ-condensing maps.

Corollary 7 ([10]). *For each* $i \in I$, *let* K_i, X_i, Y_i, D_i, T_i *and* W_i *be the same as in Corollary 4. Assume that there exist a nonempty compact subset* N *of* K *and a nonempty compact convex subset* B_i *of* K_i *for each* $i \in I$ *such that for each* $x \in K \setminus N$ *there exist* $i \in I$ *and* $\tilde{y}_i \in B_i$ *satisfying* $\tilde{y}_i \in A_i(x)$ *and* $\langle u_i, \tilde{y}_i - x_i\rangle \in -\text{int}\, C_i(x)$ *for all* $u_i \in T_i(x)$. *Then the SGVQVIP has a solution.*

Corollary 8 ([10]). *For each* $i \in I$, *let* K_i, X_i, Y_i, D_i, T_i *and* W_i *be the same as in Corollary 4. Let the multivalued map* $A = \prod_{i\in I} A_i : K \to 2^K$ *defined as* $A(x) = \prod_{i\in I} A_i(x)$ *for all* $x \in K$, *be* Φ*-condensing. Then the SGVQVIP has a solution.*

4.2 Applications

Throughout this section, unless otherwise specified, we assume that the index set I is finite, that is, $I = \{1, \ldots, n\}$. For each $i \in I$, let X_i and Y_i be finite dimensional

Euclidean spaces $\mathbb{R}^{p_i}$ and $\mathbb{R}^{q_i}$, respectively, and K_i be a nonempty convex subset of X_i. Let $K = \prod_{i\in I} K_i$. Let $K = \prod_{i=1}^{n} K_i$. For each $i \in I$, let $C_i : K \to 2^{Y_i}$ be a multivalued map such that for all $x \in K$, $C_i(x)$ is a proper, closed and convex cone with apex at the origin and $\operatorname{int} C_i(x) \neq \emptyset$ and $\mathbb{R}^{q_i}_{+} \subseteq C_i(x)$. Let the multivalued map $A = \prod_{i\in I} A_i : K \to 2^K$ be defined as $A(x) = \prod_{i\in I} A_i(x)$, for all $x \in K$. For each $i \in I$, let $\varphi_i : K \to Y_i$ be a given vector-valued function. We recall the following SVQOP which is to find $\bar{x} \in K$ such that $\bar{x} \in A(\bar{x})$ and for each $i \in I$,

$$\varphi_i(y) - \varphi_i(\bar{x}) \notin -\operatorname{int} C_i(\bar{x}) \quad \forall\, y \in A(\bar{x}),$$

where $\varphi_i(x) = (\varphi_{i_1}(x), \varphi_{i_2}(x), \ldots, \varphi_{i_{q_i}}(x))$ and for each $l \in \mathscr{L} = \{1, \ldots, q_i\}$, $\varphi_{i_l} : K \to \mathbb{R}$ is a function.

As we have seen in Sect. 3 that every solution of SVQOP is also a solution of Debreu VEP, but the converse need not be true.

We recall the following definitions.

Definition 15. A real-valued function $f : \mathbb{R}^p \to \mathbb{R}$ is said to be *locally Lipschitz* if for any $z \in \mathbb{R}^p$, there exist a neighbourhood $N(z)$ of z and a positive constant k such that

$$|f(x) - f(y)| \leq k||x - y||, \quad \forall\, x, y \in N(z).$$

The Clarke *generalized directional derivative* [34] of a locally Lipschitz function f at x in the direction d denoted by $f^0(x;d)$ is

$$f^0(x;d) = \limsup_{\substack{y \to x \\ t \downarrow 0}} \frac{f(y+td) - f(y)}{t}.$$

The Clarke *generalized gradient* [34] of a locally Lipschitz function f at x is defined as

$$\partial f(x) = \left\{\xi \in \mathbb{R}^p : f^0(x;d) \geq \langle \xi, d\rangle \text{ for all } d \in \mathbb{R}^p\right\}.$$

If f is convex, then the Clarke generalized gradient coincides with the subdifferential of f in the sense of convex analysis [90].

The generalized invex function was introduced by Craven [36] as a generalization of invex functions [51].

Definition 16. A locally Lipschitz function $f : \mathbb{R}^p \to \mathbb{R}$ is said to be *generalized invex at x w.r.t. a given function η* $: \mathbb{R}^p \times \mathbb{R}^p \to \mathbb{R}^p$ if

$$f(y) - f(x) \geq \langle \xi, \eta(y,x)\rangle, \quad \forall\, \xi \in \partial f(x) \text{ and } y \in \mathbb{R}^p.$$

For each $i \in I$, let $\phi_i : K \to \mathbb{R}$ be a locally Lipschitz function and let $x \in K$, $x_j \in K_j$. Following Clarke [34], the *generalized directional derivative at x_j in the direction $d_j \in K_j$* of the function $\phi_i(x_1, \ldots, x_{j-1}, \cdot, x_{j+1}, \ldots, x_n)$ denoted by $\phi_{ij}^0(x;d_j)$ is

$$\phi_{ij}^{0}(x;d_j) = \limsup_{\substack{y_j \to x_j \\ t \downarrow 0}} \frac{1}{t}\Big\{\phi_i(x_1,\ldots,x_{j-1},y_j+td_j,x_{j+1},\ldots,x_n) \\ -\phi_i(x_1,\ldots,x_{j-1},y_j,x_{j+1},\ldots,x_n)\Big\}.$$

The *partial generalized gradient* [34] of the function $\phi_i(x_1,\ldots,x_{j-1},\cdot,x_{j+1},\ldots,x_n)$ at x_j is defined as follows:

$$\partial_j\phi_i(x) = \left\{\xi_j \in X_j : \phi_{ij}^{0}(x;d_j) \geq \langle \xi_j, d_j\rangle \text{ for all } d_j \in K_j\right\}.$$

Lemma 4 ([34]). *For each $i \in I$, let $\phi_i : K \to \mathbb{R}$ be locally Lipschitz. Then for each $i \in I$, the multivalued map $\partial_i\phi_i$ is upper semicontinuous.*

Definition 17. For each $i \in I$, $\phi_i : K \to \mathbb{R}$ is called *generalized invex* at x w.r.t. a given function $\eta_i : K_i \times K_i \to \mathbb{R}^{p_i}$ if

$$\phi_i(y) - \phi_i(x) \geq \langle \xi_i, \eta_i(y_i,x_i)\rangle, \quad \forall\, \xi_i \in \partial_i\phi_i(x) \text{ and } \forall\, y \in K.$$

Proposition 3 ([10]). *For each $i \in I$ and for all $l \in \mathscr{L}$, let $\varphi_{i_l} : K \to \mathbb{R}$ be generalized invex w.r.t. $\eta_{i_l} : K_i \times K_i \to X_i$. Then any solution of SGVQVLIP (II) is a solution of SVQOP with $T_i(x) = \partial_i\varphi_i(x)$ for each $i \in I$ and for all $x \in K$, where $\partial_i\varphi_i(x) = (\partial_i\varphi_{i_1}(x), \partial_i\varphi_{i_2}(x),\ldots,\partial_i\varphi_{i_{q_i}}(x)) \in \mathbb{R}^{p_i \times q_i}$.*

Proof. For the sake of simplicity, we denote by $\varphi_i(x) = (\varphi_{i_1}(x),\ldots, \varphi_{i_{q_i}}(x)) \subset \mathbb{R}^{q_i}$, $u_i = (u_{i_1},\ldots,u_{i_{q_i}})$ where $u_{i_l} \in \partial_i\varphi_{i_l}(x)$ for all $l \in \mathscr{L}$, and

$$\langle u_i, \eta_i(y_i,x_i)\rangle = \Big(\langle u_{i_1}, \eta_{i_1}(y_i,x_i)\rangle,\ldots,\langle u_{i_{q_i}}, \eta_{i_{q_i}}(y_i,x_i)\rangle\Big) \in \mathbb{R}^{q_i}.$$

Assume that $\bar{x} \in K$ is a solution of the SGVQVLIP (II). Then for each $i \in I$,

$$\forall y_i \in A_i(\bar{x}),\ \exists \bar{u}_{i_l} \in \partial_i\varphi_{i_l}(\bar{x}) \text{ for all } l \in \mathscr{L} \text{ such that}$$

$$\Big(\langle \bar{u}_{i_1}, \eta_{i_1}(y_i,\bar{x}_i)\rangle,\ldots,\langle \bar{u}_{i_{q_i}}, \eta_{i_{q_i}}(y_i,\bar{x}_i)\rangle\Big) \notin -\operatorname{int} C_i(\bar{x}).$$

We can rewrite this as

$$\forall y_i \in A_i(\bar{x}),\ \exists \bar{u}_i \in \partial_i\varphi_i(\bar{x}) \ :\ \langle \bar{u}_i, \eta_i(y_i,\bar{x}_i)\rangle \notin -\operatorname{int} C_i(\bar{x}). \tag{4}$$

Since for each $i \in I$ and for all $l \in \mathscr{L}$, φ_{i_l} is generalized invex w.r.t. η_{i_l}, we have

$$\varphi_{i_l}(y) - \varphi_{i_l}(\bar{x}) \geq \langle u_{i_l}, \eta_{i_l}(y_i,\bar{x}_i)\rangle \quad \text{for all } u_{i_l} \in \partial_i\varphi_{i_l}(\bar{x}) \text{ and } y \in A(\bar{x}) = \prod_{i\in I} A_i(\bar{x}),$$

that is, for each $i \in I$

$$\varphi_i(y) - \varphi_i(\bar{x}) \geq \langle u_i, \eta_i(y_i, \bar{x}_i)\rangle \quad \text{for all } u_i \in \partial_i \varphi_i(\bar{x}) \text{ and } y \in A(\bar{x}).$$

Therefore, for each $i \in I$ and for all $u_i \in \partial_i \varphi_i(\bar{x})$, we have

$$\begin{aligned} \varphi_i(y) - \varphi_i(\bar{x}) &\in \langle \bar{u}_i, \eta_i(y_i, \bar{x}_i)\rangle + \mathbb{R}^{q_i}_+ \\ &\subseteq \langle \bar{u}_i, \eta_i(y_i, \bar{x}_i)\rangle + \operatorname{int} C_i(\bar{x}). \end{aligned} \tag{5}$$

From (4) and (5), we have $\varphi_i(y) - \varphi_i(\bar{x}) \notin -\operatorname{int} C_i(\bar{x})$. Hence $\bar{x} \in K$ is a solution of the SVQOP.

Rest of the section, unless otherwise specified, $\partial_i \varphi_i(x)$ and $\langle u_i, \eta_i(y_i, x_i)\rangle$ are the same as defined in Proposition 3.

Theorem 11 ([10]). *For each $i \in I$ and for all $l \in \mathscr{L}$, let $\varphi_{i_l} : K \to \mathbb{R}$ be generalized invex w.r.t. $\eta_{i_l} : K_i \times K_i \to X_i$ such that η_{i_l} is affine in the first argument, continuous in the second argument and $\eta_{i_l}(x_i, x_i) = 0$ for all $x_i \in K_i$. Assume that there exists $r > 0$ such that for all $x \in K$, $||x|| > r$, there exist $i \in I$ and $\tilde{y}_i \in K_i$ with $||\tilde{y}_i||_i \leq r$ satisfying $\tilde{y}_i \in A_i(x)$ and*

$$\langle u_i, \eta_i(\tilde{y}_i, x_i)\rangle \in -\operatorname{int} C_i(x), \quad \forall\, u_i \in \partial_i \varphi_i(x),$$

where $||\cdot||$ and $||\cdot||_i$ denote the norms on X and X_i, respectively. Then the SVQOP has a solution.

Theorem 12 ([10]). *For each $i \in I$ and for all $l \in \mathscr{L}$, let $\varphi_{i_l} : K \to \mathbb{R}$ be generalized invex w.r.t. $\eta_{i_l} : K_i \times K_i \to X_i$ such that η_{i_l} is affine in the first argument, continuous in the second argument and $\eta_{i_l}(x_i, x_i) = 0$ for all $x_i \in K_i$. Let the multivalued map $A = \prod_{i \in I} A_i : K \to 2^K$ defined as $A(x) = \prod_{i \in I} A_i(x)$ for all $x \in K$, be Φ-condensing. Then the SVQOP has a solution.*

The following example, provided by one of the referees, shows that if η is affine in the second argument, then it is not necessary that $\eta(x,x) = 0$.

Example 4. Consider the map $\eta : \mathbb{R}_+ \times \mathbb{R}_+ \to \mathbb{R}$ defined by

$$\eta(x,y) = (x + y + 1, 0), \quad \text{for all } x, y \in \mathbb{R}_+ = [0, \infty).$$

Then η is affine in the second argument but $\eta(x,x) \neq 0$ for all $x \in \mathbb{R}_+$.

In the next three corollaries, we set $\varphi_i(x) = (\varphi_{i_1}(x), \ldots, \varphi_{i_{q_i}}(x)) \in \mathbb{R}^{q_i}$, $u_i = (u_{i_1}, \ldots, u_{i_{q_i}})$, $\langle u_i, y_i - x_i\rangle = (\langle u_{i_1}, y_i - x_i\rangle, \ldots, \langle u_{i_{q_i}}, y_i - x_i\rangle) \in \mathbb{R}^{q_i}$ and $\partial_i \varphi_i(x) = (\partial_i \varphi_{i_1}(x), \partial_i \varphi_{i_2}(x), \ldots, \partial_i \varphi_{i_{q_i}}(x)) \in \mathbb{R}^{p_i \times q_i}$, where $\partial_i \varphi_{i_j}(x)$ $(j = 1, \ldots, q_i)$ is the partial subdifferential in the sense of convex analysis.

Corollary 9 ([10]). *For each $i \in I$ and for all $l \in \mathscr{L}$, let $\varphi_{i_l} : K \to \mathbb{R}$ be convex and lower semicontinuous Assume that there exists $r > 0$ such that for all $x \in K$, $||x|| > r$, there exist $i \in I$ and $\tilde{y}_i \in K_i$ with $||\tilde{y}_i||_i \leq r$ satisfying $\tilde{y}_i \in A_i(x)$ and*

$$\langle u_i, \tilde{y}_i - x_i\rangle \in -\operatorname{int} C_i(x), \quad \forall\, u_i \in \partial_i \varphi_i(x),$$

where $||\cdot||$ *and* $||\cdot||_i$ *denote the norms on* X *and* X_i*, respectively. Then the SVQOP has a solution.*

Corollary 10 ([10]). *For each* $i \in I$ *and for all* $l \in \mathscr{L}$*, let* $\varphi_{i_l} : K \to \mathbb{R}$ *be convex and lower semicontinuous on* K*. Let the multivalued map* $A = \prod_{i\in I} A_i : K \to 2^K$ *defined as* $A(x) = \prod_{i\in I} A_i(x)$ *for all* $x \in K$*, be* Φ*-condensing. Then the SVQOP has a solution.*

5 System of Generalized Implicit Vector Quasi-equilibrium Problems

As we have seen in the previous sections that systems of vector quasi-equilibrium problems are used as tools to establish the existence of a solution of Debreu VEP, also known as constrained Nash equilibrium problem, both for nondifferentiable and (non)convex vector-valued functions. These are also used to solve mathematical programs with equilibrium constraints [70], fixed point theory for a family of nonexpansive multivalued maps [68] and several related topics. By using different types of maximal element theorems for a family of multivalued maps and different types of fixed point theorems for a multivalued map, several authors studied the existence of solutions of different kinds of systems of vector quasi-equilibrium problems; see, for example, [5,6,9,10,12,39,68,70,71,75,98,99] and the references therein.

For each $i \in I$, let $W_i : K \to 2^{Y_i}$ be a multivalued map defined as $W_i(x) = Y_i \setminus (-\operatorname{int} C_i(x))$ for all $x \in K$ such that its graph is closed. For each $i \in I$, let $F_i : K_i \to 2^{Y_i}$ be a multivalued map with nonempty values, $A_i : K \to 2^{K_i}$ be a multivalued map with nonempty convex values such that $A(x) = \prod_{i\in I} A_i(x)$, and $\psi_i : D_i \times K_i \times K_i \to Y_i$ be a function. We consider the following *Systems of Generalized Implicit Vector Quasi-Equilibrium Problems* (SGIVQEP) [4]:

Problem 1. Find $\bar{x} \in K$ such that $\bar{x} \in A(\bar{x})$ and for each $i \in I$,

$$\forall \bar{u}_i \in F_i(\bar{x}) \quad : \quad \psi_i(\bar{u}_i, \bar{x}_i, y_i) \notin -\operatorname{int} C_i(\bar{x}), \quad \forall y_i \in A_i(\bar{x}).$$

Problem 2. Find $\bar{x} \in K$ such that $\bar{x} \in A(\bar{x})$ and for each $i \in I$,

$$\exists \bar{u}_i \in F_i(\bar{x}) \quad : \quad \psi_i(\bar{u}_i, \bar{x}_i, y_i) \notin -\operatorname{int} C_i(\bar{x}), \quad \forall y_i \in A_i(\bar{x}).$$

Problem 3. Find $\bar{x} \in K$ such that $\bar{x} \in A(\bar{x})$ and for each $i \in I$,

$$\forall y_i \in A_i(\bar{x}),\ \exists \bar{u}_i \in F_i(\bar{x})\ (\bar{u}_i \text{ depends on } y_i) \quad : \quad \psi_i(\bar{u}_i, \bar{x}_i, y_i) \notin -\operatorname{int} C_i(\bar{x}).$$

Problem 4. Find $\bar{x} \in K$ such that $\bar{x} \in A(\bar{x})$ and for each $i \in I$,

$$\forall y \in A(\bar{x}) \text{ and } \forall v_i \in F_i(y) \quad : \quad \psi_i(v_i, y_i, \bar{x}_i) \notin \operatorname{int} C_i(\bar{x}),$$

where y_i is the ith component of y.

Problem 5. Find $\bar{x} \in K$ such that $\bar{x} \in A(\bar{x})$ and for each $i \in I$,

$$\forall y \in A(\bar{x}),\ \exists v_i \in F_i(y)\ (v_i \text{ depends on } y)\ :\ \psi_i(v_i, y_i, \bar{x}_i) \notin \operatorname{int} C_i(\bar{x}),$$

where y_i is the ith component of y.

Remark 13. Problem 1 $\Rightarrow$ Problem 2 $\Rightarrow$ Problem 3 and Problem 4 $\Rightarrow$ Problem 5.

The solutions of Problems 1, 2 and 3 are called general solution, strong solution and weak solution, respectively. In view of Remark 13, every general solution is a strong solution and every strong solution is a weak solution. But the converse assertions may not be true.

When $A_i(x) = K_i$ for all $x \in K$ and for each $i \in I$, Problems 1–5 are called *systems of generalized implicit vector equilibrium problems* (SGIVEP) considered and studied in [1]. In this case, the existence results for solutions of these problems are investigated by introducing different kinds of generalized pseudomonotonicities. In this case, Nash equilibrium problem for vector-valued functions can be solved by using Problems 1–5 but not Debreu VEP.

As we have seen in Sect. 4 that Problem 3 provides a sufficient condition (which is in general not necessary) for a solution of a SVQOP that includes Debreu VEP for nondifferentiable and nonconvex functions. But, in this case, Problem 2 provides necessary and sufficient conditions for a solution of a SVQOP.

If for each $i \in I$, $A_i(x) = K_i$ for all $x \in K$, Problem 3 is called a *system of generalized implicit vector equilibrium problems* and it is introduced and studied in [15]. It is also used to give the existence of a solution of the Nash equilibrium problem for nondifferentiable and nonconvex functions. Further, if $Y_i = \mathbb{R}$ and $C_i(x) = \mathbb{R}_-$ and $A_i(x) = K_i$ for all $x \in K$, Problem 3 was studied by in [22]. As an application of our results, we established some existence results for solutions of systems of optimization problems and the Nash equilibrium problem.

When I is a singleton set, $Y_i = \mathbb{R}$ and $C_i(x) = \mathbb{R}_+$ for all $x \in K$, the existence of a solution of Problem 2 is studied in [46].

When I is a singleton set, $A_i(x) = K_i$ for all $x \in K$ and $\psi_i(u_i, x_i, y_i) = \langle u_i, \eta_i(y_i, x_i)\rangle$ (respectively, $\psi_i(u_i, x_i, y_i) = \langle u_i, y_i - x_i\rangle$), then Problem 2 provides necessary and sufficient conditions for solutions of vector optimization problems for nondifferentiable and nonconvex functions (respectively, for nondifferentiable, but convex functions). See, for example, [2, 17] and the references therein. In this case, Problem 1 is considered and studied in [2, 30, 62].

When I is a singleton set, Problems 2 and 3 are studied by Kum and Lee [58, 64]. They proved the existence of solutions of these problems under some kind of pseudomonotonicity assumptions.

In Sect. 5.1, we give some relationships among Problems 1–5 by using different kinds of generalized pseudomonotonicities. Section 5.2 is devoted to the existence results for a solution of Problem 1 under lower semi-continuity of the family of multivalued maps involved in the formulation of the problem. The existence of a solution of Problem 1 and so Problems 2 and 3 without any coercivity condition but for Φ-condensing maps is also established. In Sect. 5.3, we establish the existence of a strong solution of our SGVQEP by using $\mathscr{H}$-hemicontinuity assumption in the setting of real Banach spaces. We also present an existence result for a weak solution under generalized pseudomonotonicity and u-hemicontinuity assumptions. Basically, besides establishing existence results for solutions of Problems 1–3 without any coercivity condition but for Φ-condensing maps, we extend the results of [1] for SGIVEP to SGIVQEP. Our results provide the existence of solutions of Problems 1–5 under some kind of pseudomonotonicity assumption and under lower semicontinuity assumption which is one of main motivations of this section.

5.1 Relationships Among Problems 1–5

Throughout this section, for each $i \in I$, we assume that X_i and Y_i are locally convex Hausdorff topological vector spaces and K_i is a nonempty convex subset of X_i, and C_i is the same as defined in the previous section. We set $K = \prod_{i\in I} K_i$, $X = \prod_{i\in I} X_i$, and $Y = \prod_{i\in I} Y_i$.

We recall different kinds of generalized pseudomonotonicities introduced in [1].

Definition 18 ([1]). Let $\{\psi_i\}_{i\in I}$ be a family of mappings $\psi_i : D_i \times K_i \times K_i \longrightarrow Y_i$. A family $\{F_i\}_{i\in I}$ of multivalued maps $F_i : K \longrightarrow 2^{K_i}$ with nonempty values is called:

(i) *Generalized strongly pseudomonotone w.r.t.* $\{\psi_i\}_{i\in I}$ if for all $x, y \in K$ and for each $i \in I$,

$$\forall u_i \in F_i(x) \;:\; \psi_i(u_i, x_i, y_i) \notin -\operatorname{int} C_i(x) \Rightarrow \forall v_i \in F_i(y) \;:\; \psi_i(v_i, y_i, x_i) \notin \operatorname{int} C_i(x).$$

(ii) *Generalized pseudomonotone w.r.t.* $\{\psi_i\}_{i\in I}$ if for all $x, y \in K$ and for each $i \in I$,

$$\exists u_i \in F_i(x) \;:\; \psi_i(u_i, x_i, y_i) \notin -\operatorname{int} C_i(x) \Rightarrow \forall v_i \in F_i(y) \;:\; \psi_i(v_i, y_i, x_i) \notin \operatorname{int} C_i(x).$$

(iii) *Generalized weakly pseudomonotone w.r.t.* $\{\psi_i\}_{i\in I}$ if for all $x, y \in K$ and for each $i \in I$,

$$\exists u_i \in F_i(x) \;:\; \psi_i(u_i, x_i, y_i) \notin -\operatorname{int} C_i(x) \Rightarrow \exists v_i \in F_i(y) \;:\; \psi_i(v_i, y_i, x_i) \notin \operatorname{int} C_i(x).$$

(iv) *Generalized pseudomonotone*$^+$ *w.r.t.* $\{\psi_i\}_{i\in I}$ if for all $x, y \in K$ and for each $i \in I$,

$$\forall u_i \in F_i(x) \;:\; \psi_i(u_i, x_i, y_i) \notin -\operatorname{int} C_i(x) \Rightarrow \exists v_i \in F_i(y) \;:\; \psi_i(v_i, y_i, x_i) \notin \operatorname{int} C_i(x).$$

(v) *u-Hemicontinuous w.r.t.* $\{\psi_i\}_{i\in I}$ if for all $x, y \in K$ and $\alpha \in [0,1]$ and for each $i \in I$, the multivalued map

$$\alpha \mapsto \psi_i(F_i(x + \alpha(y - x)), x_i, y_i)$$

is upper semicontinuous at 0^+, where

$$\psi_i(F_i(x+\alpha(y-x)),x_i,y_i) = \{\psi_i(w_i,x_i,y_i) : w_i \in F_i(x+\alpha(y-x))\}.$$

Remark 14. Definition (i) ⇒ Definition (ii) ⇒ Definition (iii); Definition (iv) ⇒ Definition (iii); Definition (i) ⇒ Definition (iv); that is, Definition (i) ⇒ Definition (iv) ⇒ Definition (iii).

In the next three lemmas, we discuss the relationships among Problems 1–5.

Lemma 5 ([4]).

(a) Problem 3 ⇒ Problem 4 if $\{F_i\}_{i\in I}$ is generalized pseudomonotone w.r.t. $\{\psi_i\}_{i\in I}$.
(b) Problem 3 ⇒ Problem 5 if $\{F_i\}_{i\in I}$ is generalized weakly pseudomonotone w.r.t. $\{\psi_i\}_{i\in I}$.
(c) Problem 1 ⇒ Problem 5 if $\{F_i\}_{i\in I}$ is generalized pseudomonotone$^+$ w.r.t. $\{\psi_i\}_{i\in I}$.
(d) Problem 1 ⇒ Problem 4 if $\{F_i\}_{i\in I}$ is generalized strongly pseudomonotone w.r.t. $\{\psi_i\}_{i\in I}$.
(e) Problem 2 ⇒ Problem 4 if $\{F_i\}_{i\in I}$ is generalized pseudomonotone w.r.t. $\{\psi_i\}_{i\in I}$.

Lemma 6 ([4]). *For each $i \in I$, assume that the following conditions hold:*

(i) For all $x \in K$ and all $u_i \in F_i(x)$, $\psi_i(u_i,x_i,x_i) \in \mathscr{C}_i = \bigcap_{x\in K} C_i(x)$.
(ii) For all $x \in K$ and all $u_i \in F_i(x)$, $\psi_i(u_i,x_i,\cdot)$ is $\mathscr{C}_i$-convex, that is, for all $s_i \in L(X_i,Y_i)$, $x,y \in K$ and $\alpha \in [0,1]$,

$$\psi_i(s_i,x_i,\alpha x_i+(1-\alpha)y_i) \in \alpha\psi_i(s_i,x_i,x_i)+(1-\alpha)\psi_i(s_i,x_i,y_i)-\mathscr{C}_i.$$

(iii) For all $s_i \in L(X_i,Y_i)$, $x,y,z \in K$ and $\alpha \in [0,1]$,

$$\psi_i(s_i,x_i+\alpha(y_i-x_i),z_i) = (1-\alpha)\psi_i(s_i,x_i,z_i).$$

(iv) $\{F_i\}_{i\in I}$ is u-hemicontinuous w.r.t. $\{\psi_i\}_{i\in I}$.

Then Problem 5 ⇒ Problem 3 as well as Problem 4 ⇒ Problem 3.

Proposition 4 ([4]). *Under the conditions of Lemmas 5(a) and 6, Problems 3, 4 and 5 are equivalent.*

Lemma 7. *For each $i \in I$, let $(X_i,\|\cdot\|)$ and Y_i be real Banach spaces and K_i be a nonempty convex subset of X_i. Let $K = \prod_{i\in I} K_i$. For each $i \in I$, assume that the following conditions hold:*

(i) For all $x \in K$ and all $u_i \in F_i(x)$, $\psi_i(u_i,x_i,x_i) \in \mathscr{C}_i = \bigcap_{x\in K} C_i(x)$.
(ii) For all $x \in K$ and all $u_i \in F_i(x)$, $\psi_i(u_i,x_i,\cdot)$ is $\mathscr{C}_i$-convex, that is, for all $s_i \in L(X_i,Y_i)$, $x,y \in K$ and $t \in [0,1]$,

$$\psi_i(s_i,x_i,tx_i+(1-t)y_i) \in t\psi_i(s_i,x_i,x_i)+(1-t)\psi_i(s_i,x_i,y_i)-\mathscr{C}_i.$$

(iii) For all $s_i \in L(X_i,Y_i)$, $x,y,z \in K$ and $t \in [0,1]$,

$$\psi_i(s_i,x_i+t(y_i-x_i),z_i) = (1-t)\psi_i(s_i,x_i,z_i).$$

(iv) ψ_i *is continuous in the first argument.*
(v) F_i *is* $\mathscr{H}$*-hemicontinuous and for all* $x \in K$, $F_i(x)$ *is a nonempty compact set in* Y_i.
(vi) The family $\{F_i\}_{i\in I}$ *is generalized pseudomonotone w.r.t.* $\{\psi_i\}_{i\in I}$.

Then Problems 2 and 4 are equivalent.

5.2 Existence Results Under Lower Semicontinuity

For each $i \in I$, we assume that the graph of the multivalued map $W_i : K \to 2^{Y_i}$ defined by $W_i(x) = Y_i \setminus \{-\text{int}\, C_i(x)\}$ for all $x \in K$, is closed. For each $i \in I$, we also assume that $A_i : K \to 2^{K_i}$ is a multivalued map such that for all $x \in K$, $A_i(x)$ is nonempty and convex, $A_i^{-1}(y_i)$ is open in K for all $y_i \in K_i$ and the set $\mathscr{F}_i := \{x \in K : x_i \in A_i(x)\}$ is closed in K, where x_i is the ith component of x.

We extend and generalize Definition 14 for a family of trifunctions.

Definition 19 ([1]). For each $i \in I$, let $F_i : K \to 2^{D_i}$ be a multivalued map with nonempty values. A family $\{\psi_i\}_{i\in I}$ of functions $\psi_i : D_i \times K_i \times K_i \to Y_i$ is called *$C_i(x)$-quasiconvex-like w.r.t.* $\{F_i\}_{i\in I}$ if for all $x \in K$, $y_i', y_i'' \in K_i$ and $t \in [0,1]$, we either have $\forall u_i \in F_i(x)$,

$$\psi_i(u_i, x_i, ty_i' + (1-t)y_i'') \in \psi_i(u_i, x_i, y_i') - \text{int}\, C_i(x),$$

or

$$\psi_i(u_i, x_i, ty_i' + (1-t)y_i'') \in \psi_i(u_i, x_i, y_i'') - \text{int}\, C_i(x).$$

Definition 20 ([1]). For each $i \in I$, let $F_i : K \to 2^{D_i}$ be multivalued map with nonempty values. A family $\{\psi_i\}_{i\in I}$ of functions $\psi_i : D_i \times K_i \times K_i \to Y_i$ is called *simultaneously $C_i(x)$-quasiconvex-like w.r.t.* $\{F_i\}_{i\in I}$ if for all $x \in K$, $y_i', y_i'' \in K_i$ and $t \in [0,1]$, we either have $\forall u_i', u_i'' \in F_i(x)$,

$$\psi_i(tu_i' + (1-t)u_i'', x_i, ty_i' + (t)y_i'') \in \psi_i(u_i', x_i, y_i') - \text{int}\, C_i(x),$$

or

$$\psi_i(tu_i' + (1-t)u_i'', x_i, ty_i' + (1-t)y_i'') \in \psi_i(u_i'', x_i, y_i'') - \text{int}\, C_i(x).$$

We present an existence result for a solution of Problem 1 under lower semicontinuity of the family of multivalued maps involved in the formulation of the problem.

Theorem 13 ([4]). *For each* $i \in I$, *let* K_i *be a nonempty convex subset of a Hausdorff topological vector space* X_i. *Let* $K = \prod_{i\in I} K_i$. *For each* $i \in I$, *let* $F_i : K \to 2^{K_i}$ *be a lower semicontinuous multivalued map with nonempty convex values and* $\psi_i : D_i \times K_i \times K_i \to Y_i$ *be a function such that the following conditions are satisfied:*

(i) For all $x \in K$, *the family* $\{\psi_i\}_{i\in I}$ *of functions* ψ_i *is simultaneously* $C_i(x)$*-quasiconvex-like w.r.t.* $\{F_i\}_{i\in I}$.

(ii) *For all* $x \in K$ *and for all* $u_i \in F_i(x)$, $\psi_i(u_i, x_i, x_i) \notin -\text{int}\, C_i(x)$.
(iii) *For each fixed* y_i, *the map* $(u_i, x_i) \mapsto \psi_i(u_i, x_i, y_i)$ *is continuous on* $D_i \times K_i$.
(iv) *There exist a nonempty compact subset* M *of* K *and a nonempty compact convex subset* N_i *of* K_i *for each* $i \in I$ *such that for all* $x \in K \setminus M$, *there exist* $i \in I$ *and* $\tilde{y}_i \in N_i$ *such that* $\tilde{y}_i \in A_i(x)$ *and* $\psi_i(u_i, x_i, \tilde{y}_i) \in -\text{int}\, C_i(x)$ *for all* $u_i \in F_i(x)$.

Then Problem 1 has a solution.

The following result provides the existence of a solution of Problem 1 without any coercivity condition but for Φ-condensing maps.

Theorem 14 ([4]). *For each* $i \in I$, *let* K_i *be a nonempty, closed and convex subset of a locally convex Hausdorff topological vector space* X_i *and let the multivalued map* $A = \prod_{i \in I} A_i : K \to 2^K$ *defined as* $A(x) = \prod_{i \in I} A_i(x)$ *for all* $x \in K$, *be* Φ*-condensing. Assume that the conditions (i)–(iii) of Theorem 13 hold. Then Problem 1 has a solution.*

5.3 Existence Results Under Pseudomonotonicity

In this section, we present some existence results for a solution of the SGIVQEP under generalized pseudomonotonicity assumption.

Theorem 15 ([4]). *For each* $i \in I$, *let* K_i *be a nonempty convex subset of a Hausdorff topological vector space* X_i. *Let* $K = \prod_{i \in I} K_i$. *For each* $i \in I$, *let* $F_i : K \to 2^{K_i}$ *be a multivalued map with nonempty values and* $\psi_i : D_i \times K_i \times K_i \to Y_i$ *be a function such that the following conditions are satisfied:*

(i) *The family* $\{F_i\}_{i \in I}$ *of multivalued maps* F_i *is generalized pseudomonotone w.r.t.* $\{\psi_i\}_{i \in I}$.
(ii) *For all* $x \in K$, *the family* $\{\psi_i\}_{i \in I}$ *of functions* ψ_i *is* $C_i(x)$*-quasiconvex-like w.r.t.* $\{F_i\}_{i \in I}$.
(iii) *For all* $x \in K$ *and for all* $u_i \in F_i(x)$, $\psi_i(u_i, x_i, x_i) \notin -\text{int}\, C_i(x)$.
(iv) *For each fixed* $(v_i, y_i) \in D_i \times K_i$, *the map* $x_i \mapsto \psi_i(v_i, y_i, x_i)$ *is continuous on* K_i.
(v) *There exist a nonempty compact subset* M *of* K *and a nonempty compact convex subset* N_i *of* K_i *for each* $i \in I$ *such that for all* $x \in K \setminus M$, *there exist* $i \in I$ *and* $\tilde{y}_i \in N_i$ *such that* $\tilde{y}_i \in A_i(x)$ *and* $\psi_i(u_i, x_i, \tilde{y}_i) \in -\text{int}\, C_i(x)$ *for all* $u_i \in F_i(x)$.

Then Problem 4 has a solution.

Theorem 16 ([4]). *For each* $i \in I$, *let* K_i *be a nonempty convex subset of a Hausdorff topological vector space* X_i. *For each* $i \in I$, *let* $F_i : K \to 2^{K_i}$ *be a multivalued map with nonempty values and* $\psi_i : D_i \times K_i \times K_i \to Y_i$ *be a function such that the following conditions are satisfied:*

(i) *The family* $\{F_i\}_{i \in I}$ *of multivalued maps* F_i *is u-hemicontinuous and generalized pseudomonotone w.r.t.* $\{\psi_i\}_{i \in I}$.

(ii) *The family* $\{\psi_i\}_{i\in I}$ *of functions* ψ_i *is* $\mathscr{C}_i$*-convex in the third argument.*
(iii) *For all* $s_i \in L(X_i, Y_i)$, $x, y, z \in X$ *and* $\alpha \in [0,1]$,

$$\psi_i(s_i, x_i + \alpha(y_i - x_i), z_i) = (1-\alpha)\psi_i(s_i, x_i, z_i).$$

(iv) *For all* $x \in K$ *and for all* $u_i \in F_i(x)$, $\psi_i(u_i, x_i, x_i) \in \mathscr{C}_i$.
(v) *For each fixed* $(v_i, y_i) \in D_i \times K_i$, *the map* $x_i \mapsto \psi_i(v_i, y_i, x_i)$ *is continuous on* K_i.
(vi) *There exist a nonempty compact subset* M *of* K *and a nonempty compact convex subset* N_i *of* K_i *for each* $i \in I$ *such that for all* $x \in K \setminus M$, *there exist* $i \in I$ *and* $\tilde{y}_i \in N_i$ *such that* $\tilde{y}_i \in A_i(x)$ *and* $\psi_i(u_i, x_i, \tilde{y}_i) \in -\text{int}\, C_i(x)$ *for all* $u_i \in F_i(x)$.

Then Problem 3 has a solution.

The following result provides the existence of a strong solution of Problem 2.

Theorem 17 ([4]). *For each* $i \in I$, *let* K_i *be a nonempty convex subset of a real Banach space* X_i *and* Y_i *be a real Banach space. For each* $i \in I$, *let* $F_i : K \to 2^{K_i}$ *be a multivalued map with nonempty compact values and* $\psi_i : D_i \times K_i \times K_i \to Y_i$ *be a function such that the following conditions are satisfied:*

(i) *The family* $\{F_i\}_{i\in I}$ *of multivalued maps* F_i *is* $\mathscr{H}$*-hemicontinuous and generalized pseudomonotone w.r.t.* $\{\psi_i\}_{i\in I}$.
(ii) *The family* $\{\psi_i\}_{i\in I}$ *of functions* ψ_i *is* $\mathscr{C}_i$*-convex in the third argument.*
(iii) *For all* $s_i \in L(X_i, Y_i)$, $x, y, z \in X$ *and* $\alpha \in [0,1]$,

$$\psi_i(s_i, x_i + \alpha(y_i - x_i), z_i) = (1-\alpha)\psi_i(s_i, x_i, z_i).$$

(iv) *For all* $x \in K$ *and for all* $u_i \in F_i(x)$, $\psi_i(u_i, x_i, x_i) \in \mathscr{C}_i$.
(v) *For each fixed* $(v_i, y_i) \in D_i \times K_i$, *the map* $x_i \mapsto \psi_i(v_i, y_i, x_i)$ *is continuous on* K_i.
(vi) *There exist a nonempty compact subset* M *of* K *and a nonempty compact convex subset* N_i *of* K_i *for each* $i \in I$ *such that for all* $x \in K \setminus M$, *there exist* $i \in I$ *and* $\tilde{y}_i \in N_i$ *such that* $\tilde{y}_i \in A_i(x)$ *and* $\psi_i(u_i, x_i, \tilde{y}_i) \in -\text{int}\, C_i(x)$ *for all* $u_i \in F_i(x)$.

Then Problem 2 has a solution.

6 System of Simultaneously Generalized Vector Quasi-equilibrium Problems

Throughout the section, unless otherwise specified, I is any index set (finite or infinite). For each $i \in I$, let X_i and Y_i be two nonempty convex subsets of locally convex topological vector spaces E_i and F_i, respectively, and Z_i be a real topological vector space. Let $X = \prod_{i\in I} X_i$ and $Y = \prod_{i\in I} Y_i$. For each $i \in I$, let $C_i : X \to 2^{Z_i}$ be a multivalued map such that for all $x \in X$, $C_i(x)$ is a closed convex cone with apex at the origin $\mathbf{0}$. For each $i \in I$, let $P_i = \bigcap_{x\in X} C_i(x)$ such that P_i defines a vector ordering on Z_i. For each $i \in I$, let $S_i : X \to 2^{X_i}$ and $T_i : X \to 2^{Y_i}$ be multivalued maps with nonempty values, and $f_i : X \times Y \times X_i \to Z_i$ and $g_i : X \times Y \times Y_i \to Z_i$ be trifunctions.

We consider the following problems of *system of simultaneous generalized vector quasi-equilibrium problems* (SSGVQEP) [12]:
SSGVQEP(I): Find $(\bar{x},\bar{y}) \in X \times Y$ such that for each $i \in I$, $\bar{x}_i \in S_i(\bar{x})$, $\bar{y}_i \in T_i(\bar{x})$,

$$f_i(\bar{x},\bar{y},x_i) \in C_i(\bar{x}), \quad \forall\, x_i \in S_i(\bar{x})$$

and

$$g_i(\bar{x},\bar{y},y_i) \in C_i(\bar{x}), \quad \forall\, y_i \in T_i(\bar{x}).$$

SSGVQEP(II): Find $(\bar{x},\bar{y}) \in X \times Y$ such that for each $i \in I$, $\bar{x}_i \in S_i(\bar{x})$, $\bar{y}_i \in T_i(\bar{x})$,

$$f_i(\bar{x},\bar{y},x_i) \notin -C_i(\bar{x}) \setminus \{\mathbf{0}\}, \quad \forall\, x_i \in S_i(\bar{x})$$

and

$$g_i(\bar{x},\bar{y},y_i) \notin -C_i(\bar{x}) \setminus \{\mathbf{0}\}, \quad \forall\, y_i \in T_i(\bar{x}).$$

SSGVQEP(III): Find $(\bar{x},\bar{y}) \in X \times Y$ such that for each $i \in I$, $\bar{x}_i \in S_i(\bar{x})$, $\bar{y}_i \in T_i(\bar{x})$,

$$f_i(\bar{x},\bar{y},x_i) \notin -\operatorname{int} C_i(\bar{x}), \quad \forall\, x_i \in S_i(\bar{x})$$

and

$$g_i(\bar{x},\bar{y},y_i) \notin -\operatorname{int} C_i(\bar{x}), \quad \forall\, y_i \in T_i(\bar{x}),$$

in this case we assume that $\operatorname{int} C_i$ is nonempty for each $i \in I$.

Remark 15. For each $i \in I$ and $\forall\, x \in X$, let $C_i(x)$ be a pointed cone and $P_i = \bigcap_{x\in X} C_i(x)$, then P_i is also pointed. Indeed,

$$\begin{aligned} P_i \cap (-P_i) &= \left(\cap_{x\in X} C_i(x)\right) \bigcap \left(\cap_{x\in X} (-C_i(x))\right) \\ &= \bigcap_{x\in X} \left(C_i(x) \cap (-C_i(x))\right) = \{\mathbf{0}\}. \end{aligned}$$

Therefore, for each $i \in I$, P_i is pointed.

Remark 16. If for each $i \in I$ and $\forall\, x \in X$, $C_i(x)$ is also pointed, then every solution of SSGVQEP(I) is a solution of SSGVQEP(II) and every solution of SSGVQEP(II) is a solution of SSGVQEP(III). But the reverse implication does not hold.

Indeed, let $(\bar{x},\bar{y}) \in X \times Y$ be a solution of SSGVQEP(I), then for each $i \in I$, $\bar{x}_i \in S_i(\bar{x})$, $\bar{y}_i \in T_i(\bar{x})$,

$$f_i(\bar{x},\bar{y},x_i) \in C_i(\bar{x}), \quad \forall\, x_i \in S_i(\bar{x})$$

and

$$g_i(\bar{x},\bar{y},y_i) \in C_i(\bar{x}), \quad \forall\, y_i \in T_i(\bar{x}).$$

Since for each $i \in I$ and $\forall\, x \in X$, $C_i(x)$ is a pointed cone, we have $C_i(x) \cap (-C_i(x)) = \{\mathbf{0}\}$ and therefore

$$C_i(x) \cap \Big(-C_i(x) \setminus \{\mathbf{0}\}\Big) = \emptyset.$$

Hence

$$f_i(\bar{x},\bar{y},x_i) \notin -C_i(\bar{x}) \setminus \{\mathbf{0}\}, \quad \forall\, x_i \in S_i(\bar{x})$$

and

$$g_i(\bar{x},\bar{y},y_i) \notin -C_i(\bar{x}) \setminus \{\mathbf{0}\}, \quad \forall\, y_i \in T_i(\bar{x}).$$

The second statement follows from the fact that $-\operatorname{int} C_i(x) \subseteq -C_i(x) \setminus \{\mathbf{0}\}, \forall\, x \in X$ and for each $i \in I$.

For each $i \in I$, we denote by $L(E_i, Z_i)$ the space of all continuous linear operators from E_i into Z_i and let Y_i be a nonempty subset of $L(E_i, Z_i)$. For each $i \in I$, let $g_i \equiv 0$, then SSGVQEP(I) reduces to the following *problem of system of generalized implicit vector quasi-variational inequalities*:
SGIVQVIP(I): Find $(\bar{x},\bar{y}) \in X \times Y$ such that for each $i \in I$, $\bar{x}_i \in S_i(\bar{x})$, $\bar{y}_i \in T_i(\bar{x})$ satisfying

$$f_i(\bar{x},\bar{y},x_i) \in C_i(\bar{x}), \quad \forall\, x_i \in S_i(\bar{x}).$$

Analogous, we can define SGIVQVIP(II) and SGIVQVIP(III) (problems of system of generalized implicit vector quasi-variational inequalities) corresponding to SSGVQEP(II) and SSGVQEP(III), respectively.

The SGIVQVIP contains the problem of system of generalized vector quasi-variational-like inequalities (SGVQVLIP) as a special case. Recently, the weak formulation of SGVQVLIP is studied in [10]. We used SGVQVLIP as a tool to prove the existence of a solution of Debreu type equilibrium problem for nondifferentiable and nonconvex vector-valued functions.

When for each $i \in I$, $X_i = Y_i$, $S_i \equiv T_i$ and $f_i \equiv g_i$, then SSGVQEP is called a *system of vector quasi-equilibrium problems*. In this case, SSGVQEP(III) is considered and studied in [5] for $f_i(x,y,y_i) = h_i(x,y_i)$ with further applications to systems of generalized vector quasi-variational-like inequalities and Debreu type equilibrium problems for vector-valued functions.

When I is a singleton set and $g_i \equiv 0$, then SSGVQEP(I) is considered and studied in [46].

When I is a singleton set, $X = Y$, $S_i \equiv T_i$, $S_i(x) = X$, $f_i(x,y,x_i) = \varphi(x,y)$, $g_i(x,y,y_i) = \phi(x,y)$, then SSGVQEP(III) reduces to the problem of *simultaneous vector variational inequalities* which is considered and studied by Fu [45] for a fixed cone C. If $C = \mathbb{R}_+$, then the problem of simultaneous vector variational inequalities becomes the problem of simultaneous variational inequalities, which is introduced and studied by Husain and Tarafdar [52] with applications to optimization problems.

By making suitable choices of f_i and g_i, we can derive several systems of quasi-variational inequalities and systems of (quasi-) equilibrium problems studied in the literature; see, for example, [5, 9, 10, 14–16] and the references therein.

6.1 Existence Results for Solutions of SSGVQEP

In this section, we present an existence result for a solution of SSGVQEP and derive existence results for solutions of SGIVQVIP(I), simultaneous generalized vector quasi-equilibrium problem and a system of generalized vector quasi-variational-like inequalities.

Theorem 18 ([12]). *For each $i \in I$, let E_i, F_i and Z_i be real locally convex topological vector spaces and F_i be also quasi-complete. For each $i \in I$, let $X_i \subseteq E_i$ be a nonempty compact convex set and $Y_i \subseteq F_i$ a nonempty convex set. Let $X = \prod_{i \in I} X_i$ and $Y = \prod_{i \in I} Y_i$. For each $i \in I$, let $S_i : X \longrightarrow 2^{X_i}$ be a continuous multivalued map with nonempty closed convex values and $T_i : X \longrightarrow 2^{Y_i}$ a continuous multivalued map with nonempty compact convex values. For each $i \in I$, assume that the following conditions are satisfied:*

(i) $C_i : X \longrightarrow 2^{Z_i}$ *is a closed multivalued map such that* $\forall\, x \in X$, $C_i(x)$ *is a closed convex cone with apex at the origin, and* $P_i = \bigcap_{x \in X} C_i(x)$.
(ii) P_i^* *has a weak* compact convex base* B_i^* *and* Z_i *is ordered by* P_i.
(iii) $f_i : X \times Y \times X_i \longrightarrow Z_i$ *is a continuous function such that:*

(a) $\forall\, x \in X$ *and* $y \in Y$, $f_i(x, y, x_i) \geq_{P_i} \mathbf{0}$.
(b) $\forall\, (x, y) \in X \times Y$, *the map* $u_i \mapsto f_i(x, y, u_i)$ *is properly quasi-convex.*

(iv) $g_i : X \times Y \times Y_i \longrightarrow Z_i$ *is a continuous function such that:*

(a) $\forall\, x \in X$ *and* $y \in Y$, $g_i(x, y, y_i) \geq_{P_i} \mathbf{0}$.
(b) $\forall\, (x, y) \in X \times Y$, *the map* $v_i \mapsto g_i(x, y, v_i)$ *is properly quasi-convex.*

Then there exists a solution $(\bar{x}, \bar{y}) \in X \times Y$ *of SSGVQEP(I).*

If for each $i \in I$, $g_i \equiv 0$, then we have the following result.

Corollary 11. *For each $i \in I$, let E_i, F_i and Z_i be real locally convex topological vector spaces and F_i be also quasi-complete. For each $i \in I$, let $X_i \subseteq E_i$ be a nonempty compact convex set and $Y_i \subseteq F_i$ a nonempty convex set. Let $X = \prod_{i \in I} X_i$ and $Y = \prod_{i \in I} Y_i$. For each $i \in I$, let $S_i : X \longrightarrow 2^{X_i}$ be a continuous multivalued map with nonempty closed convex values and $T_i : X \longrightarrow 2^{Y_i}$ a continuous multivalued map with nonempty compact convex values. For each $i \in I$, assume that the following conditions are satisfied:*

(i) $C_i : X \longrightarrow 2^{Z_i}$ *is a closed multivalued map such that* $\forall\, x \in X$, $C_i(x)$ *is a closed convex cone with apex at the origin, and* $P_i = \bigcap_{x \in X} C_i(x)$.
(ii) P_i^* *has a weak* compact convex base* B_i^* *and* Z_i *is ordered by* P_i.
(iii) $f_i : X \times Y \times X_i \longrightarrow Z_i$ *is a continuous function such that:*

(a) $\forall\, x \in X$ *and* $y \in Y$, $f_i(x, y, x_i) \geq_{P_i} \mathbf{0}$.
(b) $\forall\, (x, y) \in X \times Y$, *the map* $u_i \mapsto f_i(x, y, u_i)$ *is properly quasi-convex.*

Then there exists a solution $(\bar{x}, \bar{y}) \in X \times Y$ *of SGIVQVIP(I).*

Remark 17. Corollary 11 is an extension of Theorem 1 in [46] to the system of quasi-equilibrium problems with a moving cone.

When I is a singleton set, then we have the following result.

Corollary 12. *Let E, F and Z be real locally convex topological vector spaces and F be also quasi-complete. Let $X \subseteq E$ be a nonempty compact convex set and $Y \subseteq F$ a nonempty convex set. Let $S : X \to 2^X$ be a continuous multivalued map with nonempty closed convex values and $T : X \to 2^Y$ a continuous multivalued map with nonempty compact convex values. Assume that the following conditions are satisfied:*

(i) $C : X \to 2^Z$ *is a closed multivalued map such that* $\forall\, x \in X$, $C(x)$ *is a closed convex cone with apex at the origin, and* $P = \bigcap_{x\in X} C(x)$.
(ii) P^* *has a weak* compact convex base* B^* *and* Z *is ordered by* P.
(iii) $f : X \times Y \times X \to Z$ *is a continuous function such that:*

(a) $\forall\, x \in X$ *and* $y \in Y$, $f(x,y,x) \geq_P \mathbf{0}$.
(b) $\forall\, (x,y) \in X \times Y$, *the map* $u \mapsto f(x,y,u)$ *is properly quasi-convex.*

(iv) $g : X \times Y \times Y \to Z$ *is a continuous function such that:*

(a) $\forall\, x \in X$ *and* $y \in Y$, $g(x,y,y) \geq_P \mathbf{0}$.
(b) $\forall\, (x,y) \in X \times Y$, *the map* $v \mapsto g(x,y,v)$ *is properly quasi-convex.*

Then there exists a solution $(\bar{x},\bar{y}) \in X \times Y$ *of the* simultaneous generalized vector quasi-equilibrium problem *(SGVQEP): find* $(\bar{x},\bar{y}) \in X \times Y$ *such that* $\bar{x} \in S(\bar{x})$, $\bar{y} \in T(\bar{x})$,

$$f(\bar{x},\bar{y},x) \in C(\bar{x}), \quad \forall\, x \in S(\bar{x})$$

and

$$g(\bar{x},\bar{y},y) \in C(\bar{x}), \quad \forall\, y \in T(\bar{x}).$$

In addition to the assumptions on $C_i : K \to 2^{Z_i}$, in the following corollary, we further assume that $C_i(x)$ is pointed, for each $i \in I$ and for all $x \in K$. Then the following result can be easily derived from Corollary 3.1 by setting

$$f_i(x,y,u_i) = \langle \theta_i(x,y), \eta_i(u_i,x_i) \rangle.$$

Corollary 13. *For each $i \in I$, let E_i and Z_i be real locally convex topological vector spaces and let $L(E_i,Z_i)$ be quasi-complete. For each $i \in I$, let $X_i \subseteq E_i$ be a nonempty compact convex set and $Y_i \subseteq L(E_i,Z_i)$ a nonempty convex set. Let $X = \prod_{i\in I} X_i$ and $Y = \prod_{i\in I} Y_i$. For each $i \in I$, let $S_i : X \to 2^{X_i}$ be a continuous multivalued map with nonempty closed convex values and $T_i : X \to 2^{Y_i}$ a continuous multivalued map with nonempty compact convex values. For each $i \in I$, assume that the following conditions are satisfied:*

(i) $C_i : X \to 2^{Z_i}$ *is a closed multivalued map such that* $\forall\, x \in X$, $C_i(x)$ *is a nonempty closed convex pointed cone, and* $P_i = \bigcap_{x\in X} C_i(x)$.
(ii) P_i^* *has a weak* compact convex base* B_i *and* Z_i *is ordered by* P_i.

(iii) $\theta_i : X \times Y \to Y_i$ *and* $\eta_i : X_i \times X_i \to X_i$ *are continuous bifunctions such that for each* $i \in I$*:*

(a) $\forall\, x_i \in X_i,\ \eta_i(x_i,x_i))\rangle \geq_{P_i} 0.$
(b) $\forall\, (x,y) \in X \times Y$*, the map* $u_i \mapsto \langle \theta_i(x,y), \eta_i(u_i,x_i)\rangle$ *is properly quasi-convex.*

Then there exists a solution $(\bar{x},\bar{y}) \in X \times Y$ *of the problem of system of generalized vector quasi-variational-like inequalities (SGVQVLIP)(I): find* $(\bar{x},\bar{y}) = ((\bar{x}_i)_{i\in I}, (\bar{y}_i)_{i\in I}) \in X \times Y$ *such that for each* $i \in I$*,* $\bar{x}_i \in S_i(\bar{x})$*,* $\bar{y}_i \in T_i(\bar{x})$ *and*

$$\langle \theta_i(\bar{x},\bar{y}), \eta_i(x_i,\bar{x}_i)\rangle \in C_i(\bar{x}), \quad \forall\, x_i \in S_i(\bar{x}).$$

Remark 18. It is worth to mention that the weak formulation of SGVQVLIP(III) is considered and studied in [10]. Corollary 13 provides the existence of a solution of a more general problem than SGVQVLIP(III).

6.2 Systems of Vector Quasi-saddle Point Problems

In this section, we define systems of quasi-saddle point problems and systems of quasi-minimax inequalities. As application of the results of previous section, we derive existence results for solutions of these problems.

Let X, Y, X_i, Y_i, Z_i and C_i be the same as defined in the formulations of SSGVQEP. Let $\ell_i : X_i \times Y_i \to Z_i$ be a bifunction. We consider the following *systems of quasi-saddle point problems.*

SVQSPP(I): Find $\bar{x} = (\bar{x}_i)_{i\in I} \in X$ and $\bar{y} = (\bar{y}_i)_{i\in I} \in Y$ such that for each $i \in I$, $\bar{x}_i \in S_i(\bar{x})$, $\bar{y}_i \in T_i(\bar{x})$,

$$\ell_i(x_i,\bar{y}_i) - \ell_i(\bar{x}_i,\bar{y}_i) \in C_i(\bar{x}), \quad \forall\, x_i \in S_i(\bar{x})$$

and

$$\ell_i(\bar{x}_i,\bar{y}_i) - \ell_i(\bar{x}_i,y_i) \in C_i(\bar{x}), \quad \forall\, y_i \in T_i(\bar{x}).$$

SVQSPP(II): Find $\bar{x} = (\bar{x}_i)_{i\in I} \in X$ and $\bar{y} = (\bar{y}_i)_{i\in I} \in Y$ such that for each $i \in I$, $\bar{x}_i \in S_i(\bar{x})$, $\bar{y}_i \in T_i(\bar{x})$,

$$\ell_i(x_i,\bar{y}_i) - \ell_i(\bar{x}_i,\bar{y}_i) \notin -C_i(\bar{x}) \setminus \{\mathbf{0}\}, \quad \forall\, x_i \in S_i(\bar{x})$$

and

$$\ell_i(\bar{x}_i,\bar{y}_i) - \ell_i(\bar{x}_i,y_i) \notin -C_i(\bar{x}) \setminus \{\mathbf{0}\}, \quad \forall\, y_i \in T_i(\bar{x}).$$

SVQSPP(III): Find $\bar{x} = (\bar{x}_i)_{i\in I} \in X$ and $\bar{y} = (\bar{y}_i)_{i\in I} \in Y$ such that for each $i \in I$, $\bar{x}_i \in S_i(\bar{x})$, $\bar{y}_i \in T_i(\bar{x})$,

$$\ell_i(x_i, \bar{y}_i) - \ell_i(\bar{x}_i, \bar{y}_i) \notin -\text{int}\, C_i(\bar{x}), \quad \forall\, x_i \in S_i(\bar{x})$$

and

$$\ell_i(\bar{x}_i, \bar{y}_i) - \ell_i(\bar{x}_i, y_i) \notin -\text{int}\, C_i(\bar{x}), \quad \forall\, y_i \in T_i(\bar{x}).$$

Remark 19. If for each $i \in I$ and $\forall\, x \in X$, $C_i(x)$ is a convex pointed cone, then every solution of SVQSPP(I) is a solution of SVQSPP(II) and every solution of SVQSPP(II) is a solution of SVQSPP(III). But the converse implication is not true.

If I is a singleton set and $Z = \mathbb{R}$ then SVQSPP(I), SVQSPP(II) and SVQSPP(III) are called a *quasi-saddle point problem* (QSPP). Of course, if I is a singleton set, $S_i(x) = X_i$ and $T_i(x) = Y_i$, $\forall\, x \in X$ and $Z_i = \mathbb{R}$, then above-mentioned SVQSPPs reduce to the classical saddle point problem. A study of saddle point for set-valued maps can be found in [77].

For each $i \in I$, let ℓ_i be a real-valued bifunction. We also consider the following *problem of system of quasi-minimax inequalities* (SQMIP): find $\bar{x} = (\bar{x}_i)_{i\in I} \in X$ and $\bar{y} = (\bar{y}_i)_{i\in I} \in Y$ such that for each $i \in I$, $\bar{x}_i \in S_i(\bar{x})$, $\bar{y}_i \in T_i(\bar{x})$ and

$$\min_{u_i \in S_i(\bar{x}_i)} \max_{v_i \in T_i(\bar{x}_i)} \ell_i(u_i, v_i) = \ell_i(\bar{x}_i, \bar{y}_i) = \max_{v_i \in T_i(\bar{x}_i)} \min_{u_i \in S_i(\bar{x}_i)} \ell_i(u_i, v_i).$$

When I is a singleton set, SQMIP is called *quasi-minimax inequality problem* (QMIP). A study of a minimax type inequality for vector-valued functions can be found in [60, 66].

As application of Theorem 18, we derive the following existence result for a solution of SGVQSPP(I).

Theorem 19. *For each $i \subset I$, let E_i, F_i and Z_i be real locally convex topological vector spaces and also F_i be quasi-complete. For each $i \in I$, let $X_i \subseteq E_i$ be a nonempty compact convex set and $Y_i \subseteq F_i$ a nonempty convex set. Let $X = \prod_{i\in I} X_i$ and $Y = \prod_{i\in I} Y_i$. For each $i \in I$, let $S_i : X \to 2^{X_i}$ be a continuous multivalued map with nonempty closed convex values and $T_i : X \to 2^{Y_i}$ a continuous multivalued map with nonempty compact convex values. For each $i \in I$, assume that the following conditions are satisfied:*

(i) *$C_i : X \to 2^{Z_i}$ is a closed multivalued map such that $\forall\, x \in X$, $C_i(x)$ is a closed convex cone with apex at the origin, and $P_i = \bigcap_{x\in X} C_i(x)$.*
(ii) *P_i^* has a weak* compact convex base B_i^* and Z_i is ordered by P_i.*
(iii) *$\ell_i : X_i \times Y_i \to Z_i$ is a continuous function such that:*

   (a) *For each fixed $y_i \in Y_i$, $x_i \mapsto \ell_i(x_i, y_i)$ is properly quasi-convex.*
   (b) *For each fixed $x_i \in X_i$, $y_i \mapsto \ell_i(x_i, y_i)$ is properly quasi-concave.*

Then the SVQSPP(I) has a solution.

If I is a singleton set and $Z = \mathbb{R}$, then we have following existence result for a solution of the quasi-saddle point problem.

Corollary 14. *Let E and F be real locally convex topological vector spaces and also F be quasi-complete. Let $X \subseteq E$ be a nonempty compact convex set and $Y \subseteq F$*

a nonempty convex set. Let $S : X \longrightarrow 2^X$ be a continuous multivalued map with nonempty closed convex values and $T : X \longrightarrow 2^Y$ a continuous multivalued map with nonempty compact convex values. Assume that $\ell : X \times Y \longrightarrow Z$ is a continuous function such that:

(a) *For each fixed $y \in Y$, $x \mapsto \ell(x,y)$ is quasi-convex.*
(b) *For each fixed $x \in X$, $y \mapsto \ell(x,y)$ is quasi-concave.*

Then the QSPP has a solution.

As a consequence of Theorem 19, we have the following existence result for a solution of the system of quasi-minimax inequalities.

Theorem 20. *Let E_i, F_i, X_i, Y_i, X, Y, S_i and T_i be the same as in Theorem 3.1. For each $i \in I$, assume that $\ell_i : X_i \times Y_i \longrightarrow \mathbb{R}$ is a continuous function satisfying the following conditions:*

(i) *For each fixed $y_i \in Y_i$, $x_i \mapsto \ell_i(x_i, y_i)$ is quasi-convex.*
(ii) *For each fixed $x_i \in X_i$, $y_i \mapsto \ell_i(x_i, y_i)$ is quasi-concave.*

Then the SQMIP has a solution.

If for each $i \in I$, X_i and Y_i are nonempty compact convex sets, and $S_i(x) = X_i$ and $T_i(x) = Y_i$, $\forall\, x \in X$, then from Theorem 20 we derive the following corollary which can be seen as an extension of Sion's minimax theorem [92] for a family of continuous bifunctions.

Corollary 15. *For each $i \in I$, let X_i and Y_i be nonempty compact convex subsets of E_i and F_i, respectively. For each $i \in I$, assume that $\ell_i : X_i \times Y_i \longrightarrow \mathbb{R}$ is a continuous function satisfying the following conditions:*

(i) *For each fixed $y_i \in Y_i$, $x_i \mapsto \ell_i(x_i, y_i)$ is quasi-convex.*
(ii) *For each fixed $x_i \in X_i$, $y_i \mapsto \ell_i(x_i, y_i)$ is quasi-concave.*

Then there exist $\bar{x} = (\bar{x}_i)_{i \in I} \in X$ and $\bar{y} = (\bar{y}_i)_{i \in I} \in Y$ such that for each $i \in I$,

$$\min_{u_i \in X_i} \max_{v_i \in Y_i} \ell_i(u_i, v_i) = \ell_i(\bar{x}_i, \bar{y}_i) = \max_{v_i \in Y_i} \min_{u_i \in X_i} \ell_i(u_i, v_i).$$

If I is a singleton, then Theorem 20 reduces to the Corollary 3.2 in [73].

Corollary 16. *Let E, F, X, Y, S and T be the same as in Corollary 14. Assume that $\ell : X \times Y \longrightarrow \mathbb{R}$ is a continuous function satisfying the following conditions:*

(i) *For each fixed $y \in Y$, $x \mapsto \ell(x,y)$ is quasi-convex.*
(ii) *for each fixed $x \in X$, $y \mapsto \ell(x,y)$ is quasi-concave.*

Then there exists $(\bar{x}, \bar{y}) \in X \times Y$ such that $\bar{x} \in S(\bar{x})$, $\bar{y} \in T(\bar{y})$ and

$$\min_{u \in S(\bar{x})} \max_{v \in T(\bar{x})} \ell(u,v) = \ell(\bar{x}, \bar{y}) = \max_{v \in T(\bar{x})} \min_{u \in S(\bar{x})} \ell(u,v).$$

6.3 Debreu Type Equilibrium Problem

In this section, we give another application of Corollary 11 to prove the existence of a solution of the Debreu VEP.

Let X, X_i, Z_i and C_i be the same as defined in the formulations of SSGVQEP. For each $i \in I$, let $\varphi_i : X \to Z_i$ be a vector-valued function and let $X^i = \prod_{j\in I,\ j\neq i} X_j$ and we write $X = X^i \times X_i$. For $x \in X$, x^i denotes the projection of x onto X^i and hence we write $x = (x^i, x_i)$. We consider the following Debreu VEP:

Debreu VEP(I): Find $\bar{x} \in X$ such that for each $i \in I$, $\bar{x}_i \in S_i(\bar{x})$ and

$$\varphi_i(\bar{x}^i, y_i) - \varphi_i(\bar{x}) \in C_i(\bar{x}), \quad \forall\, y_i \in S_i(\bar{x}).$$

Debreu VEP(II): Find $\bar{x} \in X$ such that for each $i \in I$, $\bar{x}_i \in S_i(\bar{x})$ and

$$\varphi_i(\bar{x}^i, y_i) - \varphi_i(\bar{x}) \notin -C_i(\bar{x}) \setminus \{\mathbf{0}\}, \quad \forall\, y_i \in S_i(\bar{x}).$$

Debreu VEP(III): Find $\bar{x} \in X$ such that for each $i \in I$, $\bar{x}_i \in S_i(\bar{x})$ and

$$\varphi_i(\bar{x}^i, y_i) - \varphi_i(\bar{x}) \notin -\operatorname{int} C_i(\bar{x}), \quad \forall\, y_i \in S_i(\bar{x}),$$

in this case we assume that $\operatorname{int} C_i$ is nonempty for each $i \in I$.

Of course, if for each $i \in I$, φ_i is a scalar-valued function, then Debreu VEPs are the same as the one introduced and studied by Debreu in [38], see also [80–82]. In this case, a large number of papers have already been appeared in the literature; see, for example, [9,98] and the references therein. In [5], we introduced and studied Debreu VEP(III) and established several existence results for its solution with or without involving Φ-condensing maps. It is the first paper in the literature in which the Debreu type equilibrium problem for vector-valued functions is considered.

As in the case of SSGVQEPs, if for each $i \in I$ and $\forall\, x \in X$, $C_i(x)$ is also pointed, then every solution of Debreu VEP(I) is a solution of Debreu VEP(II) and every solution of Debreu VEP(II) is a solution of Debreu VEP(III). But the reverse implication does not hold.

Let $\mathscr{Z}^*$ be the dual of a locally convex topological vector space $\mathscr{Z}$, $P^* \subseteq \mathscr{Z}^*$ the polar cone of P, that is, $P^* = \{z^* \in \mathscr{Z}^* : \langle z^*, z\rangle \geq 0,\ \forall z \in P\}$. We assume that P^* has a weak* compact convex base B^*. This means that $B^* \subseteq P^*$ is a weak* compact convex set such that $0 \notin B^*$ and $P^* = \bigcup_{\lambda\geq 0} \lambda B^*$; see, for example, [53].

Theorem 21. *For each $i \in I$, let E_i and Z_i be real locally convex topological vector spaces and E_i be also quasi-complete. For each $i \in I$, let $X_i \subseteq E_i$ be a nonempty compact convex set and let $X = \prod_{i\in I} X_i$. For each $i \in I$, let $S_i : X \to 2^{X_i}$ be a continuous multivalued map with nonempty closed convex values. For each $i \in I$, assume that the following conditions are satisfied:*

(i) *$C_i : X \to 2^{Z_i}$ is a closed multivalued map such that $\forall\, x \in X$, $C_i(x)$ is a closed convex cone with apex at the origin, and $P_i = \bigcap_{x\in X} C_i(x)$.*
(ii) *P_i^* has a weak* compact convex base B_i^* and Z_i is ordered by P_i.*
(iii) *$\varphi_i : X \to Z_i$ is continuous and properly quasi-convex in each argument.*

Then there exists a solution $\bar{x} \in X$ of the Debreu VEP(I).

Acknowledgements Authors are grateful to the referees for their valuable comments and suggestions.

References

1. S. Al-Homidan, Q. H. Ansari and S. Schaible, Existence of solutions of systems of generalized implicit vector variational inequalities, *J. Optim. Theory Appl.*, **134** (2007), 515–531.
2. Q. H. Ansari, A note on generalized vector variational-like inequalities, *Optimization*, **41** (1997), 197–205.
3. Q. H. Ansari, Vector equilibrium problems and vector variational inequalities, In [48], pp. 1–17, 2000.
4. Q. H. Ansari, Existence of solutions of systems of generalized implicit vector quasi-equilibrium problems, *J. Math. Anal. Appl.*, **341** (2008), 1271–1283.
5. Q. H. Ansari, W. K. Chan and X. Q. Yang, The systems of vector quasi-equilibrium problems with applications, *J. Global Optim.*, **29** (2004), 45–57.
6. Q. H. Ansari, W. K. Chan and X. Q. Yang, Weighted quasi-variational inequalities and constrained Nash equilibrium problems, *Taiwanese J. Math.*, **10** (2006), 361–380.
7. Q. H. Ansari and F. Flores-Bazán, Generalized vector quasi-equilibrium problems with applications, *J. Math. Anal. Appl.*, **277** (2003), 246–256.
8. Q. H. Ansari and F. Flores-Bazan, Recession method for generalized vector equilibrium problems, *J. Math. Anal. Appl.*, **321**(1) (2006), 132–146.
9. Q. H. Ansari, A. Idzik and J. C. Yao, Coincidence and fixed point theorems with applications, *Topolgical Meth. Nonlinear Anal.*, **15**(1) (2000), 191–202.
10. Q. H. Ansari and Z. Khan, System of generalized vector quasi-equilibrium problems with applications, in *Mathematical Analysis and Applications*, Edited by S. Nanda and G. P. Rajasekhar, Narosa, New Delhi, pp. 1–13, 2004.
11. Q. H. Ansari, I. V. Konnov and J. C. Yao, On generalized vector equilibrium problems, *Nonlinear Anal.*, **47** (2001), 543–554.
12. Q. H. Ansari, L. J. Lin and L. B. Su, Systems of simultaneous generalized vector quasiequilibrium problems and their applications, *J. Optim. Theory Appl.*, **127**(1) (2005), 27–44.
13. Q. H. Ansari, W. Oettli and D. Schläger, A generalization of vectorial equilibria, *Math. Meth. Oper. Res.*, **46** (1997), 147–152.
14. Q. H. Ansari, S. Schaible and J. C. Yao, System of vector equlibrium problems and its applications, *J. Optim. Theory Appl.*, **107** (2000), 547–557.
15. Q. H. Ansari, S. Schaible and J. C. Yao, The system of generalized vector equilibrium problems with applications, *J. Global Optim.*, **22** (2002), 3–16.
16. Q. H. Ansari, S. Schaible and J. C. Yao, Generalized vector quasi-variational inequality problems over product sets, *J. Global Optim.*, **32**(4) (2005), 437–449.
17. Q. H. Ansari and A. H. Siddiqi, A generalized vector variational-like inequality and optimization over an efficient set, in *Functional Analysis with Current Applications in Science, Technology and Industry*, Edited by M. Brokate and A. H. Siddiqi, Pitman Research Notes in Mathematics, Vol. 377, Addison Wesley Longman, Essex, UK, pp. 177–191, 1998.
18. Q. H. Ansari, A. H. Siddiqi and S. Y. Wu, Existence and duality of generalized vector equilibrium problems, *J. Math. Anal. Appl.*, **259** (2001), 115–126.
19. Q. H. Ansari, X. Q. Yang and J. C. Yao, Existence and duality of implicit vector variational problems, *Numer. Funct. Anal. Optim.*, **22**(7–8) (2001), 815–829.
20. Q. H. Ansari and J. C. Yao, A fixed point theorem and its applications to the system of variational inequalities, *Bull. Austral. Math. Soc.*, **59** (1999), 433–442.
21. Q. H. Ansari and J. C. Yao, An existence result for the generalized vector equilibrium problem, *Appl. Math. Lett.*, **12**(8) (1999), 53–56.

22. Q. H. Ansari and J. C. Yao, Systems of generalized variational inequalities and their applications, *Appl. Anal.*, **76** (2000), 203–217.
23. Q. H. Ansari and J. C. Yao, On nondifferentiable and nonconvex vector optimization problems, *J. Optim. Theory Appl.*, **106** (2000), 487–500.
24. Q. H. Ansari and J. C. Yao, On vector quasi-equilibrium problems, in *Equilibrium Problems and Variational Models*, Edited by P. Daniele, F. Giannessi and A. Maugeri, Kluwer, Dordrecht, 2002.
25. J.-P. Aubin, *Mathematical Methods of Game and Economic Theory*, North-Holland, Amsterdam, 1979.
26. J.-P. Aubin and H. Frankowska, *Set-Valued Analysis*, Birkhäuser, Boston, 1990.
27. C. Berge, *Topological Space*, Oliver&Boyd, Edinburgh, 1963.
28. M. Bianchi, N. Hadjisavvas and S. Schaible, Vector equilibrium problems with generalized monotone bifunctions, *J. Optim. Theory Appl.*, **92** (1997), 527–542.
29. S. Chebbi and M. Florenzano, Maximal elements and equilibria for condensing correspondences, *Nonlinear Anal.*, **38** (1999), 995–1002.
30. G. Y. Chen and B. D. Craven, A vector variational inequality and optimization over an efficient set, *ZOR-Math. Meth. Oper. Res.*, **34** (1990), 1–12.
31. G. Y. Chen, C. J. Goh and X. Q. Yang, Existence of a solution for generalized vector variational inequalities, *Optimization*, **50**(1) (2001), 1–15.
32. G. Y. Chen and X. Q. Yang, Vector complementarity problem and its equivalence with weak minimal element in ordered spaces, *J. Math. Anal. Appl.*, **153** (1990), 136–158.
33. G. Y. Chen, X. Q. Yang and H. Yu, A nonlinear scalarization function and generalized vector quasi-equilibrium problems, *J. Global Optim.*, **32** (2005), 451–466.
34. F. H. Clarke, *Optimization and Nonsmooth Analysis*, SIAM, Philadelphia, PA, 1990.
35. G. Cohen and F. Chaplais, Nested monotony for variational inequalities over a product of spaces and convergence of iterative algorithms, *J. Optim. Theory Appl.*, **59** (1988), 360–390.
36. B. D. Craven, Nondifferentiable optimization by smooth approximation, *Optimization*, **17** (1986), 3–17.
37. A. Daniilidis and N. Hadjisavvas, Existence theorems for vector variational inequalities, *Bull. Austral. Math. Soc.*, **54** (1996), 473–481.
38. G. Debreu, A social equilibrium existence theorem, *Proc. Natl. Acad. Sci. U.S.A.*, **38** (1952), 886–893.
39. P. Deguire, K. K. Tan and G. X.-Z. Yuan, The study of maximal element, fixed point for L_S-majorized mappings and their applications to minimax and variational inequalities in the product topological spaces, *Nonlinear Anal.*, **37** (1999), 933–951.
40. N. H. Dien, Some remarks on variational-like and quasivariational-like inequalities, *Bull. Austral. Math. Soc.*, **46** (1992), 335–342.
41. X. P. Ding, Existence of solutions for quasi-equilibrium problems in noncompact topological spaces, *Comput. Math. Appl.*, **39** (2000), 13–21.
42. X. P. Ding and E. Tarafdar, Generalized vector variational-like inequalities without monotonicity, in [48], pp. 113–124, 2000.
43. F. Ferro, A minimax theorem for vector-valued functions, *J. Optim. Theory Appl.*, **60**(1) (1989), 19–31.
44. F. Flores-Bazán and F. Flores-Bazán, Vector equilibrium problems under asymptotic analysis, *J. Global Optim.*, **26** (2003), 141–166.
45. J. Fu, Simultaneous vector variational inequalities and vector implicit complementarity problem, *J. Optim. Theory Appl.*, **93** (1997), 141–151.
46. J. Y. Fu, Generalized vector quasi-equilibrium problems, *Math. Meth. Oper. Res.*, **52** (2000), 57–64.
47. F. Giannessi, Theorems of the alternative, quadratic programs and complementarity problems, in *Variational Inequalities and Complementarity Problems*, Edited by R. W. Cottle, F. Giannessi and J. L. Lions, Wiley, New York, pp. 151–186, 1980.
48. F. Giannessi (Ed.), *Vector Variational Inequalities and Vector Equilibria. Mathematical Theories*, Kluwer, Dordrecht, 2000.

49. X. H. Gong, Efficiency and Henig efficiency for vector equilibrium problems, *J. Optim. Theory Appl.*, **108** (2001), 139–154.
50. N. Hadjisavvas and S. Schaible, From scalar to vector equilibrium problems in the quasi-monotone case, *J. Optim. Theory Appl.*, **96** (1998), 297–309.
51. M. A. Hanson, On sufficiency of the Kuhn–Tucker conditions, *J. Math. Anal. Appl.*, **80** (1982), 545–550.
52. T. Husain and E. Tarafdar, Simultaneous variational inequalities, minimization problems and related results, *Math. Japonica*, **39** (1994), 221–231.
53. V. Jeykumar, W. Oettli and M. Natividad, A solvability theorem for a class of quasiconvex mappings with applications to optimization, *J. Math. Anal. Appl.*, **179** (1993), 537–546.
54. S. Kakutani, A generalization of Brouwer's fixed point theorem, *Duke Math. J.*, **8** (1941), 457–459.
55. W. K. Kim and K. K. Tan, On generalized vector quasi-variational inequalities, *Optimization*, **46** (1999), 185–198.
56. I. V. Konnov and J.-C. Yao, On the generalized vector variational inequality problem, *J. Math. Anal. Appl.*, **206** (1997), 42–58.
57. I. V. Konnov and J.-C. Yao, Existence of solutions for generalized vector equilibrium problems, *J. Math. Anal. Appl.*, **233** (1999), 328–335.
58. S. Kum and G. M. Lee, Remarks on implicit vector variational inequalities, *Taiwanese J. Math.*, **6**(3) (2002), 369–382.
59. G. M. Lee, On relations between vector variational inequality and vector optimization problem, in *Progress in Optimization, II: Contributions from Australasia*, Edited by X. Q. Yang, A. I. Mees, M. E. Fisher and L. S. Jennings, Kluwer, Dordrecht, pp. 167–179, 2000.
60. G. M. Lee and D. S. Kim, Existence of solutions for vector optimization problems, *J. Math. Anal. Appl.*, **220** (1998), 90–98.
61. G. M. Lee, D. S. Kim and B. S. Lee, On noncooperative vector equilibrium, *Indian J. Pure Appl. Math.*, **27** (1996), 735–739.
62. G. M. Lee, D. S. Kim and B. S. Lee, Generalized vector variational inequality, *Appl. Math. Lett.*, **9**(1) (1996), 39–42.
63. G. M. Lee, D. S. Kim and B. S. Lee, On vector quasivariational-like inequality, *Bull. Korean Math. Soc.*, **33** (1996), 45–55.
64. G. M. Lee and S. Kum, On implicit vector variational inequalities, *J. Optim. Theory Appl.*, **104**(2) (2000), 409–425.
65. G. M. Lee, B. S. Lee and S.-S. Change, On vector quasivariational inequalities, *J. Math. Anal. Appl.*, **203** (1996), 626–638.
66. Z. F. Li and S. Y. Wang, A type of minimax inequality for vector-valued mappings, *J. Math. Anal. Appl.*, **227** (1998), 68–80.
67. L. J. Lin, Existence theorems of simultaneous equilibrium problems and generalized vector quasi-saddle points, *J. Global Optim.*, **32** (2005), 613–632.
68. L.-J. Lin, System of generalized vector quasi-equilibrium problems with applications to fixed point theorems for a family of nonexpansive multivalued Maps, *J. Global Optim.*, **34** (2006), 15–32.
69. L.-J. Lin and Q. H. Ansari, Collective fixed points and maximal elements with applications to abstract economies, *J. Math. Anal. Appl.*, **296** (2004), 455–472.
70. L.-J. Lin and H.-W. Hsu, Existence theorems for systems of generalized vector quasiequilibrium problems and optimization problems, *J. Optim. Theory Appl.*, **130** (2006), 461–475.
71. L.-J. Lin and Y. H. Liu, Existence theorems for systems of generalized vector quasi-equilibrium problems and mathematical programs with equilibrium constraint, *J. Optim. Theory Appl.*, **130** (2006), 461–475.
72. L.-J. Lin and S. Park, On some generalized quasi-equilibrium problems, *J. Math. Anal. Appl.*, **224** (1998), 167–181.
73. L. J. Lin and Y. L. Tasi, On vector quasi-saddle points of set-valued maps, in *Generalized Convexity/Monotonicity*, Edited by D. T. Luc and N. Hadjisavvs, Kluwer, Dordrecht, pp. 311–319, 2004.

74. L. J. Lin and Z. T. Yu, On some equilibrium problems for multimaps, *J. Comput. Appl. Math.*, **129** (2001), 171–183.
75. Z. Lin and J. Yu, The existence of solutions for the systems of generalized vector quasi-equilibrium problems, *Appl. Math. Lett.*, **18** (2005), 415–422.
76. D. T. Luc, *Theory of Vector Optimization*, Lecture Notes in Economics and Mathematical Systems, Vol. 319, Springer, Berlin, 1989.
77. D. T. Luc and C. A. Vargas, A saddle point theorem for set-valued maps, *Nonlinear Anal. Theory Meth. Appl.*, **18** (1992), 1–7.
78. S. R. Mohan and S. K. Neogy, On invex sets and preinvex functions, *J. Math. Anal. Appl.*, **189** (1995), 901–908.
79. S. B. Nadler, Jr., Multi-valued contraction mappings, *Pacific J. Math.*, **30** (1969), 475–488.
80. J. F. Nash, Equilibrium point in n-person games, *Proc. Natl. Acad. Sci. U.S.A.*, **36** (1950), 48–49.
81. J. F. Nash, Non-cooperative games, *Ann. Math.*, **54** (1951), 286–295.
82. J. F. Nash, Two-persons cooperative games, *Econometrica*, **21** (1953), 128–140.
83. W. Oettli, A remark on vector-valued equilibria and generalized monotonicity, *Acta Math. Vietnam.*, **22**(1) (1997), pp.213–221.
84. W. Oettli and D. Schläger, Generalized vectorial equilibria and generalized monotonicity, in *Functional Analysis with current Applications in Science, Technology and Industry*, Edited by M. Brokate and A. H. Siddiqi, Pitman Research Notes in Mathematics Series, Vol. 377, Longman, Essex, pp. 145–154, 1998.
85. W. Oettli and D. Schläger, Existence of equilibria for monotone multivalued mappings, *Math. Meth. Oper. Res.*, **48** (1998), 219–228.
86. J. S. Pang, Asymmetric variational inequality problems over product sets: Applications and iterative methods, *Math. Prog.*, **31** (1985), 206–219.
87. M. Patriksson, *Nonlinear Programming and Variational Inequality Problems*, Kluwer, Dordrecht, 1999.
88. W. V. Petryshyn and P. M. Fitzpatrick, Fixed point theorems of multivalued noncompact inward maps, *J. Math. Anal. Appl.*, **46** (1974), 756–767.
89. W. V. Petryshyn and P. M. Fitzpatrick, Fixed point theorems of multivalued noncompact acyclic mappings, *Pacific J. Math.*, **54** (1974), 17–23.
90. R. T. Rockafellar, *Convex Analysis*, Princeton University Press, Princeton, 1970.
91. H. H. Schaefer, *Topological Vector Spaces*, Springer, Berlin, 1971.
92. M. Sion, On general minimax theorem, *Pacific J. Math.*, **8** (1958), 171–176.
93. W. Song, Vector equilibrium problems with set-valued mappings, in [48], pp. 403–422, 2000.
94. N. X. Tan and P. N. Tinh, On the existence of equilibrium points of vector functions, *Numer. Funct. Anal. Optim.*, **19** (1998), 141–156.
95. T. Tanaka, Generalized semicontinuity and existence theorems for cone saddle points, *Appl. Math. Optim.*, **36** (1997), 313–322.
96. T. Weir and V. Jeyakumar, A class of nonconvex functions and mathematical programming, *Bull. Austral. Math. Soc.*, **38** (1988), 177–189.
97. X. Q. Yang, Vector variational inequality and its duality, *Nonlinear Anal. Theory Meth. Appl.*, **21** (1993), 869–877.
98. G. X.-Z. Yuan, *KKM Theory and Applications in Nonlinear Analysis*, Marcel Dekker, New York, 1999.
99. G. X.-Z. Yuan, G. Isac, K.-K. Tan and J. Yu, The study of minimax inequalities, abstract economies and applications to variational inequalities and Nash equilibra, *Acta Appl. Math.*, **54** (1998), 135–166.
100. E. Zeidler, *Nonlinear Functional Analysis and Its Applications, I: Fixed Point Theorems*, Springer, New York, 1986.
101. L. C. Zeng and J. C. Yao, An existence result for generalized vector equilibrium problems without pseudomonotonicity, *Appl. Math. Lett.*, **19** (2006) 1320–1326.

Properties of Derivates and Some Applications

Michael McAsey and Libin Mou

Abstract In this chapter, we generalize the concept of derivates, defined recently in the literature, to maps defined on a topological space. The derivate of a map has some interesting properties and applications to optimization problems. For example, it is closely related to various notions of tangent spaces of the range of the map. It strengthens the necessary condition (Fermat's theorem) for an extremum point to a sufficient condition.

1 Introduction

One motivation for the notion of differentiation and subsequent generalizations is to solve the optimization problem: $\min f(x)$, $x \in W$, where W is a metric space or topological space and $f : W \to \mathbb{R}$. Such a consideration is quite natural because the solution(s) of the problem should be independent of the structure of W (algebraic or topological). Of course, in most applications, W is endowed with certain natural structures, which makes it possible to develop a variational analysis on W for solving the problem. On the other hand, as pointed out in [4] and [7], there are many optimization problems that are naturally formulated on metric spaces or topological spaces. In [4] and [7], W is assumed to be a metric space. In this chapter, we allow W to be a topological space and define the derivate and strict derivate of a map f from W to a Banach space Y. The derivate in [4] and [7] is generalized herein that it depends only on the topology of W. In other words, two metrics on W defining the same topology induce the same derivate for a map f.

The rest of the chapter is arranged as follows. In Sect. 2, the notion of derivate is defined and some basic properties of derivates are proved, including a strengthened version of Fermat's Theorem and general rules of differentiation for derivates.

Michael McAsey, Libin Mou
Bradley University, Peoria, IL 61625, USA
e-mail: mcasey@bradley.edu, mou@bradley.edu

R.S. Burachik and J.-C. Yao (eds.), *Variational Analysis and Generalized Differentiation in Optimization and Control*, Springer Optimization and Its Applications 47,
DOI 10.1007/978-1-4419-0437-9_2,

In Sect. 3, we explore the relationship between derivate and several versions of tangent spaces. We show, for example, that the set of zero derivates of f is a subspace of the Bouligand tangent space of the range $f(W) \subset Y$. For a map f with the RTR (restrictive topological regularity) property, these two spaces are actually the same. Under additional conditions, such as when W is a subset of a Banach space and f is strictly differentiable in the sense of Gateaux or Fréchet, then the derivates of f can be described more specifically. In the final section, we derive some necessary conditions for solutions to typical optimization problems by using derivates; these results generalize some of the results in [2] and [3] and extend the Lagrange multiplier rules in [4] and [7].

2 Derivates and Their Properties

Definition 2.1. (1) Let f be a map from a Hausdorff topological space W to a Banach space $(Y, \|\cdot\|)$ and $\varepsilon \geq 0$ a number. We say that $y \in Y$ is an *ε-derivate* of f at $\bar{x}$ if there are sequences $x_i \in W$, $x_i \to \bar{x}$ and $d_i \downarrow 0$ such that

$$\lim_{i\to\infty} \left\| \frac{f(x_i) - f(\bar{x})}{d_i} - y \right\| \leqslant \varepsilon. \tag{1}$$

We denote by $D^{\varepsilon} f(\bar{x})$ or $D^{\varepsilon} f(\bar{x}, W)$ the set of all ε-derivates of f at $\bar{x}$.
(2) We say that $y \in Y$ is a *strict derivate* of f at $\bar{x}$ if for every sequence $x^k \to \bar{x}$ there is a sequence $\varepsilon_k \to 0$ such that $y \in D^{\varepsilon_k} f(\bar{x})$ for all k. The set of all strict derivates of f at $\bar{x}$ is denoted as $D_s f(\bar{x})$ or $D_s f(\bar{x}, W)$.

Note that if two metrics d', d'' are equivalent, that is, for any sequence $x_i \in W$ and $\bar{x} \in W$, $d'(x_i, \bar{x}) \to 0$ if and only if $d''(x_i, \bar{x}) \to 0$, then the metrics define the same notion of derivate. So the notion of derivate depends only on the topology of W induced by the metric.

Remark 2.2. Our notion of derivate slightly generalizes those defined in papers [4] and [7], where the sequence x_i is also required to satisfy $d(x_i, \bar{x}) \leqslant d_i$ for all i. It should also be noted that an individual derivate is not a mapping as the usual derivatives are, but is closer in spirit to a derivative evaluated at a point.

The following proposition contains some basic properties of derivates of scalar functions.

Proposition 2.3. *Let $f : W \to \mathbb{R}$ be a real-valued function and $\varepsilon \geq 0$ be a real number.*
(a) *If $l = \lim_{i\to\infty} \frac{f(x_i) - f(\bar{x})}{d_i} \in \mathbb{R}$ for some sequences $x_i \to \bar{x}$ and $d_i \downarrow 0$, then $l \in D^0 f(\bar{x})$ and $[l - \varepsilon, l + \varepsilon] \subset D^{\varepsilon} f(\bar{x})$.*
(b) *If there is a sequence $x_i \to \bar{x}$ such that $f(x_i) \downarrow f(\bar{x})$ then $[-\varepsilon, \infty) \subset D^{\varepsilon} f(\bar{x})$.*
(c) *If there is a sequence $x_i \to \bar{x}$ such that $f(x_i) \uparrow f(\bar{x})$ then $(-\infty, \varepsilon] \subset D^{\varepsilon} f(\bar{x})$.*
(d) *If $\varepsilon \leq \delta$, then $D^{\varepsilon} f(\bar{x}) \subseteq D^{\delta} f(\bar{x})$.*

(e) *If* $y \in D^{\varepsilon} f(\bar{x})$ *and* $\alpha > 0$, *then* $\alpha y \in D^{\alpha\varepsilon} f(\bar{x})$. *In particular, if* $y \in D^0 f(\bar{x})$ *and* $\alpha \geq 0$, *then* $\alpha y \in D^0 f(\bar{x})$

Proof. (a) In this case, every $y \in [l-\varepsilon, l+\varepsilon]$ satisfies the definition of an ε-derivate.

(b) In this case, we can take $d_i = (f(x_i) - f(\bar{x}))^s / r$, where $r > 0$ and $s \in (0,1]$. Then $d_i \to 0$. Note that if $s = 1$, then $\lim_{i\to\infty} \frac{f(x_i)-f(\bar{x})}{d_i} = r$. This implies that $[r-\varepsilon, r+\varepsilon] \subset D^{\varepsilon} f(\bar{x})$ by (a). Since $r > 0$ is arbitrary, we have $(-\varepsilon, \infty) \subset D^{\varepsilon} f(\bar{x})$. If $s \in (0,1)$, then $\lim_{i\to\infty} \frac{f(x_i)-f(\bar{x})}{d_i} = 0$. This shows that $[-\varepsilon, \varepsilon] \subset D^{\varepsilon} f(\bar{x})$ by (a). Together we see that the conclusion holds.

(c) is similar to (b). (d) is obvious from the definition of derivate.

(e) If $y \in D^{\varepsilon} f(\bar{x})$, then $|l - y| \leqslant \varepsilon$, where $l = \lim_{i\to\infty} \frac{f(x_i)-f(\bar{x})}{d_i}$ for some sequences as in the definition. Then $\alpha l = \lim_{i\to\infty} \frac{f(x_i)-f(\bar{x})}{d_i/\alpha}$ and $|\alpha l - \alpha y| \leqslant \alpha\varepsilon$. So $\alpha y \in D^{\alpha\varepsilon} f(\bar{x})$. If $\varepsilon = 0$, then $y = \lim_{i\to\infty} \frac{f(x_i)-f(\bar{x})}{d_i}$ for some sequences as in the definition. Replace by d_i by $\sqrt{d_i} \downarrow 0$, then we see that $0 = \lim_{i\to\infty} \frac{f(x_i)-f(\bar{x})}{\sqrt{d_i}}$. So $0 \in D^0 f(\bar{x})$. □

Let us look at some examples.

Example 2.4 (General real-valued functions). Let $f(x)$ be a function on a closed interval $[a,b]$ and $\varepsilon \geq 0$ a given number. Let $\bar{x} \in [a,b]$. A *neighborhood* of $\bar{x}$ refers a set of the form $[a,b] \cap (\bar{x} - \delta, \bar{x} + \delta)$ for some number $\delta > 0$. The following is a collection of elementary computations of derivates and strict derivates for functions with simple properties:

1. $D^{\varepsilon} f(\bar{x}) = [-\varepsilon, \varepsilon]$ if f is constant in a neighborhood of $\bar{x}$. This is because for all sequences $x_i \in W$, $x_i \to \bar{x}$ and $d_i \downarrow 0$, we have $\lim_{i\to\infty} \frac{f(x_i)-f(\bar{x})}{d_i} = 0$.
2. $D^{\varepsilon} f(\bar{x}) = [-\varepsilon, \infty)$ if $f(\bar{x})$ is a local minimum of f and f is not constant in any neighborhood of $\bar{x}$. This follows from Proposition 2.3(b) and the fact that there exists a sequence $x_i \to \bar{x}$ such that $f(x_i) \downarrow f(\bar{x})$.
3. $D^{\varepsilon} f(\bar{x}) = (-\infty, \varepsilon]$ if $f(\bar{x})$ is a local maximum point of f and f is not constant in any neighborhood of $\bar{x}$. In this case, there exists a sequence $x_i \to \bar{x}$ such that $f(x_i) \uparrow f(\bar{x})$ and the result follows from Proposition 2.3(c).
4. $D^{\varepsilon} f(x) = (-\infty, \infty)$ if $f(x)$ is not a local maximum nor minimum. This follows from (2) and (3) above.
5. $D_s f(x) = \{0\}$ if f is constant in a neighborhood of x, or there exists two sequences $x_i \to x$ and $x_i' \to x$ such that x_i are local maximum points while x_i' are local minimum points. In the latter cases, by (2) and (3) above, we have that $D^{\varepsilon_i} f(x_i) = (-\infty, \varepsilon_i]$ and $D^{\varepsilon_i} f(x_i') = [-\varepsilon_i, \infty)$ for any $\varepsilon_i \downarrow 0$, which imply the conclusion by the definition of strict derivate.
6. $D_s f(x) = (-\infty, \infty)$ if every point in a neighborhood of x is not a local maximum nor local minimum point. This follows from (4) and the definition of $D_s f$. For example, this is the case if f is strictly increasing or decreasing in a neighborhood of x.

Example 2.5 (A continuous function). Let $f(0) = 0$ and $f(x) = |x| \sin^2 \frac{1}{x}$ for $x \neq 0$. Then by Example 2.4(2) and (5), $D^{\varepsilon} f(0) = [-\varepsilon, \infty)$ and $D_s f(0) = \{0\}$.

Example 2.6 (A discontinuous function). Let

$$f = \begin{cases} ax & x \text{ is rational}, \\ ax+b & x \text{ is irrational}, \end{cases}$$

where $a > 0, b \neq 0$. Then $D^{\varepsilon} f(x) = D_s f(x) = (-\infty, \infty)$ for every x. This is a special case of Example 2.4(4) and (6).

Various derivatives or differentials can be used to give necessary conditions for a local extremum as in classical Fermat-type theorems. Typically, the necessary conditions of Fermat's theorem are not sufficient. However, derivates give both a necessary and sufficient condition for a local extremum, as shown next.

Theorem 2.7 (Fermat's theorem). *Let $f : W \to \mathbb{R}$ be a function on a topological space W and $\bar{x} \in W$. Then $\bar{x}$ is a local minimum (maximum) point of f on W iff $y \geq 0$ ($y \leqslant 0$, respectively) for all $y \in D^0 f(\bar{x})$.*

Proof. Suppose that $\bar{x}$ is a local minimum point and $y \in D^0 f(\bar{x})$. Then we have $y = \lim_{i\to\infty} \frac{f(x_i)-f(\bar{x})}{d_i} \geq 0$ for some sequences $x_i \in W$, $x_i \to \bar{x}$ and $d_i \to 0$. Because $f(x_i) \geq f(\bar{x})$, $y = \lim_{i\to\infty} \frac{f(x_i)-f(\bar{x})}{d_i} \geq 0$. Conversely, if there is a $y \in D^0 f(\bar{x})$ such that $y < 0$, then $y = \lim_{i\to\infty} \frac{f(x_i)-f(\bar{x})}{d_i} < 0$. This shows that $\frac{f(x_i)-f(\bar{x})}{d_i} < 0$ for all i sufficiently large, and this implies that $\bar{x}$ is not a local minimum point of f. The proof for maximum points is similar. □

As an example, we can apply Theorem 2.7 to the functions in the two examples above. The function in Example 2.5 has a minimum at 0 because $D^0 f(0) = [0, \infty)$. The function in Example 2.6 does not have a minimum because $D^0 f(x) = (-\infty, \infty)$ everywhere.

A simple example shows how derivates will find a minimizer while the usual elementary calculus derivative will sometimes provide candidates for minima that satisfy the necessary condition $f'(x) = 0$ but the condition is not sufficient. Consider the function $f(x) = x^3$ for $x \in W = [-1, 1]$. Elementary calculus suggests there are three numbers of significance: the two endpoints and $x = 0$ since $f'(0) = 0$. But, of course, $x = 0$ satisfies a necessary condition and fails to be the minimizer of f on W. Using Proposition 2.3, however, the zero derivates are: $D^0 f(-1, W) = [0, \infty)$, $D^0 f(+1, W) = (-\infty, 0]$, and $D^0 f(x, W) = (-\infty, \infty)$. Theorem 2.7 tells us immediately that the minimum occurs at $x = -1$.

To calculate derivates, the following version of the chain rule will be useful. The "outside" function G in the chain rule will be assumed to be Fréchet differentiable in part (a) of the theorem and strictly (Fréchet) differentiable in part (b). Recall that Fréchet differentiability of $G : Y \to Z$ at $\bar{y} \in Y$ means that there is bounded linear map $\nabla G(\bar{y})$ and a map $\alpha(\cdot) = \alpha(\cdot; G, \bar{y})$ with $\alpha : Y \to Z$ so that $G(y) - G(\bar{y}) - \nabla G(\bar{y})(y - \bar{y}) = \alpha(y - \bar{y}) \|y - \bar{y}\|$ and $\|\alpha(h)\| \to 0$ as $\|h\| \to 0$. We recall the definition of strict differentiability for convenience. See also [5, p. 19]. (We use the same notation for Fréchet derivative and strict Fréchet derivative.)

Definition 2.8. For Banach spaces Y and Z, the mapping $G: Y \to Z$ is *strictly Fréchet differentiable* at $\bar{y} \in Y$ in case there is a bounded linear operator $\nabla G(\bar{y}):$ $Y \to Z$ satisfying $r(G,\bar{y},\eta) \to 0$ as $\eta \to 0$, where

$$r(G,\bar{y},\eta) = \sup\left\{ \frac{\|G(y_1) - G(y_2) - \nabla G(\bar{y})(y_1 - y_2)\|}{\|y_1 - y_2\|} : y_1 \neq y_2, \|y_i - \bar{y}\| < \eta \ (i = 1,2) \right\}.$$

The definition implies that for all $y_1, y_2 \in B(\bar{y},\eta)$, the following holds

$$\|G(y_1) - G(y_2) - \nabla G(\bar{y})(y_1 - y_2)\| \leqslant r(G,\bar{y}, \|y_1 - \bar{y}\| + \|y_2 - y_1\|)\ \|y_1 - y_2\|. \quad (2)$$

Remark 2.9. Note that the definition of strict differentiability is equivalent to (2) for some function $r(G,\bar{y},\eta) \to 0$ as $\eta \to 0$. In (2), we may assume that $r(G,\bar{y},\eta)$ is continuous in η from the right by replacing it by $s(G,\bar{y},\eta) = \lim_{\tau \to \eta^+} r(G,\bar{y},\tau)$ if r is not continuous from the right. Here $s(G,\bar{y},\eta)$ is continuous from the right and $s(G,\bar{y},\eta) \leqslant r(G,\bar{y},2\eta) \to 0$ as $\eta \to 0$.

The concept of strict differentiability can be generalized/localized by requiring the approximating vectors y_1 and y_2 in the definition to be elements of a prescribed subset. For this, let $W \subset Y$ be a subset of the Banach space Y, $\bar{y} \in W$ and $G: W \to Z$. The next definition deviates a bit from the setting in which W is a general topological space, of course, but the restriction of approximating elements to be elements in the set W can be used as a replacement for the hypothesis on G in Theorem 2.11(b); see Remark 2.12.

Definition 2.10. We say that G is *strictly Fréchet differentiable at* $\bar{y}$ *in* W if $r(G,\bar{y},\eta,W) \to 0$, where $r(G,\bar{y},\eta,W)$ is defined as $r(G,\bar{y},\eta)$ with the additional requirement that $y_1, y_2 \in W$. (The notation $\nabla G(\bar{x})$ will sometimes be written as $\nabla G(\bar{x},W)$ to emphasize the dependence on W.)

For example, $G(y) = -|y|$ is not differentiable at $y = 0$. However, for $W = (-\infty, 0]$, $G|_W = y$ is strictly differentiable at $0 \in W$ with $\nabla G(0) = 1$.

Now we prove a chain rule for derivates.

Theorem 2.11 (Chain rules for derivates). *Let $\varepsilon \geq 0$, W a topological space, Y,Z Banach spaces, and $\bar{x} \in W$. Suppose $F: W \to Y$ and $G: Y \to Z$.*

(a) *If $y \in D^{\varepsilon}F(\bar{x})$ and G is Fréchet differentiable at $\bar{y} = F(\bar{x})$, then $\nabla G(\bar{y})y \in D^{\|\nabla G(\bar{y})\|\varepsilon}(G \circ F)(\bar{x})$. In particular, if $y \in D^0 F(\bar{x})$, then $\nabla G(\bar{y})y \in D^0(G \circ F)(\bar{x})$.*

(b) *If $y \in D_s F(\bar{x})$ and G is strictly differentiable at $\bar{y} = F(\bar{x})$, then $\nabla G(\bar{y})y \in D_s(G \circ F)(\bar{x})$.*

Proof. (a) Since $y \in D^{\varepsilon}F(\bar{x})$, by assumption, there are $x_i \to \bar{x}$ and $d_i \downarrow 0$ such that

$$\lim_{i \to \infty} \left\| \frac{F(x_i) - F(\bar{x})}{d_i} - y \right\| \leqslant \varepsilon. \quad (3)$$

This implies that $\limsup_{i\to\infty}\left\|\frac{F(x_i)-F(\bar{x})}{d_i}\right\| \leqslant \|y\|+\varepsilon$. In particular, $F(x_i)\to F(\bar{x})$. Write

$$\begin{aligned}
&\frac{1}{d_i}[G(F(x_i))-G(F(\bar{x}))]-\nabla G(F(\bar{x}))y \\
&\quad=\nabla G(F(\bar{x}))\left(\frac{F(x_i)-F(\bar{x})}{d_i}-y\right) \\
&\qquad+\frac{G(F(x_i))-G(F(\bar{x}))-\nabla G(F(\bar{x}))(F(x_i)-F(\bar{x}))}{d_i}.
\end{aligned}$$

By Definition (2), the norm of the second term is bounded by

$$r(G,\bar{y},\|F(x_i)-F(\bar{x})\|)\|\,\|F(x_i)-F(\bar{x})\|/d_i.$$

So it follows from (3) that

$$\lim_{i\to\infty}\left\|\frac{G(F(x_i))-G(F(\bar{x}))}{d_i}-\nabla G(F(\bar{x}))y\right\| \leqslant \|\nabla G(F(\bar{x}))\|\varepsilon.$$

This shows the conclusion of part (a).

(b) To show $\nabla G(\bar{y})y$ is a strict derivate of $G\circ F$, let $x^k\in W$ be a sequence converging to $\bar{x}$. Since $y\in D_sF(\bar{x})$, there is a sequence $\varepsilon_k\to 0$ so that $y\in D^{\varepsilon_k}F(x^k)$ for all k. This, in turn, means that for each k there exist sequences $x_i^k\to x^k$ and $t_i^k\downarrow 0$ as $i\to\infty$ such that

$$\limsup_{i\to\infty}\left\|\frac{F(x_i^k)-F(x^k)}{t_i^k}-y\right\| \leqslant \varepsilon_k.$$

This implies

$$\limsup_{i\to\infty}\left\|\frac{F(x_i^k)-F(x^k)}{t_i^k}\right\| \leqslant \|y\|+\varepsilon_k. \tag{4}$$

Using the strict differentiability of G, we get

$$\begin{aligned}
&\frac{1}{t_i^k}\left\|G(F(x_i^k))-G(F(x^k))-\nabla G(F(\bar{x}))(F(x_i^k)-F(x^k))\right\| \\
&\quad\le r(G,F(\bar{x}),\|F(x_i^k)-F(x^k)\|+\|F(x^k)-F(\bar{x})\|)\frac{\|F(x_i^k)-F(x^k)\|}{t_i^k}.
\end{aligned}\tag{5}$$

Finally, to see that $\nabla G(\bar{y})y$ is a strict derivate of $G\circ F$, add and subtract appropriately to get

$$\begin{aligned}
&\frac{1}{t_i^k}[G(F(x_i^k))-G(F(x^k))]-\nabla G(F(\bar{x}))y \\
&\quad=\frac{1}{t_i^k}[G(F(x_i^k))-G(F(x^k))]-\frac{1}{t_i^k}\nabla G(F(\bar{x}))(F(x_i^k)-F(x^k))
\end{aligned}$$

$$+\frac{1}{t_i^k}\nabla G(F(\bar{x}))(F(x_i^k)-F(x^k))-\nabla G(F(\bar{x}))y$$

$$=\frac{1}{t_i^k}[G(F(x_i^k))-G(F(x^k))-\nabla G(F(\bar{x}))(F(x_i^k)-F(x^k))] \tag{6}$$

$$+\nabla G(F(\bar{x}))\left(\frac{F(x_i^k)-F(x^k)}{t_i^k}-y\right).$$

Using (4) and (5) in (6), we obtain

$$\limsup\left\|\frac{G(F(x_i^k))-G(F(x^k))}{t_i^k}-\nabla G(F(\bar{x}))y\right\|\leqslant\sigma_k,$$

where $\sigma_k = r(G,F(\bar{x}),||F(x^k)-F(\bar{x})||)(||y||+\varepsilon_k)+||\nabla G(F(\bar{x}))||\varepsilon_k \to 0$ as $k\to\infty$. This shows that $\nabla G(F(\bar{x}))y\in D^{\sigma_k}(G\circ F,x^k)$ for each k. Consequently, $\nabla G(F(\bar{x}))y\in D_s(G\circ F)(\bar{x})$. □

Remark 2.12. It is obvious that we need only to assume the weaker condition that G is Fréchet or strictly Fréchet differentiable at $\bar{y}$ in $F(W)$, the range of F.

As special cases of Theorem 2.11, we have

Corollary 2.13 (Sum and product rules). *Let $\varepsilon\geq 0$.*

(a) *Suppose that $f,g:W\to Y$ and $(y,z)\in D^{\varepsilon}(f,g)(\bar{x})$, c is a constant, then $y\pm z\in D^{2\varepsilon}(f\pm g)(\bar{x})$, and $cy\in D^{|c|\varepsilon}(cf)(\bar{x})$.*

(b) *Suppose that $f:W\to\mathbb{R}$ and $g:W\to Y$ and $(y,z)\in D^{\varepsilon}(f,g)(\bar{x})$. Then $yg(\bar{x})+f(\bar{x})z\in D^{\varepsilon(||g(\bar{x})||+|f(\bar{x})|)}(fg)(\bar{x})$.*

Proof. Part (a) follows from the chain rule with $F=(f,g):W\to Y\times Y$ and $G: Y\times Y\to Y$ define by $G(y,z)=y+z$. Part (b) is a special case of the chain rule with $F=(f(x),g(x)):W\to\mathbb{R}\times Y$ and $G:\mathbb{R}\times Y\to Y$ defined by $G(r,y)=ry$. □

3 Relationship Between Derivates and Tangents

In this section, we focus on the relationship between derivates and several types of tangent spaces. The applications of the concepts and results in this section will be discussed in Sect. 4. First we recall some definitions.

Let Q be a subset of the Banach space Y and $\bar{y}\in Q$. The *Bouligand tangent space* $T(\bar{y},Q)$ is the set of $v\in Y$ such that there are sequences $v_i\in Y$, $v_i\to v$ and $t_i\downarrow 0$ such that $\bar{y}+t_iv_i\in Q$ for all i (see [5]). The following definition generalizes the notion of restrictive metric regularity (RMR) in [8, Definition 2.1].

Definition 3.1. Let W be a topological space. Let $f:W\to Y$ be a map, $\bar{x}\in W$ and $\bar{y}=f(\bar{x})$. We say that f has the RTR property around $\bar{x}$ if there are neighborhoods U and V of $\bar{x}$ and $\bar{y}$, respectively, such that for all $x\in U$ and a sequence $y_i\to f(x)$ (as $i\to\infty$), there exists a sequence $x_i\in f^{-1}(y_i)$ such that $x_i\to x$.

As shown by the following proposition, RTR is much weaker than RMR. Essentially, RTR requires that f has a "right" inverse function that is continuous, while RMR requires that the inverse be Lipschitz continuous.

Proposition 3.2. *If $f : W \longrightarrow Y$ is one-to-one, continuous and W is compact, then f has RTR at each $\bar{x} \in W$.*

Proof. It suffices to show the inverse $f^{-1} : f(W) \longrightarrow W$ is continuous. If f^{-1} is discontinuous at $\bar{y} = f(\bar{x})$, then there exists a sequence $y_i \longrightarrow \bar{y}$ such that $x_i = f^{-1}(y_i)$ does not converge to $\bar{x}$. By compactness of W, we may assume (by passing to a subnet) that $x_i \longrightarrow \bar{x}'$ as $i \longrightarrow \infty$. Consequently, $\bar{x}' \neq \bar{x}$ and by continuity of f, $f(\bar{x}') = \lim_{i\to\infty} y_i = \bar{y} = f(\bar{x})$, a contradiction to the assumption that f is one-to-one. So f has RTR at $\bar{x}$. □

First we prove a relationship with Bouligand tangent space.

Theorem 3.3. (a) *Let $f : W \longrightarrow Y$ be a map. Let $\bar{x} \in W$ and $\bar{y} = f(\bar{x})$. Then $D^0 f(\bar{x}) \subset T(\bar{y}, f(W))$.*

(b) *If f has the RTR property around $\bar{x}$, then $T(\bar{y}, f(W)) = D^0 f(\bar{x})$.*

Proof. (a) Let $y \in D^0 f(\bar{x})$. Then $y_i = \frac{f(x_i)-f(\bar{x})}{d_i} \longrightarrow y$ for some sequences $x_i \in W$ and $d_i \downarrow 0$. This implies that $\bar{y} + d_i y_i = f(x_i) \in f(W)$. So $y \in T(\bar{y}, f(W))$ by definition of Bouligand tangent.

(b) Let $y \in T(\bar{y}, f(W))$. We show $y \in D^0 f(\bar{x})$. Let $y_i \in Y$, $y_i \longrightarrow y$ and $d_i \downarrow 0$ be such that $\bar{y} + d_i y_i \in f(W)$. Since f has the RTR property around $\bar{x}$, there exists x_i' such that $f(x_i') = \bar{y} + d_i y_i$ and $x_i' \longrightarrow \bar{x}$. Since $\bar{y} = f(\bar{x})$, $\frac{f(x_i')-f(\bar{x})}{d_i} = y_i \longrightarrow y$. This shows that $y \in D^0 f(\bar{x})$. □

Next we generalize the definition of Gateaux differentiability of a map on X to a map $f : W \longrightarrow Y$, where W is a subset of a Banach space $(X, ||\cdot||)$.

Definition 3.4. We say that $f : W \longrightarrow Y$ is *restrictively Gateaux differentiable at $\bar{x} \in W$* if there exists a bounded and linear operator $\nabla f(\bar{x}) : X \longrightarrow Y$ such that for every $v \in T(\bar{x}, W)$ and sequences $t_i \downarrow 0$ and $v_i \in X$ with $v_i \longrightarrow v$, $\bar{x} + t_i v_i \in W$, the following holds

$$\lim_{i\to\infty} \frac{f(\bar{x}+t_i v_i) - f(\bar{x})}{t_i} = \nabla f(\bar{x}) v. \tag{7}$$

Remark 3.5. Recall that in the case $\bar{x}$ is an interior point of W, we say that f is Gateaux differentiable at $\bar{x}$ if

$$\lim_{i\to\infty} \frac{f(\bar{x}+t_i v) - f(\bar{x})}{t_i} = \nabla g(\bar{x}) v$$

for every $v \in X$ and $t_i \downarrow 0$. So restrictive Gateaux differentiability implies Gateaux differentiability. However, when $\bar{x}$ is not an interior point, the sequence $\bar{x} + t_i v$ may not be in W and so $f(\bar{x}+t_i v)$ and, hence the Gateaux derivative, may not be defined. That is why we replace $\bar{x}+t_i v$ by $\bar{x}+t_i v_i \in W$ in the definition of restrictive Gateaux differentiability.

We also note that the limit in (7) is assumed to exist and be independent of the sequence v_i. However, the linear and bounded operator $\nabla f(\bar{x})$ may not be uniquely defined on the whole space X. For example, if W is a subspace of X and W has a complemented subspace V so that $X = W \oplus V$ then $\nabla f(\bar{x})$ can be extended arbitrarily on V. Finally, note that strict Fréchet differentiability (Definition 2.8) implies restrictive Gateaux differentiability.

To further generalize restrictive Gateaux differentiability by localizing to a subspace, let $Z \subset X$ be a subspace. Denote by $T(\bar{x},W,Z)$ the set of all $v \in X$ such that there are sequences $z_i \in Z$, $z_i \to 0$ and $t_i \downarrow 0$ such that $\bar{x}+t_i(z_i+v) \in W$. Clearly, $T(\bar{x},W,Z) \subset T(\bar{x},W,X) = T(\bar{x},W)$ – the Bouligand tangent space. Note that $z_i+v \in Z_v$, where Z_v is the linear space spanned by Z and the vector v. So each $v \in T(\bar{x},W,Z)$ is a Bouligand tangent vector of $(W-\bar{x}) \cap Z_v$ at 0 (in the space Z_v).

Definition 3.6. Let W be a subset of the Banach space X and Z a subspace of X. We say that $f : W \to Y$ is *restrictively Gateaux differentiable at* $\bar{x} \in W$ *with respect to* Z if there exists a bounded and linear operator $\nabla f(\bar{x}) : X \to Y$ such that for every $v \in T(\bar{x},W,Z)$ and all sequences $t_i \downarrow 0$ and $z_i \in Z \to 0$ with $\bar{x}+t_i(z_i+v) \in W$, the following holds

$$\lim_{i\to\infty} \frac{f(\bar{x}+t_i z_i+t_i v)-f(\bar{x})}{t_i} = \nabla f(\bar{x})v.$$

Definition 3.6 generalizes the notion "f is (X,Z)-differentiable" as defined in [2, p. 443]. In particular, we require f be defined only on a subset W, not on all of X. If $W = X$, then our notion is the same as the same as "f is (X,Z)-differentiable" at $\bar{x}$. As noted in [2, Remarks 2.1 and 2.3], a restrictively Gateaux differentiable map at $\bar{x} \in W$ with respect to Z does not require the map be continuous at $\bar{x} \in W$. The idea here is to restrict the possible directions for approaches to $\bar{x}$.

The next theorem shows that derivates can be calculated in terms of these generalized derivatives defined above.

Theorem 3.7. (a) *If* $\bar{x}$ *is an interior point of* W *and* f *is Gateaux differentiable at* $\bar{x}$, *then* $\nabla f(\bar{x})v \in D^0 f(\bar{x})$ *for every* $v \in X$.

(b) *If* f *is restrictively Gateaux differentiable at* $\bar{x}$ *with respect to* Z *then* $\nabla f(\bar{x})v \in D^0 f(\bar{x})$ *for every* $v \in T(\bar{x},W,Z)$.

Proof. This follows directly from Definitions 3.4 and 3.6 and the definition of derivate. □

Next we prove a relationship between the Clarke tangent space $T_C(\bar{x},W)$ of W and the strict derivate $D_s f(W)$. Recall that $v \in T_C(\bar{x},W)$ if and only if for every pair of sequences $x_i \in W$, $x_i \to \bar{x}$ and $t_i \to 0$ there exists $v_i \in X$, $v_i \to v$ such that $\bar{x}+t_i v_i \in W$ for all i. A vector v is called a *hypertangent* of W at $\bar{x}$ if there exists $\varepsilon > 0$ such that $y+tw \in W$ for all $t \in (0,\varepsilon)$, $y \in B(\bar{x},\varepsilon) \cap W$ and $w \in B(v,\varepsilon)$; see [1, p. 57]. A well-known theorem of Rockafellar (see Theorem 2.4.8 in [1]) says that if W has a hypertangent vector at $\bar{x}$, then every vector in $T_C(\bar{x},W)$ is a hypertangent vector.

Theorem 3.8. *Let W be a subset of a Banach space X. If $f : W \to Y$ is strictly Fréchet differentiable at $\bar{x}$ and W has a hypertangent vector at $\bar{x}$, then $\nabla f(\bar{x})v \in D_s f(\bar{x})$ for all $v \in T_C(\bar{x}, W)$.*

Proof. By Rockafellar's theorem, every tangent vector v in $T_C(\bar{x}, W)$ is a hypertangent vector. Let $x^k \in W$, $x^k \to \bar{x}$ and $t_i \downarrow 0$, then by the definition of hypertangent, $x^k + t_i v \in W$ for all i, k that are sufficiently large. By strict Fréchet differentiability of f in (2), for each fixed $k > 0$, we have

$$||f(x^k + t_i v) - f(x^k) - t_i \nabla f(\bar{x})(v)|| \leqslant t_i ||v|| r(f, \bar{x}, t_i ||v|| + ||x^k - \bar{x}||).$$

Dividing the inequality by t_i, letting $i \to \infty$, and using that r is continuous in η from the right, we get

$$\lim_{i \to \infty} \left\| \frac{f(x^k + t_i v) - f(x^k)}{t_i} - \nabla f(\bar{x}) v \right\| \leqslant \varepsilon_k,$$

where $\varepsilon_k = ||v|| r(f, \bar{x}, ||x^k - \bar{x}||)$. So $\nabla f(\bar{x})(v) \in D^{\varepsilon_k} f(x^k)$. Since $\varepsilon_k \to 0$, we see that $\nabla f(\bar{x}) v \in D_s f(\bar{x})$. □

To generalize Theorem 3.8, we introduce the following concepts, which generalize the concept of hypertangent and are in the spirit of the construction of derivates.

Definition 3.9. (1) A vector $v \in X$ is called a *strict tangent vector* at $\bar{x}$ if for every sequence $x^k \in W$, $x^k \to \bar{x}$, there exists a sequence $v^k \in T(x^k, W)$, $v^k \to v$ as $k \to \infty$. We denote by $T^s(\bar{x}, W)$ the set of all strict tangent vectors at $\bar{x}$.

(2) The *strict normal* $N^s(\bar{x}, W)$ is defined as the set of $x^* \in X$ such that

$$\langle x^*, v \rangle \leqslant 0$$

for all $v \in T^s(\bar{x}, W)$. (The symbol $\langle x^*, v \rangle$ is used here to denote the value of the functional x^* acting on the vector v.)

Note that $T^s(\bar{x}, W)$ may be nonempty even though W has no hypertangent vectors. For example, $W = \{(x, 0) : x \in [0, 1]\} \subset \mathbb{R}^2$ has no hypertangent but $T^s(\bar{x}, W)$, for any $\bar{x} \in W$ is the linear subspace $\mathbb{R} \times \{0\}$.

The following example shows that $T^s(0, W)$ may be significantly larger than the Clarke tangent cone $T_C(0, W)$.

Example 3.10. Let $W = \{0\} \cup_{k=0}^{\infty} [a_k, \frac{5}{4} a_k)$, where $a_k = 2^{-k}$. We claim that

$$T_C(0, W) = \{0\}; \quad T(0, W) = T^s(0, W) = [0, \infty).$$

To show $T_C(0, W) = \{0\}$, it suffices to show that $1 \notin T_C(0, W)$ (since $T_C(0, W)$ is convex). Indeed, if $1 \in T_C(0, W)$, then for $x_k = a_k \to 0$ and $t_k = a_k / 2 \to 0$ there should exist $v_k \to 1$ such that

$$a_k + \frac{1}{2} a_k v_k \in W.$$

This means either $a_k + \frac{1}{2}a_k v_k \leqslant \frac{5}{4}a_k$ or $a_k + \frac{1}{2}a_k v_k \geq a_{k-1}$. In the first case, $v_k \leqslant 1/2$. In the second case, $v_k \geq 2$. We see that in either case, v_k does not approach 1, a contradiction.

To show that $T(0,W) = T^s(0,W) = [0,\infty)$, we need only show that 1 is in both $T(0,W)$ and $T^s(0,W)$ because they are cones. For $1 \in T(0,W)$, we need $v_k \to 1$ and $t_k \downarrow 0$ so that $0 + t_k v_k \in W$. Let $v_k \equiv 1$ and $t_k = 2^{-k}$. Then $t_k v_k = 2^{-k} \cdot 1 \in [2^{-k}, (5/4)2^{-k})$. So $0 + t_k v_k \in W$, which implies $v = 1 \in T(0,W)$. In fact, it is easy to see that $1 \in T(x,W)$ for every $x \in W$. So $1 \in T^s(0,W)$.

The next proposition clarifies the relationships among the hypertangents, the Bouligand tangent $T(\bar{x},W)$, and the strict tangent $T^s(\bar{x},W)$.

Proposition 3.11. (1) $T^s(\bar{x},W) \subset T(\bar{x},W)$, *that is, each strict tangent is a Bouligand tangent.*

(2) *Each hypertangent vector of W at $\bar{x}$ is a strict tangent vector.*

Proof. (1) Let $v \in T^s(\bar{x},W)$. Then apply the definition to the constant sequence $x^k = \bar{x}$ to get a sequence $v^k \in T^{\varepsilon_k}(\bar{x},W)$, $v^k \to v$. So, for every k, there exists $x_i^k \to x^k$ and $t_i^k \to 0$ as $i \to \infty$ such that $\limsup_{i\to\infty} \frac{x_i^k - x^k}{t_i^k} = v^k$. Consequently, there exists a subsequence i_k such that $x_k \equiv x_{i_k}^k \to \bar{x}$ and $t_k = t_{i_k} \to 0$ with $\lim_{k\to\infty} \frac{x_k - \bar{x}}{t_k} = v$. This shows that $v \in T(\bar{x},W)$.

(2) Let $v \in X$ be a hypertangent vector and $x^k \to \bar{x}$ be a sequence. By definition of hypertangent, there exists $\varepsilon > 0$ such that $y + tw \in W$ for all $t \in (0,\varepsilon)$, $y \in B(\bar{x},\varepsilon) \cap W$ and $w \in B(v,\varepsilon)$. Let K be an integer such that $x^k \in B(\bar{x},\varepsilon)$ for all $k \geq K$. Let $t_i = 1/i$ for $i = 1,2,\ldots$. Then $x^k + t_i v \in W$ for all $i \geq 1/\varepsilon$. Consequently, $v \in T(x^k,W)$, which implies that $v \in T^s(\bar{x},W)$. □

The next theorem describes certain strict derivates of a strictly differentiable map.

Theorem 3.12. *If $f : W \to Y$ is strictly Fréchet differentiable at $\bar{x}$ then $\nabla f(\bar{x},W)v \in D_s f(\bar{x})$ for all $v \in T^s(\bar{x},W)$.*

Proof. Let $x^k \in W$, $x^k \to \bar{x}$ and $v \in T^s(\bar{x},W)$. By definition, there exists a sequence $v^k \in T^{\varepsilon^k}(x^k,W)$, $v^k \to v$. So, for each $k \geq 1$, there exists a sequence $x_i^k \to x^k$ as $i \to \infty$ and a sequence $t_i^k \downarrow 0$ such that $\lim_{i\to\infty} v_i^k = v^k$, where $v_i^k = \frac{x_i^k - x^k}{t_i^k}$. Therefore, $x^k + t_i^k v_i^k = x_i^k \in W$ for all i,k that are sufficiently large. By strict Fréchet differentiability of f, for each fixed $k > 0$, we have

$$||f(x^k + t_i^k v_i^k) - f(x^k) - t_i^k \nabla f(\bar{x}) v_i^k|| \leqslant t_i^k \, ||v_i^k|| \, r(f,\bar{x},t_i^k ||v_i^k|| + ||x^k - \bar{x}||).$$

This implies that

$$\limsup_{i\to\infty} \left\| \frac{f(x^k + t_i^k v_i^k) - f(x^k)}{t_i^k} - \nabla f(\bar{x}) v_i^k \right\| \leqslant \delta_k,$$

where $\delta_k = ||v||r(f,\bar{x},||x^k - \bar{x}||)$. Since

$$\limsup_{i\to\infty} ||v_i^k - v|| \leqslant ||v^k - v||,$$

we conclude that

$$\limsup_{i\to\infty} \left\| \frac{f(x^k + t_i^k v_i^k) - f(x^k)}{t_i^k} - \nabla f(\bar{x})v \right\| \leqslant \varepsilon_k,$$

where $\varepsilon_k = \delta_k + ||\nabla f(\bar{x})|| \cdot ||v^k - v|| \to 0$ as $k \to \infty$. This implies that $\nabla f(\bar{x}, W)v \in D^{\varepsilon_k} f(x^k)$ and $\nabla f(\bar{x})v \in D_s f(\bar{x})$. □

4 Some Applications of Derivates

Next we apply the concept of derivate to derive necessary conditions for solutions to typical optimization problems. There is an abundant literature on this problem; see [6] and [9], for example.

Let $f : W \to \mathbb{R}$ be a function and $g : W \to Y$ be a map, where W is a topological space and Y a Banach space. Let $Q \subset Y$ be a closed subset. Consider

$$\min f(x), \quad \text{for } g(x) \in Q. \tag{8}$$

One approach is to rewrite the operator constraint as part of a geometric constraint. That is, the problem is equivalent to minimizing $f(x)$ for $x \in \Omega = W \cap g^{-1}(Q)$. By Fermat's Theorem 2.7, we have

Theorem 4.1. *The minimum of f on W subject to $g(x) \in Q$ occurs at $\bar{x}$ iff $D^0 f(\bar{x}, W \cap g^{-1}(Q)) \subset [0,\infty)$.*

The following example shows that Theorem 4.1 can be used to exclude candidates for the minimum which are proposed as candidates for solutions by other necessary conditions.

Example 4.2. Consider problem (8) with the functions

$$f(x) = \begin{cases} x & \text{if } x \leqslant 0, \\ x\sin^2(\ln x)) & \text{if } x > 0, \end{cases}$$

and $g(x) = x$, $W = (-\infty,\infty)$ and $Q = (-\infty, 0]$. (We could also phrase this with the same functions and $W = Q = (-\infty, 0]$ and no operator constraint.) Obviously, the minimization problem (8), which is the same as minimizing $f(x) = x$ on $(-\infty, 0]$, has no solution. Now let us apply Theorem 4.1 to this problem. Note that $W = g^{-1}(Q) = (-\infty, 0]$ and $D^0 f(\bar{x}, (-\infty, 0]) = (-\infty, \infty)$ for all $\bar{x} < 0$ and

$D^0 f(0,(-\infty,0]) = (-\infty,0]$. By Theorem 4.1, no point $\bar{x}$ can be a minimum point. However, if we apply Theorem 5.21 in [6], which is a general necessary condition for the problem (5.23) in [6, p. 22], we would get

$$0 \in \lambda_0 \partial f(\bar{x}) + \lambda_1 \partial g(\bar{x}), \tag{9}$$

where ∂f is the limiting subdifferential of f. Let us check this condition at $\bar{x} = 0$. It is easy see that $\partial g(0) = \{1\}$. To find $\partial f(0)$, by Theorem 1.89 in [5, p. 90], $\partial f(0) = \limsup_{x\to 0} \hat{\partial} f(x)$, where $\hat{\partial} f(x)$ is the Fréchet subdifferential of f at x. For $x \neq 0$, f is differentiable. So $\hat{\partial} f(x) = \{1\}$ for $x < 0$, $\hat{\partial} f(x) = \sin^2(\ln x) + \sin(2\ln x) \in [\frac{1-\sqrt{5}}{2}, \frac{1+\sqrt{5}}{2}]$ for $x > 0$ and $\hat{\partial} f(0) = \{0\}$. It follows that $\partial f(0) = [\frac{1-\sqrt{5}}{2}, \frac{1+\sqrt{5}}{2}]$ and the condition (9) is satisfied with $\lambda_0 = \lambda_1 = 1$ at $\bar{x} = 0$, even though 0 is not a minimum point, as mentioned above.

Now let us consider the case where W is an open subset of a Banach space X and $Q \subset Y$ is a closed subset. Although the open set W is a metric space, this case is distinct from the setting in [4] and [7], where W is a complete metric space. Let f and g be defined on W and $\bar{x} \in W$. The problem can be equivalently stated as

$$\min f(x), \quad x \in g^{-1}(Q). \tag{10}$$

First we derive a necessary condition for a solution $\bar{x}$ of (8) in terms of f, g, and Q. The proof is based on Theorem 3.7.

Theorem 4.3. *Let $\bar{x} \in g^{-1}(Q)$ be a solution of Problem (10). Suppose that there is a closed subspace $Z \subset X$ such that f and g are continuous and restrictively Gateaux differentiable at $\bar{x} \in g^{-1}(Q)$ with respect to Z and*

(a) *The map $\nabla g(\bar{x}) : Z \to Y_g = \nabla g(\bar{x})(X)$ is one-to-one and onto.*

(b) *$Ker(\nabla g(\bar{x})) \subset T(\bar{x}, g^{-1}(Q), Z)$.*

(c) *For each $w \in T(\bar{y}, Q) \cap Y_g$, there exists $v \in T(\bar{x}, g^{-1}(Q), Z)$ such that $\nabla g(\bar{x})v = w$.*

Then there exists $y^* \in Y^*$ such that

$$\nabla f(\bar{x}) = \langle y^*, \nabla g(\bar{x}) \rangle \tag{11}$$

and $\langle y^*, w \rangle \geq 0$ for all $w \in T(\bar{y}, Q) \cap Y_g$.

The notation $\langle y^*, \nabla g(\bar{x}) \rangle$ denotes the linear functional in X^* whose value at $v \in X$ is $\langle y^*, \nabla g(\bar{x}) \rangle (v) = y^*(\nabla g(\bar{x})v) = \langle y^*, \nabla g(\bar{x})v \rangle$.

Proof. Since $\nabla g(\bar{x})$ is a bijection between Z and Y_g, by the open mapping theorem, it has a bounded inverse $\Lambda : Y_g \to Z$. In particular, $\nabla g(\bar{x})\Lambda = id$ on Y_g. So any $v \in X$, we have that, $v - \Lambda \nabla g(\bar{x})v \in Ker(\nabla g(\bar{x}))$. By the assumption (b),

$$v - \Lambda \nabla g(\bar{x})v \in T(\bar{x}, g^{-1}(Q), Z).$$

By Theorem 3.7(b), $\nabla f(\bar{x})(v - \Lambda \nabla g(\bar{x})v) \in D^0 f(\bar{x}, g^{-1}(Q))$. Since $\bar{x}$ is a solution of Problem (10), by Theorem 4.1, we have that

$$\nabla f(\bar{x})v - [\nabla f(\bar{x})\Lambda]\nabla g(\bar{x})v \geq 0.$$

Since $v \in X$ is arbitrary, we have $\nabla f(\bar{x}) = [\nabla f(\bar{x})\Lambda]\nabla g(\bar{x}) = \langle y^*, \nabla g(\bar{x})\rangle$ with $y^* = \nabla f(\bar{x})\Lambda$.

Now let $w \in T(\bar{y}, Q) \cap Y_g$ and we prove $\langle y^*, w\rangle \geq 0$. By assumption (c), there is a $v \in T(\bar{x}, g^{-1}(Q), Z)$ such that $w = \nabla g(\bar{x})v$. By (11) and Theorem 3.7

$$\langle y^*, w\rangle = \langle y^*, \nabla g(\bar{x})v\rangle = \langle y^*, \nabla g(\bar{x})\rangle(v) = \nabla f(\bar{x})v \in D^0 f(\bar{x}, g^{-1}(Q)).$$

Since $\bar{x}$ is a minimum point, by Theorem 4.1, $\langle y^*, w\rangle \geq 0$. □

The main difference between Theorem 4.3 and the other necessary conditions is we only assume that f and g are differentiable in $g^{-1}(Q)$ with respect to some subspace Z plus the assumptions (a)–(c).

For problems with finite constraints, conditions (b) and (c) are automatic. In this case, Theorem 4.3 generalizes the relevant results in [2] and [3]. Let W be an open neighborhood of $\bar{x}$ in X, $f : W \to \mathbb{R}$ and $g = (g_1, \ldots, g_{n+m}) : W \to \mathbb{R}^{n+m}$. Let $Q \subset \mathbb{R}^{n+m}$ be any closed subset. We obtain the following corollary.

Corollary 4.4. *Suppose there is an $n+m$-dimensional subspace $Z \subset X$ such that f and g are continuous in a neighborhood of $\bar{x}$ in Z restrictively Gateaux differentiable at $\bar{x} \in g^{-1}(Q)$ with respect to Z and the map $\nabla g(\bar{x}) : Z \to \mathbb{R}^{n+m}$ is one-to-one and onto. If $\bar{x}$ is a minimum point of f subject to $g(x) \in Q$, then there exists $y^* \in Y^*$ such that*

$$\nabla f(\bar{x}) = \langle y^*, \nabla g(\bar{x})\rangle$$

and $\langle y^*, w\rangle \leqslant 0$ for all $w \in T(\bar{y}, Q)$.

Proof. We need only verify conditions (b) and (c) in Theorem 4.3. Condition (b) is proved in [2, Lemma 2.1]. To prove (c), let $w \in T(\bar{y}, Q)$. Then there exist sequences $t_i \downarrow 0$ and $w_i \in \mathbb{R}^{n+m}$, $w_i \to w$ such that $\bar{y} + t_i w_i \in Q$. Since $\nabla g(\bar{x})$ is onto $\mathbb{R}^{n+m}$, there exists $v_i \in Z$ such that $w_i = \nabla g(\bar{x})v_i$. Because $v_i = \Lambda w_i$, and the inverse Λ of $\nabla g(\bar{x})$ is bounded, we have that $v_i \to \Lambda w$. Apply Theorem F in [3] to the function $g(\bar{x}+z)$ for $z \in Z$ near $z = 0$. We get a function $\xi : Z \to Z$ such that $\xi(z) = o(||z||)$ and

$$g(\bar{x} + z + \xi(z)) = g(\bar{x}) + \nabla g(\bar{x})z$$

for z near 0 in Z. In particular,

$$g(\bar{x} + t_i v_i + \xi(t_i v_i)) = g(\bar{x}) + \nabla g(\bar{x})t_i v_i = \bar{y} + t_i w_i \in Q.$$

This shows that $\bar{x} + t_i v_i + \xi(t_i v_i) \in g^{-1}(Q)$. Since $\xi(t_i v_i) = o(t_i)$, $v = \lim_{i\to\infty} v_i \in T(\bar{x}, g^{-1}(Q)Z)$. Therefore, $v = \Lambda w$, that is, $w = \nabla g(\bar{x})v$. This shows (c). □

We note that Corollary 4.4 generalizes the main theorem in [2, Theorem 2.2] and the multiplier rule in [3, p. 235], where Q is the "cube"

$$Q = \{y \in \mathbb{R}^{n+m} : y_1 \leqslant 0, \ldots, y_n \leqslant 0, y_{n+1} = \cdots = y_{n+m} = 0\}.$$

Finally, we mention that the Lagrange multiplier rule in [4] and [7] can be applied to the problem (8) to obtain necessary conditions in terms of the strict normal introduced earlier.

Theorem 4.5 (Necessary conditions). *Suppose that Y^* is strictly convex and Q is closed, convex and finite codimensional. Then there exists a non-zero pair $(\psi^0, \psi) \in \mathbb{R}^+ \times Z^*$ such that*

$$z^0\psi^0 + \langle z, \psi\rangle \geq 0 \quad \textit{for } (z^0, z) \in D_s(f, g)(\bar{x}, W)$$

and $\langle \psi, \eta - g(\bar{x})\rangle \leqslant 0$ for all $\eta \in Q$. In particular, if f and g are strictly differentiable at $\bar{x}$, then

$$\psi^0\nabla f(\bar{x}, W) + \langle \psi, \nabla g(\bar{x}, W)\rangle \in -N^s(\bar{x}, W).$$

Proof. The first part is a restatement of Theorem 6 in [4]. In the case f and g are strictly differentiable at $\bar{x}$ on W, we have that $(\nabla f(\bar{x}, W)v, \nabla g(\bar{x}, W)v) \in D_s(f, g)(\bar{x}, W)$ for all $v \in T^s(\bar{x}, W)$. The second conclusion follows from the first by the definition of strict normal in (3.9). □

More general but similar conditions can be obtained by applying Theorem 5.4 in [7] to Problem (8), where the set Q can be nonconvex and Y can be a Gateaux smooth Banach space.

References

1. Clarke, Frank H.: Optimization and Nonsmooth Analysis, Wiley, New York (1983)
2. Duc, Duong Minh; Hoang, Nguyen Dinh; and Nguyen, Lam Hoang: Lagrange multipliers theorem and saddle point optimality criteria in mathematical programming, J. Math. Anal. Appl. **323**, 441–455 (2006)
3. Halkin, H.: Implicit functions and optimization problems without continuous differentiability of the data, SIAM J. Control **12**, 229–236 (1974)
4. McAsey, Michael and Mou, Libin: A multiplier rule on a metric space, J. Math. Anal. Appl. **337**, 1064–1071 (2008)
5. Mordukhovich, Boris S.: Variational Analysis and Generalized Differentiation, I: Basic Theory, Springer, Berlin (2006)
6. Mordukhovich, Boris S.: Variational Analysis and Generalized Differentiation, II: Applications, Springer, Berlin (2006)
7. Mordukhovich, Boris S. and Mou, Libin: Necessary conditions for nonsmooth optimization problems with operator constraints in metric spaces. J. Convex Anal. **16**, 913–937 (2009)
8. Mordukhovich, Boris S. and Wang, Bingwu: Restrictive metric regularity and generalized calculus in Banach spaces, Int. J. Math. Math. Sci. **50**, 2653–2680 (2004)
9. Rockafellar, R. Tyrell and Wets, Roger J.-B.: Variational Analysis, Springer, Berlin (1998)

On Computing the Mordukhovich Subdifferential Using Directed Sets in Two Dimensions

Robert Baier, Elza Farkhi, and Vera Roshchina

Dedicated to Boris Mordukhovich on his sixtieth birthday

Abstract The Mordukhovich subdifferential, being highly important in variational and nonsmooth analysis and optimization, often happens to be hard to calculate. We propose a method for computing the Mordukhovich subdifferential of differences of sublinear (DS) functions via the directed subdifferential of differences of convex (DC) functions. We restrict ourselves to the two-dimensional case mainly for simplicity of the proofs and for the visualizations.

The equivalence of the Mordukhovich symmetric subdifferential (the union of the corresponding subdifferential and superdifferential) to the Rubinov subdifferential (the visualization of the directed subdifferential) is established for DS functions in two dimensions. The Mordukhovich subdifferential and superdifferential are identified as parts of the Rubinov subdifferential. In addition, it is possible to construct the directed subdifferential in a way similar to the Mordukhovich one by considering outer limits of Fréchet subdifferentials. The results are extended to the case of DC functions. Examples illustrating the obtained results are presented.

2010 Mathematics Subject Classification. Primary 49J52; Secondary 26B25, 49J50, 90C26

Robert Baier
Chair of Applied Mathematics, University of Bayreuth, 95440 Bayreuth, Germany
e-mail: robert.baier@uni-bayreuth.de

Elza Farkhi
School of Mathematical Sciences, Sackler Faculty of Exact Sciences, Tel Aviv University, 69978 Tel Aviv, Israel
e-mail: elza@post.tau.ac.il

Vera Roshchina
Centro de Investigação em Matemática e Aplicações, Universidade de Évora, Colégio Luís Verney, Rua Romão Ramalho 59, 7000-671 Évora, Portugal
e-mail: vera.roshchina@gmail.com

R.S. Burachik and J.-C. Yao (eds.), *Variational Analysis and Generalized Differentiation in Optimization and Control*, Springer Optimization and Its Applications 47, DOI 10.1007/978-1-4419-0437-9_3,

1 Introduction

The Mordukhovich subdifferential is a highly important notion in variational analysis, closely related to optimality conditions, metric regularity, Lipschitzness and other fundamental concepts of modern optimization theory (see [23, 24]). This subdifferential is a closed subset of the Clarke subdifferential (see e.g., [25, Theorem 9.2]), and may be nonconvex for nonconvex functions, thus achieving sharper optimality conditions. In contrast to the Fréchet subdifferential (cf. [18, Example 1.1]), the Mordukhovich subdifferential of a locally Lipschitz function is always nonempty (see e.g., [22, (2.17)]).

Along with these essential advantages, there comes a substantial drawback: the Mordukhovich subdifferential is difficult to calculate even for relatively simple examples, as such computation normally involves finding the Painlevé–Kuratowski outer limit (see Sect. 2). For most known subdifferentials, the sum rule only has the form of an inclusion – the subdifferential of a sum is a subset of the sum of the subdifferentials [23, Theorem 3.36]. This rule applied in calculations only provides a superset of the subdifferential of the sum.

We propose a method for computing the Mordukhovich subdifferential of differences of sublinear (DS) functions, which are positively homogeneous DC (difference of convex) functions, applying directed sets [2] and the directed subdifferential of DC functions [4]. The DC functions represent a large family of functions. They are dense in the space of continuous functions [16] and constitute an important subclass of the quasidifferentiable functions [10]. Various aspects of calculus and optimality conditions for this class of functions are discussed, for example, in [1, 8, 10–12, 14, 20].

The class of positively homogeneous DC functions is important enough since it contains differences of support functions and directional derivatives of DC functions. Many interesting examples of nonconvex DC functions in the literature are in this class (see, e.g., [4]). All results in Sect. 3 obtained first for DS functions can be formulated as a corollary for the directional derivative of DC functions.

The main advantage of directed subdifferentials based on directed sets is the sum rule: the directed subdifferential of a sum is equal to the sum of the directed subdifferentials [4, Proposition 4.2]. This rule applied for directed subdifferentials provides the exact result.

We restrict ourselves to the two-dimensional case mainly for simplicity of the proofs and for the visualizations. Furthermore, the visualization of the directed subdifferential is essentially more complicated in dimensions higher than two, since lower dimensional mixed-type parts missing in the two-dimensional case would emerge in higher dimensions.

In this chapter, the equivalence of the Mordukhovich symmetric subdifferential, the union of the corresponding subdifferential and superdifferential, to the Rubinov subdifferential (the visualization of the directed subdifferential), is established in Theorem 3.14 for the special class of DS functions in two dimensions.

While the Mordukhovich subdifferential is based on the corresponding normal cone and can be calculated by outer limits of the Fréchet subdifferential, the directed subdifferential for DC functions is essentially based on the subtraction of convex subdifferentials embedded in the Banach space of directed sets. Although these two concepts differ substantially, there are many interesting links between them.

In Theorem 3.13, we prove that certain parts of the Rubinov subdifferential comprise the Mordukhovich subdifferential. The remaining parts coincide with the Mordukhovich superdifferential (see Theorem 3.14). Furthermore, Theorem 3.11 links outer limits of the Fréchet subdifferential to the directed subdifferential. The assumption on positive homogeneity of the DC functions is dropped in Theorems 3.16 and 3.17 yielding the connection of the Rubinov subdifferential to the Mordukhovich symmetric subdifferential of the directional derivative for the broader class of DC functions.

This chapter is organized as follows. In the next section, we recall necessary definitions, notation and results on Fréchet subdifferential. In Sect. 3, the relation between the Mordukhovich and the directed subdifferential is discussed. We illustrate our results with several examples in Sect. 4. In the last section, we sketch directions for future research.

2 Preliminaries

Recall that $f : \mathbb{R}^n \to \mathbb{R}$ is called *positively homogeneous*, if $f(\lambda x) = \lambda f(x)$ for all $x \in \mathbb{R}^n$ and $\lambda > 0$. Clearly, $f(0) = 0$ for positively homogeneous functions. A function is *sublinear* if it is convex and positively homogeneous. Recall that support functions of compact sets are sublinear. We denote by $\mathscr{S}_{n-1}$ the unit sphere in $\mathbb{R}^n$, and by $\mathrm{cl}(A), \mathrm{co}(A)$ the closure and the convex hull of the set A, respectively. The following operations on sets $A, B \subset \mathbb{R}^n$ are well known:

$$\begin{aligned} A + B &:= \{a + b \,|\, a \in A,\ b \in B\} && \text{(Minkowski addition)},\\ \ominus A &:= \{-a \,|\, a \in A\} && \text{(the pointwise negative of the set } A). \end{aligned}$$

The last operation is used in the definition of the Mordukhovich superdifferential and in the negative part of the visualization of the directed subdifferential.

For the sets $A, B \subset \mathbb{R}^n$ the operation

$$A \stackrel{*}{-} B = \{x \in \mathbb{R}^n \,|\, x + B \subset A\} = \bigcap_{b \in B} (A - b)$$

is called the *geometric difference* of the sets A and B. This difference is introduced by Hadwiger in [13] as well as in [28] and is also called Minkowski–Pontryagin difference.

Let $C \subset \mathbb{R}^n$ be nonempty, convex, compact and $l \in \mathbb{R}^n$. Then, the *support function* and respectively the *supporting face* of C in direction l are defined by

$$\delta^*(l,C) = \max_{c \in C} \langle l, c \rangle ,$$
$$Y(l,C) = \{y \in C \,|\, \langle l, y \rangle = \delta^*(l,C)\} = \arg\max_{c \in C} \langle l, c \rangle .$$

Note that for $l = 0$, $Y(l,C) = C$. By $y(l,C)$, we denote any point of the set $Y(l,C)$, and if the latter is a singleton (i.e., there is a unique supporting point), then $Y(l,C) = \{y(l,C)\}$.

The supporting face $Y(l,C)$ equals the subdifferential of the support function of C at l [29, Corollary 23.5.3].

We denote by Lim sup the *Painlevé–Kuratowski outer limit* and by Lim inf the *inner limit* of sets (see [30, Chap. 4]). Intuitively, the outer limit of a sequence of sets consists of the limiting points of all converging subsequences of points from these sets. In contrast, the inner limit consists of limiting points of all sequences constructed from points taken from almost every set in a way that only a finite number of sets can be missed out. For a more rigorous definition (see [30, Sect. 4.A]), first consider the set $\mathscr{N}_\infty^\sharp$ of all infinite subsequences in the set of natural numbers $\mathscr{N}_\infty^\sharp := \{N \subset \mathbb{N} \,|\, N \text{ infinite}\}$, and the set $\mathscr{N}_\infty$ of all the sequences of natural numbers which include all numbers beyond a certain value, that is, $\mathscr{N}_\infty := \{N \subset \mathbb{N} \,|\, \mathbb{N} \setminus N \text{ finite}\}$. Given a sequence $\{C_k\}$ of sets in $\mathbb{R}^n$, we set

$$\operatorname*{Lim\,sup}_{k \to \infty} C_k = \{x \in \mathbb{R}^n \,|\, \exists N \in \mathscr{N}_\infty^\sharp ,\ \exists x_k \in C_k (k \in N) \text{ with } x_k \to x\} ,$$
$$\operatorname*{Lim\,inf}_{k \to \infty} C_k = \{x \in \mathbb{R}^n \,|\, \exists N \in \mathscr{N}_\infty ,\ \exists x_k \in C_k (k \in N) \text{ with } x_k \to x\} .$$

For a set-valued mapping $F : \mathbb{R}^n \to \mathbb{R}^m$ and $\bar{x} \in \mathbb{R}^n$, the outer and inner limit of F as $x \to \bar{x}$ is naturally defined as

$$\operatorname*{Lim\,sup}_{x \to \bar{x}} F(x) := \{y \in \mathbb{R}^m \,|\, \exists x_k \to \bar{x} ,\ y_k \to y \text{ with } y_k \in F(x_k)\ \forall k \in \mathbb{N}\} , \tag{1}$$
$$\operatorname*{Lim\,inf}_{x \to \bar{x}} F(x) := \{y \in \mathbb{R}^m \,|\, \forall x_k \to \bar{x} ,\ \exists N \in \mathscr{N}_\infty ,\ \exists y_k \to y \text{ with } y_k \in F(x_k)\ \forall k \in N\} . \tag{2}$$

Clearly, the inner limit is a subset of the outer limit. If they are equal, this set is called the *Painlevé–Kuratowski limit* and is denoted by $\operatorname{Lim}_{k \to \infty} C_k$, respectively, $\operatorname{Lim}_{x \to \bar{x}} F(x)$.

Remark 2.1. Let $F(\cdot)$ be a uniformly bounded mapping defined in a neighborhood of the point $\bar{x} \in \mathbb{R}^n$ with nonempty images in a finite-dimensional space. It is easy to show that if the Painlevé–Kuratowski outer limit is a singleton $\operatorname{Lim\,sup}_{x \to \bar{x}} F(x) = \{\bar{y}\}$, it is equal to the Painlevé–Kuratowski limit. Indeed, by the assumption, for any sequence $x_n \to \bar{x}$, there is a converging subsequence $y_{n_k} \in F(x_{n_k})$ and any such subsequence may have only the point $\bar{y}$ as the limit.

The classical Moreau–Rockafellar *subdifferential* of a convex function $f : \mathbb{R}^n \to \mathbb{R}$ at $x \in \mathbb{R}^n$ is

$$\partial f(x) := \{s \in \mathbb{R}^n \,|\, \forall y \in \mathbb{R}^n :\ \langle s, y - x \rangle + f(x) \le f(y)\} . \tag{3}$$

It is well-known (see, e.g., [15, Chap. V, Definition 1.1.4]) that

$$\delta^*(l, \partial f(x)) = f'(x;l)\,, \tag{4}$$

where $f'(x;l)$ is the *directional derivative* of f at x in direction l.

In the sequel, the Moreau–Rockafellar *subdifferential of a sublinear function g* at zero is denoted by ∂g instead of $\partial g(0)$.

Also, for the *unique supporting point of a supporting face* we denote

$$d_h(l;l') = y(l', Y(l, \partial h)) \quad (l, l' \in \mathscr{S}_1 \text{ with } l \perp l')\,. \tag{5}$$

The *Dini subdifferential* (see [5,17,26,27]) of a directionally differentiable function $f : \mathbb{R}^n \to \mathbb{R}$ at $x \in \mathbb{R}^n$ is

$$\partial_{\mathrm{D}} f(x) = \{v \in \mathbb{R}^n \,|\, f'(x;d) \geq \langle v, d\rangle \quad \forall d \in \mathbb{R}^n\}\,.$$

The *Fréchet subdifferential* and the *superdifferential/upper subdifferential* (see [5,6,18,23]) of a function $f : \mathbb{R}^n \to \mathbb{R}$ at a point $\bar{x} \in \mathbb{R}^n$ are defined as follows:

$$\partial_{\mathrm{F}} f(\bar{x}) = \left\{ v \in \mathbb{R}^n \;\middle|\; \liminf_{x \to \bar{x}} \frac{f(x) - f(\bar{x}) - \langle v, x - \bar{x}\rangle}{\|x - \bar{x}\|} \geq 0 \right\}\,,$$

$$\partial_{\mathrm{F}}^{+} f(\bar{x}) = \left\{ v \in \mathbb{R}^n \;\middle|\; \limsup_{x \to \bar{x}} \frac{f(x) - f(\bar{x}) - \langle v, x - \bar{x}\rangle}{\|x - \bar{x}\|} \leq 0 \right\}\,.$$

The Fréchet subdifferential coincides with the Fréchet gradient for a Fréchet differentiable function, and with the subdifferential for a convex function. One can think of $\partial_{\mathrm{F}} f(\bar{x})$ and $\partial_{\mathrm{F}}^{+} f(\bar{x})$ as of the set of gradients of linear functions "supporting" f from below resp. above at $\bar{x}$. While the Fréchet subdifferential is defined for a vast class of functions, and can be used to check optimality conditions, in many cases it happens to be an empty set, which is a serious drawback for applications.

The Fréchet subdifferential possesses several useful properties summarized in the following two lemmas.

Lemma 2.2. *Let $f : \mathbb{R}^n \to \mathbb{R}$ be positively homogeneous and $l \in \mathbb{R}^n$. Then*

$$\partial_{\mathrm{F}} f(0) = \{v \in \mathbb{R}^n \,|\, f(d) \geq \langle v, d\rangle \quad \forall d \in \mathscr{S}_{n-1}\} \tag{6}$$

and $f(\cdot)$ is the support function of the Fréchet subdifferential, that is

$$f'(0;l) = f(l)\,. \tag{7}$$

Furthermore,

$$\partial_{\mathrm{F}} f(l) = \partial_{\mathrm{F}} f(\lambda l), \qquad \lambda > 0\,. \tag{8}$$

Proof. The relation (6) is obtained easily from the positive homogeneity of f and $f(0) = 0$ (see, e.g., [18, Proposition 1.9(a)]), and (8) follows from [18, Proposition 1.9(b)]. □

The following result, which is an immediate consequence of [9, Theorem 2], is used for evaluating Fréchet and Mordukhovich subdifferentials in the examples.

Lemma 2.3. *Let $f : \mathbb{R}^n \to \mathbb{R}$ be directionally differentiable.*
(i) If the directional derivative of f at x can be represented as

$$f'(x;g) = \inf_{t \in T} \varphi_t(g) ,$$

where φ_t are sublinear functions for every $t \in T$ and T is an arbitrary index set, then

$$\partial_{\mathrm{F}} f(x) = \bigcap_{t \in T} \partial \varphi_t(x) . \tag{9}$$

(ii) Analogously, if

$$f'(x;g) = -\inf_{t \in T} \varphi_t(g) ,$$

where φ_t are sublinear functions for every $t \in T$, then

$$\partial_{\mathrm{F}}^{+} f(x) = \ominus \bigcap_{t \in T} \partial \varphi_t(x) . \tag{10}$$

The next lemma states that the Fréchet subdifferential coincides with the Dini one for DC functions.

Lemma 2.4. *If $f = g - h$ is DC with convex functions g and h, then*

$$\partial_{\mathrm{F}} f(x) = \partial_{\mathrm{D}} f(x) = \{ v \in \mathbb{R}^n \,|\, f'(x;l) \geq \langle v,l \rangle \quad \forall l \in \mathscr{S}_{n-1} \} . \tag{11}$$

Proof. Since each convex function $g,h : \mathbb{R}^n \to \mathbb{R}$ is locally Lipschitz (see [15, Chap. IV, Theorem 3.1.2]), each DC function $f = g - h$ is also locally Lipschitz. Hence, we can apply Proposition 1.16 from [18], which yields

$$df(x)(l) = \liminf_{t \downarrow 0} \frac{f(x+tl) - f(x)}{t} ,$$

where we use the notation in [18]. In our setting, $df(x)(l)$ corresponds to the lower Hadamard directional derivative of f at x in the direction l.

Since each convex function (and hence, each DC function) is directionally differentiable, the limit inferior is indeed a limit with $df(x)(l) = f'(x;l)$. As we are dealing with finite-dimensional spaces, $d_w f(x;l) = df(x;l)$ holds, and [18, Proposition 1.17] yields

$$\begin{aligned} \partial_{\mathrm{F}} f(x) &= \{ v \in \mathbb{R}^n \,|\, d_w f(x;l) \geq \langle v,l \rangle \quad \forall l \in \mathbb{R}^n \} \\ &= \{ v \in \mathbb{R}^n \,|\, f'(x;l) \geq \langle v,l \rangle \quad \forall l \in \mathbb{R}^n \} = \partial_{\mathrm{D}} f(x) . \end{aligned}$$

□

Clearly, for convex functions it follows that

$$\partial_{\mathrm{F}} g(x) = \partial_{\mathrm{D}} g(x) = \partial g(x) . \tag{12}$$

3 The Mordukhovich and the Directed Subdifferential in $\mathbb{R}^2$

For a continuous function $f : \mathbb{R}^n \to \mathbb{R}$, the *Mordukhovich (lower) subdifferential* and *superdifferential (upper subdifferential)* can be defined as a corresponding outer limit of Fréchet subdifferentials [23, Theorem 1.89]:

$$\partial_{\mathrm{M}} f(\bar{x}) = \operatorname*{Lim\,sup}_{x \to \bar{x}} \partial_{\mathrm{F}} f(x) , \tag{13}$$

$$\partial_{\mathrm{M}}^{+} f(\bar{x}) = \operatorname*{Lim\,sup}_{x \to \bar{x}} \partial_{\mathrm{F}}^{+} f(x) . \tag{14}$$

The *Mordukhovich symmetric subdifferential* is defined as

$$\partial_{\mathrm{M}}^{0} f(x) = \partial_{\mathrm{M}} f(x) \cup \partial_{\mathrm{M}}^{+} f(x) .$$

Here, the limits are in the *Painlevé–Kuratowski* sense. Furthermore, the connection between the Fréchet/Mordukhovich superdifferential to the corresponding subdifferential is given by the following formulas

$$\partial_{\mathrm{F}}^{+} f(\bar{x}) = \ominus \partial_{\mathrm{F}}(-f)(\bar{x}) , \quad \partial_{\mathrm{M}}^{+} f(\bar{x}) = \ominus \partial_{\mathrm{M}}(-f)(\bar{x}), \tag{15}$$

which involve the negative function and the pointwise inverse of sets, see [18, remarks following Proposition 1.3] and [23, remarks below Definition 1.78].

Directed sets, offering a visualization of differences of two compact convex sets, are introduced and studied in [2, 3]. Here, we only sketch the main ideas and notations on directed sets in $\mathbb{R}^2$.

The directed sets, as well as the *embedding* J_n of convex compact sets in $\mathbb{R}^n$ into the Banach space of directed sets, are defined recursively in the space of dimension n. In one dimension, the *directed embedded intervals* are defined by the values of the support function in the two unit directions ± 1,

$$\overrightarrow{[a,b]} = J_1([a,b]) = (\delta^*(\eta, [a,b]))_{\eta=\pm 1} = (-a, b) \quad (a \le b) .$$

A general *directed interval* $\overrightarrow{A}_1 = \overrightarrow{[c,d]} = (-c,d)$ allows that c,d are arbitrary real numbers, even $c > d$ is possible (see references in [2, 3]). A *two-dimensional directed set* $\overrightarrow{A}_2$ is a pair of a uniformly bounded map $\overrightarrow{A}_1(\cdot)$ having one-dimensional directed intervals [2] as its values (the *directed "supporting face"*), and a continuous function $a_2(\cdot) : \mathbb{R}^2 \to \mathbb{R}$ (the *directed "support function"*). This pair is parametrized by the unit vectors $l \in \mathbb{R}^2$:

$$\overrightarrow{A_2} = (\overrightarrow{A}_1(l), a_2(l))_{l \in \mathscr{S}_1} . \tag{16}$$

A convex compact set $A \subset \mathbb{R}^2$ is embedded into the space of two-dimensional directed sets via the embedding map J_2 composed from the natural projection $\pi_{1,2}$

from $\mathbb{R} \times \{0\} \subset \mathbb{R}^2$ onto $\mathbb{R}$, and the rotation R_{2l} which for any unit vector $l \in \mathbb{R}^2$ maps the pair (l', l) (with l' orthonormal to l) to the standard basis (e^1, e^2) in $\mathbb{R}^2$:

$$J_2(A) = (\overrightarrow{Y(l,A)}, \delta^*(l,A))_{l \in \mathscr{S}_1} \quad \text{with}$$
$$\overrightarrow{Y(l,A)} = J_1(\pi_{1,2} R_{2l}(Y(l,A) - \delta^*(l,A)l)) . \tag{17}$$

For a directed set $\overrightarrow{A}$, its *visualization* $V_2(\overrightarrow{A}) \subset \mathbb{R}^2$ has three parts – *positive* $P_2(\overrightarrow{A})$, *negative* $N_2(\overrightarrow{A})$ and *mixed-type part* $M_2(\overrightarrow{A})$:

$$V_2(\overrightarrow{A}) = P_2(\overrightarrow{A}) \cup N_2(\overrightarrow{A}) \cup M_2(\overrightarrow{A}) , \tag{18}$$
$$M_2(\overrightarrow{A}) = \bigcup_{l \in \mathscr{S}_1} Q_{2,l} V_1(\overrightarrow{A_1}(l)) \setminus \left(\partial P_2(\overrightarrow{A}) \cup \partial N_2(\overrightarrow{A}) \right). \tag{19}$$

The last part is formed by reprojections $Q_{2,l}$ of one-dimensional visualizations from $\mathbb{R}$ onto the supporting lines $\langle x, l \rangle = a_2(l)$ for any unit vector $l \in \mathbb{R}^2$.

Equipped with a norm and operations acting separately on the components of the directed sets, the space of directed sets is a Banach space. The subtraction in this space is inverse to the Minkowski addition for embedded convex compact sets.

The *directed subdifferential* for DC functions and its visualization, the *Rubinov subdifferential*, are introduced in [4] for a DC function $f = g - h$ as

$$\overrightarrow{\partial} f(x) = J_2(\partial g(x)) - J_2(\partial h(x)), \qquad \partial_{\mathrm{R}} f(x) = V_2(\overrightarrow{\partial} f(x)) ,$$

that is, it is the difference of the two embedded subdifferentials.

An explicit formula for the Mordukhovich subdifferential of a positively homogeneous function as a union of Fréchet subdifferentials is obtained in the next statement.

Proposition 3.1. *Let* $f : \mathbb{R}^2 \to \mathbb{R}$ *be a positively homogeneous function. Then*

$$\partial_{\mathrm{M}} f(0) = \partial_{\mathrm{F}} f(0) \cup \bigcup_{l \in \mathscr{S}_1} \left(\partial_{\mathrm{F}} f(l) \cup \bigcup_{\substack{l' \in \mathscr{S}_1, \\ l \perp l'}} \operatorname*{Lim\,sup}_{t \downarrow 0} \partial_{\mathrm{F}} f(l + tl') \right) . \tag{20}$$

Proof. Denote by D the right-hand side of (20). We first show that $D \subseteq \partial_{\mathrm{M}} f(0)$. Observe that $\partial_{\mathrm{F}} f(0) \subset \partial_{\mathrm{M}} f(0)$ holds by (13). Further, for any $l \in \mathscr{S}_1$ and $\lambda > 0$ we have $\partial_{\mathrm{F}} f(\lambda l) = \partial_{\mathrm{F}} f(l)$ by Lemma 2.2 and

$$\partial_{\mathrm{F}} f(l) = \operatorname*{Lim\,sup}_{\lambda \downarrow 0} \partial_{\mathrm{F}} f(\lambda l) \subset \operatorname*{Lim\,sup}_{x \to 0} \partial_{\mathrm{F}} f(x) = \partial_{\mathrm{M}} f(0) .$$

It remains to show that for any $l, l' \in \mathscr{S}_1$, $l \perp l'$ we have

$$\operatorname*{Lim\,sup}_{t \downarrow 0} \partial_{\mathrm{F}} f(l + tl') \subset \partial_{\mathrm{M}} f(0) .$$

Again, by Lemma 2.2 for any $t > 0$

$$\partial_{\mathrm{F}} f(t(l+tl')) = \partial_{\mathrm{F}} f(l+tl') \,.$$

Therefore,

$$\operatorname*{Lim\,sup}_{t\downarrow 0} \partial_{\mathrm{F}} f(l+tl') = \operatorname*{Lim\,sup}_{t\downarrow 0} \partial_{\mathrm{F}} f(t(l+tl')) \subset \operatorname*{Lim\,sup}_{x\to 0} \partial_{\mathrm{F}} f(x) = \partial_{\mathrm{M}} f(0) \,.$$

Now we will show that $\partial_{\mathrm{M}} f(0) \subseteq D$. Let us consider an arbitrary element $v \in \partial_{\mathrm{M}} f(0)$. By (13), there exist $\{v_n\}$ and $\{x_n\}$ such that $v_n \to v$, $x_n \to 0$ and $v_n \in \partial_{\mathrm{F}} f(x_n)$. Without loss of generality, either $x_n = 0$ for all n, or $x_n \neq 0$ for all n. In the former case, we have $v_n \in \partial_{\mathrm{F}} f(0)$, and by the closedness of $\partial_{\mathrm{F}} f(0)$

$$v \in \operatorname*{Lim\,sup}_{n\to\infty} \partial_{\mathrm{F}} f(0) = \partial_{\mathrm{F}} f(0) \subset D \,.$$

In the latter case, without loss of generality suppose that $l_n = \frac{x_n}{\|x_n\|} \to l \in \mathscr{S}_1$. Observe that by Lemma 2.2

$$\partial_{\mathrm{F}} f(x_n) = \partial_{\mathrm{F}} f\left(\frac{1}{\|x_n\|} x_n\right) = \partial_{\mathrm{F}} f(l_n) \,. \tag{21}$$

There are two possibilities again. Without loss of generality, either $l_n = l$ for all n, or $l_n - \langle l_n, l\rangle \cdot l \neq 0$ and $\langle l, l_n\rangle \neq 0$ for all n. In the first case, by (21)

$$v \in \operatorname*{Lim\,sup}_{n\to\infty} \partial_{\mathrm{F}} f(l_n) = \partial_{\mathrm{F}} f(l) \subset D \,.$$

In the second case, let $l'_n = \frac{l_n - \langle l_n, l\rangle \cdot l}{\|l_n - \langle l_n, l\rangle \cdot l\|}$ and $t_n = \frac{\|l_n - \langle l_n, l\rangle \cdot l\|}{\langle l_n, l\rangle}$. Observe that $l'_n \perp l$, and $\|l'_n\| = 1$. Since in $\mathbb{R}^2$ there are only two unit vectors perpendicular to l, we can assume $l'_n = l' \in \mathscr{S}_1$ for all n, where l' is one of such two vectors. We have by (21) and Lemma 2.2

$$v \in \operatorname*{Lim\,sup}_{n\to\infty} \partial_{\mathrm{F}} f\left(\frac{l_n}{\langle l_n, l\rangle}\right) = \operatorname*{Lim\,sup}_{n\to\infty} \partial_{\mathrm{F}} f(l+t_n l') \subset \operatorname*{Lim\,sup}_{t\downarrow 0} \partial_{\mathrm{F}} f(l+tl') \subset D \,.$$

□

The following result about the Fréchet subdifferential of a DC function follows from (11) and [14, Sect. 4] resp. [10, Chap. III, Proposition 4.1]. The following lemma will be used to explicitly calculate the first term appearing in the right-hand side of (20) in Proposition 3.1.

Lemma 3.2. *Let $f = g - h$, where $g, h : \mathbb{R}^n \to \mathbb{R}$ are convex. Then*

$$\partial_{\mathrm{F}} f(x) = \partial_{\mathrm{D}} f(x) = \partial g(x) \stackrel{*}{-} \partial h(x) \,, \tag{22}$$

where $\partial g(x)$ and $\partial h(x)$ are the subdifferentials of g and h, respectively.

To obtain a formula for the second term in the right-hand side of (20) for sublinear functions, we show now that the subdifferential of a sublinear function in a point $l \neq 0$ is a lower dimensional supporting face.

Lemma 3.3. *Let $h : \mathbb{R}^n \to \mathbb{R}$ be convex. Then for any $l \in \mathbb{R}^n$,*

$$\partial[h'(x;\cdot)](l) = Y(l, \partial h(x)) \ . \tag{23}$$

If, in addition, h is sublinear, then

$$\partial h(l) = Y(l, \partial h) \ . \tag{24}$$

Proof. The equality (24) is trivial for $l = 0$. It follows from [15, Chap. VI, Proposition 2.1.5] that for $l \neq 0$ and every convex function

$$\partial[h'(x;\cdot)](l) = Y(l, \partial h(x)) \ .$$

Setting $x = 0$, the equality follows immediately, since (7) holds for the positively homogeneous function $h(\cdot)$. □

In the next two lemmas, we study the last term in the right-hand side of (20) for DS functions.

Lemma 3.4. *Let $f = g - h$, where $g, h : \mathbb{R}^2 \to \mathbb{R}$ are sublinear. Then for every $l, l' \in \mathscr{S}_1$ with $l \perp l'$,*

$$\operatorname*{Lim\,sup}_{t \downarrow 0} \partial_{\mathrm{F}} f(l + t l') \neq \emptyset \ .$$

Proof. The function f is locally Lipschitz as a difference of sublinear functions. Hence, f is Fréchet differentiable almost everywhere, and there exists a sequence $\{x_n\}_n \subset \mathbb{R}^2$ such that $\langle x_n, l' \rangle > 0$ for all n, $x_n \to 0$ and f is Fréchet differentiable at $l + x_n$. The Fréchet subdifferential of f at $l + x_n$ is nonempty and coincides with the Fréchet derivative (see [18, Proposition 1.1]). Therefore, we have

$$\partial_{\mathrm{F}} f(l + x_n) = \{\nabla f(l + x_n)\} \quad (n \in \mathbb{N}) \ .$$

Observe that for sufficiently large n we have $1 + \langle x_n, l \rangle > 0$ and

$$l + x_n = l + \langle x_n, l \rangle \cdot l + \langle x_n, l' \rangle \cdot l' = (1 + \langle x_n, l \rangle) \left(l + \frac{\langle x_n, l' \rangle}{1 + \langle x_n, l \rangle} l' \right) \ .$$

The positive homogeneity of f together with (8) yields

$$\partial_{\mathrm{F}} f \left(l + \frac{\langle x_n, l' \rangle}{1 + \langle x_n, l \rangle} l' \right) = \partial_{\mathrm{F}} f(l + x_n) = \{\nabla f(l + x_n)\} \ .$$

Let $t_n = \frac{\langle x_n, l' \rangle}{1 + \langle x_n, l \rangle}$. Observe that $t_n > 0$ and also $t_n \to 0$, that is, $t_n \downarrow 0$. Since f is locally Lipschitz, the sequence $\{\nabla f(l + x_n)\}$ is bounded, hence, has a converging subsequence. This subsequence satisfies

$$\operatorname*{Lim\,sup}_{n\to\infty} \partial_F f(l+x_n) = \operatorname*{Lim\,sup}_{n\to\infty} \partial_F f(l+t_n l') \subset \operatorname*{Lim\,sup}_{t\downarrow 0} \partial_F f(l+tl'),$$

which yields the nonemptiness of $\operatorname{Lim\,sup}_{t\downarrow 0} \partial_F f(l+tl')$. □

The following result establishes that the set limit (i.e., the limit of the sequence) of the subdifferentials $\partial h(l+tl')$ evaluated at small orthogonal disturbances of the direction l is a singleton. This fact is needed later in the representation theorem for directed subdifferentials.

Lemma 3.5. *Let $h : \mathbb{R}^2 \to \mathbb{R}$ be sublinear. Then for any $l, l' \in \mathscr{S}_1$ with $l \perp l'$, the set $Y(l', Y(l, \partial h))$ is a singleton, and*

$$\operatorname*{Lim}_{t\downarrow 0} \partial h(l+tl') = Y(l', Y(l, \partial h)) = \{y(l', Y(l, \partial h))\}\,. \tag{25}$$

Proof. First, we will prove the claimed equality for the outer limit $\operatorname{Lim\,sup}_{t\downarrow 0} \partial h(l+tl')$, and then apply Remark 2.1.
Let $\bar{v} \in Y(l, \partial h)$. Assume that $t_n \downarrow 0$ and $\{v_n\}_n$ is a sequence of points, each one in $\partial h(l+t_n l')$, and converging to a point in $\operatorname{Lim\,sup}_{t\downarrow 0} \partial h(l+tl')$. Lemma 3.3 shows that

$$v_n \in \partial h(l+t_n l') = Y(l+t_n l', \partial h) \quad (n \in \mathbb{N})\,.$$

By the definition of supporting face and by (7), we have

$$\langle v_n, l+t_n l'\rangle \geq \langle \bar{v}, l+t_n l'\rangle = \langle \bar{v}, l\rangle + t_n\langle \bar{v}, l'\rangle = h(l) + t_n\langle \bar{v}, l'\rangle \tag{26}$$

and

$$\langle l, v_n\rangle \leq \delta^*(l, Y(l+t_n l', \partial h)) \leq \delta^*(l, \partial h) = h'(0; l) = h(l)\,. \tag{27}$$

Taking limits as $n \to \infty$ ($t_n \downarrow 0$) on both sides of (26) and (27), we obtain

$$\lim_{n\to\infty} \langle v_n, l\rangle = h(l)\,. \tag{28}$$

Let $\tilde{v} \in Y(l', Y(l, \partial h))$. Observe that $\tilde{v} \in Y(l, \partial h) \subset \partial h$, $v_n \in Y(l+t_n l', \partial h)$ and

$$\langle v_n, l+t_n l'\rangle = \langle v_n, l\rangle + t_n\langle v_n, l'\rangle \leq \langle \tilde{v}, l\rangle + t_n\langle v_n, l'\rangle\,, \tag{29}$$

$$\langle v_n, l+t_n l'\rangle \geq \langle \tilde{v}, l+t_n l'\rangle = \langle \tilde{v}, l\rangle + t_n\langle \tilde{v}, l'\rangle\,. \tag{30}$$

Subtracting (30) from (29), we have $\langle v_n, l'\rangle \geq \langle \tilde{v}, l'\rangle$. Thus for any cluster point $\hat{v} \in \operatorname{Lim\,sup}_{t\downarrow 0} \partial h(l+tl')$ of the sequence $\{v_n\}_n$, we have

$$\langle \hat{v}, l'\rangle \geq \langle \tilde{v}, l'\rangle\,. \tag{31}$$

Since $Y(\cdot, \partial h)$ is upper semicontinuous and has closed values, it follows from (24) and $v_n \in Y(l+t_n l', \partial h)$ that $\hat{v} \in Y(l, \partial h)$. Hence, $\hat{v} \in Y(l', Y(l, \partial h))$ by (31) and the inclusion "⊂" in (25) is proved with the outer limit in the left-hand side.

Assume now that $Y(l', Y(l, \partial h))$ contains two different points $\tilde{v}_1, \tilde{v}_2$. Clearly,

$$\langle l', \tilde{v}_1 \rangle = \langle l', \tilde{v}_2 \rangle = \delta^*(l', Y(l, \partial h)) ,$$
$$\langle l, \tilde{v}_1 \rangle = \langle l, \tilde{v}_2 \rangle = \delta^*(l, \partial h) .$$

For any $\eta \in \mathbb{R}^2$, the representation $\eta = \langle \eta, l \rangle \cdot l + \langle \eta, l' \rangle \cdot l'$ is valid, therefore

$$\langle \eta, \tilde{v}_1 - \tilde{v}_2 \rangle = \langle \eta, l \rangle \cdot \langle l, \tilde{v}_1 - \tilde{v}_2 \rangle + \langle \eta, l' \rangle \cdot \langle l', \tilde{v}_1 - \tilde{v}_2 \rangle = 0 ,$$

which contradicts the assumption that the points are different.

Hence, the right-hand side in (25) is just a singleton and the equality follows by the nonemptiness of the left-hand side guaranteed by Lemma 3.4, (12) and Remark 2.1. □

Thus, (25) in the above lemma can be reformulated with the notation (5) as

$$\operatorname*{Lim}_{t \downarrow 0} \partial h(l + tl') = \{d_h(l; l')\} . \tag{32}$$

The previous lemma will be generalized to DC functions. The following lemma states an explicit formula for the third term appearing in the right-hand side of (20) in Proposition 3.1.

Lemma 3.6. *Let $f = g - h$, where $g, h : \mathbb{R}^2 \to \mathbb{R}$ are sublinear. Then for every $l, l' \in \mathscr{S}_1$, $l' \perp l$ the outer limit $\operatorname{Lim\,sup}_{t \downarrow 0} \partial_{\mathrm{F}} f(l + tl')$ is a singleton, and*

$$\operatorname*{Lim\,sup}_{t \downarrow 0} \partial_{\mathrm{F}} f(l + tl') = \{y(l', Y(l, \partial g)) - y(l', Y(l, \partial h))\} . \tag{33}$$

Proof. Let

$$u \in \operatorname*{Lim\,sup}_{t \downarrow 0} \partial_{\mathrm{F}} f(l + tl') .$$

Then there exist sequences $\{u_n\}$, $\{t_n\}$, $u_n \to u$, $t_n \downarrow 0$ such that $u_n \in \partial_{\mathrm{F}} f(l + t_n l')$. By Lemma 3.2 we have

$$\partial_{\mathrm{F}} f(l + t_n l') = \partial g(l + t_n l') \stackrel{*}{-} \partial h(l + t_n l') \quad (n \in \mathbb{N}) .$$

Therefore, for all $n \in \mathbb{N}$ there are

$$v_n \in \partial g(l + t_n l') \quad \text{and} \quad w_n \in \partial h(l + t_n l')$$

such that $u_n = v_n - w_n$.

Since $\{v_n\}$ and $\{w_n\}$ are bounded (as they belong to the corresponding upper semicontinuous subdifferentials of g and h), the sets $\operatorname{Lim\,sup}_{n \to \infty} \{v_n\}$ and $\operatorname{Lim\,sup}_{n \to \infty} \{w_n\}$ of cluster points of the corresponding sequences are nonempty. Moreover, by Lemma 3.5 we have

$$\operatorname*{Lim\,sup}_{n \to \infty} \{v_n\} \subset \operatorname*{Lim\,sup}_{n \to \infty} \partial g(l + t_n l') = \operatorname*{Lim}_{n \to \infty} \partial g(l + t_n l') = \{d_g(l; l')\} ,$$
$$\operatorname*{Lim\,sup}_{n \to \infty} \{w_n\} \subset \operatorname*{Lim\,sup}_{n \to \infty} \partial h(l + t_n l') = \operatorname*{Lim}_{n \to \infty} \partial h(l + t_n l') = \{d_h(l; l')\} ,$$

where we have used the notation (5).

Hence the sequences $\{v_n\}$ and $\{w_n\}$ converge and have unique cluster points. Therefore

$$u = \lim_{n\to\infty} u_n = \lim_{n\to\infty} v_n - \lim_{n\to\infty} w_n = d_g(l;l') - d_h(l;l') .$$

Since u is arbitrary, we have

$$\operatorname*{Lim\,sup}_{t\downarrow 0} \partial_{\mathrm{F}} f(l+tl') \subset \{d_g(l;l') - d_h(l;l')\} . \qquad (34)$$

Applying Lemma 3.4,

$$\operatorname*{Lim\,sup}_{t\downarrow 0} \partial_{\mathrm{F}} f(l+tl') \neq \emptyset$$

holds and we obtain (33) from (34). □

For the convenience of the reader, we include a full proof for the explicit formula of the subdifferential of a sublinear function with the help of two collinear directions orthogonal to the supporting face in Lemma 3.3, although this geometric fact is rather obvious.

Lemma 3.7. *Let $h : \mathbb{R}^2 \to \mathbb{R}$ be a sublinear function. Then for every $l, l' \in \mathscr{S}_1$ with $l' \perp l$,*

$$\partial h(l) = \mathrm{co}\{d_h(l;-l'), d_h(l;l')\} ,$$

where we used again the notation (5).

Proof. From Lemma 3.3, we know that

$$\partial h(l) = Y(l, \partial h) .$$

Obviously, $\mathrm{co}\{d_h(l;-l'), d_h(l;l')\} \subset Y(l,\partial h)$, and we only need to show the opposite inclusion. Assume that there exists $v \in Y(l,\partial h)$ such that

$$v \notin \mathrm{co}\{d_h(l;-l'), d_h(l;l')\} .$$

Then by the separation theorem there exists $\tilde{l} \in \mathbb{R}^2$ such that

$$\langle v, \tilde{l}\rangle > \max\{\langle d_h(l;-l'), \tilde{l}\rangle, \langle d_h(l;l'), \tilde{l}\rangle\} . \qquad (35)$$

Since the representation $v = \langle v, l\rangle \cdot l + \langle v, l'\rangle \cdot l'$ holds, we can use $v \in Y(l,\partial h)$ as well as (4) and (7) to observe that

$$\langle \tilde{l}, v\rangle = \langle \tilde{l}, l\rangle \cdot \langle v, l\rangle + \langle \tilde{l}, l'\rangle \cdot \langle v, l'\rangle = \langle \tilde{l}, l\rangle \cdot h(l) + \langle \tilde{l}, l'\rangle \cdot \langle v, l'\rangle .$$

Using $d_h(l;l') \in Y(l,\partial h)$ twice, the equality $h(l) = \langle d_h(l,l'), l\rangle$ follows, if $\langle \tilde{l}, l'\rangle \geq 0$, as well as

$$\begin{aligned} \langle \tilde{l}, v\rangle &\leq \langle \tilde{l}, l\rangle \cdot h(l) + \langle \tilde{l}, l'\rangle \cdot \langle d_h(l;l'), l'\rangle = \langle d_h(l;l'), \tilde{l}\rangle \\ &\leq \max\{\langle d_h(l;-l'), \tilde{l}\rangle, \langle d_h(l;l'), \tilde{l}\rangle\} . \end{aligned} \qquad (36)$$

Analogously, if $\langle \tilde{l}, l' \rangle < 0$, the following estimate is valid due to $h(l) = \langle d_h(l, -l'), l \rangle$:

$$\begin{aligned} \langle \tilde{l}, v \rangle &\le \langle \tilde{l}, l \rangle \cdot h(l) - \langle \tilde{l}, l' \rangle \cdot \langle d_h(l; -l'), -l' \rangle = \langle d_h(l; -l'), \tilde{l} \rangle \\ &\le \max\{ \langle d_h(l; -l'), \tilde{l} \rangle, \langle d_h(l; l'), \tilde{l} \rangle \} \end{aligned} \tag{37}$$

Clearly, (36) resp. (37) contradict (35), hence our assumption is wrong. □

The next two lemmas will be used in the further theorems. The first one connects the first component of the embedding (17) of convex sets into the space of directed sets to the interval which coincides with the projection of the line segment from Lemma 3.7. In the embedding, the natural projection $\pi_{1,2}$ and the rotation $R_{2,l}$ in [2] are used.

Lemma 3.8. *Let* $h : \mathbb{R}^2 \to \mathbb{R}$ *be sublinear,* $l \in \mathscr{S}_1$ *and* $l' = R_{2,l}^\top e^1$*. Then, the embedding in (17) satisfies*

$$\pi_{1,2} R_{2,l}(Y(l, \partial h) - h(l) l) = [\langle d_h(l; -l'), l' \rangle, \langle d_h(l; l'), l' \rangle] \,,$$

where we used again the notation (5).

Proof. Observe that $l \perp l'$, so that Lemmas 3.3 and 3.7 apply with

$$Y(l, \partial h) = \operatorname{co}\{ d_h(l; -l'), d_h(l; l') \} \,. \tag{38}$$

Since $h(l) = \langle d_h(l; \pm l'), l \rangle$, the following representation holds:

$$d_h(l; \pm l') = \langle d_h(l; \pm l'), l' \rangle \cdot l' + \langle d_h(l; \pm l'), l \rangle \cdot l = \langle d_h(l; \pm l'), l' \rangle \cdot l' + h(l) l. \tag{39}$$

Therefore,

$$\begin{aligned} \pi_{1,2} R_{2,l}(Y(l, \partial h) - h(l) l) &= \pi_{1,2} R_{2,l}(\operatorname{co}\{ d_h(l; -l'), d_h(l; l') \} - h(l) l) \quad \text{(by (38))} \\ &= \pi_{1,2} R_{2,l}(\operatorname{co}\{ d_h(l; -l') - h(l) l, d_h(l; l') - h(l) l \}) \\ &= \pi_{1,2} R_{2,l}(\operatorname{co}\{ \langle d_h(l; -l'), l' \rangle l', \langle d_h(l; l'), l' \rangle l' \}) \quad \text{(by (39))} \\ &= \operatorname{co}\{ \pi_{1,2} R_{2,l} \langle d_h(l; -l'), l' \rangle l', \pi_{1,2} R_{2,l} \langle d_h(l; l'), l' \rangle l' \} \\ &= \operatorname{co}\{ \langle d_h(l; -l'), l' \rangle \cdot \pi_{1,2} R_{2,l} l', \langle d_h(l; l'), l' \rangle \cdot \pi_{1,2} R_{2,l} l' \} \\ &= \operatorname{co}\{ \langle d_h(l; -l'), l' \rangle, \langle d_h(l; l'), l' \rangle \} \\ &= [\langle d_h(l; -l'), l' \rangle, \langle d_h(l; l'), l' \rangle] \\ &\qquad \text{(as } \langle d_h(l; -l'), l' \rangle \le \langle d_h(l; l'), l' \rangle) \,. \end{aligned}$$

□

The following lemma generalizes Lemma 3.8 to DS functions. To study the result of the embedded difference of subdifferentials, the convex sets in the first component of the embedding (17) can be calculated with the help of the two endpoints of the interval.

Lemma 3.9. *Let $f = g - h$, where $g,h : \mathbb{R}^2 \to \mathbb{R}$ are sublinear. Consider $l \in \mathscr{S}_1$ and the orthogonal vector $l' = R_{2,l}^\top e^1$. Then*

$$\pi_{1,2}R_{2,l}(D^-(l) - f(l)l) = \langle d_g(l;-l') - d_h(l;-l'), l' \rangle ,$$
$$\pi_{1,2}R_{2,l}(D^+(l) - f(l)l) = \langle d_g(l;l') - d_h(l;l'), l' \rangle ,$$

where the notation (5) and

$$D^-(l) := \operatorname*{Lim\,sup}_{t \downarrow 0} \partial_{\mathrm{F}} f(l - tl') \quad \text{and} \quad D^+(l) := \operatorname*{Lim\,sup}_{t \downarrow 0} \partial_{\mathrm{F}} f(l + tl')$$

are used.

Proof. Clearly, $l \perp l'$. By Lemma 3.6, the two sets

$$D^-(l) = d_g(l;-l') - d_h(l;-l') , \quad D^+(l) = d_g(l;l') - d_h(l;l')$$

are singletons and $d_g(l;\pm l') \in Y(l,\partial g)$, $d_h(l;\pm l') \in Y(l,\partial h)$. Therefore,

$$\begin{aligned}
\pi_{1,2}R_{2,l}(D^-(l) - f(l)l) &= \pi_{1,2}R_{2,l}(D^-(l) - f(l)l) \\
&= \pi_{1,2}R_{2,l}(d_g(l;-l') - d_h(l;-l') - f(l)l) \\
&= \pi_{1,2}R_{2,l}(\langle d_g(l;-l') - d_h(l;-l'), l' \rangle l' \\
&\quad + \langle d_g(l;-l') - d_h(l;-l'), l \rangle l - f(l)l) \\
&= \pi_{1,2}R_{2,l}(\langle d_g(l;-l') - d_h(l;-l'), l' \rangle l' \\
&\quad + (g(l) - h(l))l - f(l)l) \\
&= \pi_{1,2}R_{2,l}(\langle d_g(l;-l') - d_h(l;-l'), l' \rangle l') \\
&= \langle d_g(l;-l') - d_h(l;-l'), l' \rangle \cdot \pi_{1,2}R_{2,l}l' \\
&= \langle d_g(l;-l') - d_h(l;-l'), l' \rangle \cdot \pi_{1,2}R_{2,l}R_{2,l}^\top e^1 \\
&= \langle d_g(l;-l') - d_h(l;-l'), l' \rangle
\end{aligned}$$

and analogously

$$\pi_{1,2}R_{2,l}(D^+(l) - f(l)l) = \langle d_g(l;l') - d_h(l;l'), l' \rangle .$$

□

We apply the two lemmas above to represent the directed subdifferential of a positively homogeneous DC function in $\mathbb{R}^2$ with the help of outer limits of Fréchet subdifferential. The unique supporting points calculated in Lemma 3.9 are used to determine the (one-dimensional) first component of the directed subdifferential.

Lemma 3.10. *$f = g - h$, $g,h : \mathbb{R}^2 \to \mathbb{R}$, sublinear. Then, using the notation (5),*

$$\overrightarrow{\partial} f(0) = (\overrightarrow{[\langle d_g(l;-l') - d_h(l;-l'), l' \rangle, \langle d_g(l;l') - d_h(l;l'), l' \rangle]}, f(l))_{l \in \mathscr{S}_1}$$

with $l' = l'(l) = R_{2,l}^\top e^1$.

Proof. Observe that $\delta^*(l,\partial g) = g'(0;l) = g(l)$ by (4) and Lemma 2.2 and therefore,

$$\begin{aligned}
\overrightarrow{\partial} f &= J_2(\partial g) - J_2(\partial h) \quad \text{(by definition)}\\
&= (J_1(\pi_{1,2}R_{2,l}(Y(l,\partial g) - g(l)l)), g(l))_{l\in\mathcal{S}_1}\\
&\quad -(J_1(\pi_{1,2}R_{2,l}(Y(l,\partial h) - h(l)l)), h(l))_{l\in\mathcal{S}_1} \quad \text{(by definition)}\\
&= (J_1([\langle d_g(l;-l'),l'\rangle, \langle d_g(l;l'),l'\rangle])\\
&\quad -J_1([\langle d_h(l;-l'),l'\rangle, \langle d_h(l;l'),l'\rangle]), g(l)-h(l))_{l\in\mathcal{S}_1} \quad \text{(by Lemma 3.8)}\\
&= (\overrightarrow{[\langle d_g(l;-l'),l'\rangle, \langle d_g(l;l'),l'\rangle]} - \overrightarrow{[\langle d_h(l;-l'),l'\rangle, \langle d_h(l;l'),l'\rangle]}, f(l))_{l\in\mathcal{S}_1}\\
&= (\overrightarrow{[\langle d_g(l;-l'),l'\rangle - \langle d_h(l;-l'),l'\rangle, \langle d_g(l;l'),l'\rangle - \langle d_h(l;l'),l'\rangle]}, f(l))_{l\in\mathcal{S}_1}\,.
\end{aligned}$$

□

As a first main result, we connect the representation of the directed subdifferential to outer limits of Fréchet subdifferentials.

Theorem 3.11. *Let $g,h : \mathbb{R}^2 \to \mathbb{R}$ be sublinear functions, and let $f = g - h$. Then the directed subdifferential of f at zero $\overrightarrow{A} = (\overrightarrow{A_1}(l), a_2(l))_{l\in\mathcal{S}_1}$ can be constructed via limits of Fréchet normals as follows: for every $l \in \mathcal{S}_1$ let*

$$f_2(l) := f(l), \quad \overrightarrow{F_1}(l) := \overrightarrow{[\pi_{1,2}R_{2,l}\left(D^-(l) - f(l)l\right), \pi_{1,2}R_{2,l}\left(D^+(l) - f(l)l\right)]}\,, \tag{40}$$

where $D^-(l) := \operatorname{Lim\,sup}_{t\downarrow 0} \partial_F f(l - tl')$, $D^+(l) := \operatorname{Lim\,sup}_{t\downarrow 0} \partial_F f(l + tl')$, *and* $l' := R_{2,l}^\top e^1$.

Then, $\overrightarrow{F} = (\overrightarrow{F_1}(l), f_2(l))_{l\in\mathcal{S}_1}$ *coincides with* $\overrightarrow{A} = \overrightarrow{\partial} f(0)$.

Proof. By Lemma 3.9

$$\begin{aligned}
\pi_{1,2}R_{2,l}(D^-(l) - f(l)l) &= \langle d_g(l;-l') - d_h(l;-l'), l'\rangle\,,\\
\pi_{1,2}R_{2,l}(D^+(l) - f(l)l) &= \langle d_g(l;l') - d_h(l;l'), l'\rangle\,,
\end{aligned}$$

where we used again the notation (5). Therefore,

$$\overrightarrow{F} = (\overrightarrow{[\langle d_g(l;-l') - d_h(l;-l'),l'\rangle, \langle d_g(l;l') - d_h(l;l'),l'\rangle]}, f(l))_{l\in\mathcal{S}_1}\,,$$

which coincides with the directed subdifferential $\overrightarrow{A}$ of f by Lemma 3.10. □

The equality for the Fréchet subdifferential in the next lemma will be used to explicitly calculate the second term appearing in the right-hand side of (20) in Proposition 3.1. Geometrically, this fact is easy to believe so the reader may skip the technical proof.

Lemma 3.12. *Let $f = g - h$, where $g,h : \mathbb{R}^2 \to \mathbb{R}$ are sublinear. Then for every $l \in \mathcal{S}_1$*

$$\begin{aligned}
\partial_F f(l) &= \partial g(l) \stackrel{*}{-} \partial h(l) = \operatorname{co}\{d_g(l;-l'), d_g(l;l')\} \stackrel{*}{-} \operatorname{co}\{d_h(l;-l'), d_h(l;l')\}\\
&= \begin{cases} \operatorname{co}\{d_g(l;-l') - d_h(l;-l'), d_g(l;l') - d_h(l;l')\}\,, & \textit{if case 1 holds},\\ \emptyset\,, & \textit{if case 2 holds}, \end{cases}
\end{aligned}$$

where we used again the notation (5) and $l' = R_{2,l}^{\top} e^1$. Case 1 holds, if

$$\langle d_g(l;-l') - d_h(l;-l'), l' \rangle \leq \langle d_g(l;l') - d_h(l;l'), l' \rangle$$

and case 2 is given, if the inequality ">" holds.

Proof. Lemmas 3.3 and 3.7 show that

$$\begin{aligned} \partial g(l) &= Y(l, \partial g) = \operatorname{co}\{d_g(l;-l'), d_g(l;l')\} , \\ \partial h(l) &= Y(l, \partial h) = \operatorname{co}\{d_h(l;-l'), d_h(l;l')\} , \end{aligned}$$

since $l \perp l'$. Clearly, for all $v \in \partial g(l)$ and $w \in \partial h(l)$, (4) and (7) apply, that is

$$\begin{aligned} \langle l, v \rangle &= \delta^*(l, \partial g) = g'(0;l) = g(l) , \\ \langle l, w \rangle &= \delta^*(l, \partial h) = h'(0;l) = h(l) \end{aligned}$$

and especially,

$$\langle l, d_g(l;\pm l') \rangle = g(l) , \quad \langle l, d_h(l;\pm l') \rangle = h(l) . \tag{41}$$

It holds that

$$\begin{aligned} \partial g(l) &= \operatorname{co}\{d_g(l;-l') - g(l)l, d_g(l;l') - g(l)l\} + g(l)l \\ &= \operatorname{co}\{\langle d_g(l;-l'), l' \rangle \cdot l', \langle d_g(l;l'), l' \rangle \cdot l'\} + g(l)l , \\ \partial h(l) &= \operatorname{co}\{d_h(l;-l') - h(l)l, d_h(l;l') - h(l)l\} + h(l)l \\ &= \operatorname{co}\{\langle d_h(l;-l'), l' \rangle \cdot l', \langle d_h(l;l'), l' \rangle \cdot l'\} + h(l)l , \\ \partial g(l) - g(l)l &= \operatorname{co}\{\langle d_g(l;-l'), l' \rangle \cdot l', \langle d_g(l;l'), l' \rangle \cdot l'\} , \\ \partial h(l) - h(l)l &= \operatorname{co}\{\langle d_h(l;-l'), l' \rangle \cdot l', \langle d_h(l;l'), l' \rangle \cdot l'\} . \end{aligned}$$

Let us denote for abbreviation

$$\begin{aligned} \mu_1 &:= \langle d_g(l;-l'), l' \rangle , \quad & \mu_2 &:= \langle d_g(l;l'), l' \rangle , \\ \nu_1 &:= \langle d_h(l;-l'), l' \rangle , \quad & \nu_2 &:= \langle d_h(l;l'), l' \rangle . \end{aligned}$$

Since $d_g(l;l') \in Y(l', Y(l, \partial g))$ and $d_h(l;l') \in Y(l', Y(l, \partial h))$, we have the ordering

$$\mu_1 \leq \mu_2 \quad \text{and} \quad \nu_1 \leq \nu_2 .$$

Let us study the scalar product of $u \in (\partial g(l) - g(l)l) \overset{*}{-} (\partial h(l) - h(l)l)$ and $\eta \in \mathbb{R}^n$:

$$\begin{aligned} \langle \eta, u \rangle &\leq \delta^*(\eta, \operatorname{co}\{\mu_1 l', \mu_2 l'\}) - \delta^*(\eta, \operatorname{co}\{\nu_1 l', \nu_2 l'\}) \\ &= \max\{\langle \eta, \mu_1 l' \rangle, \langle \eta, \mu_2 l' \rangle\} - \max\{\langle \eta, \nu_1 l' \rangle, \langle \eta, \nu_2 l' \rangle\} \\ &= \max\{\mu_1 \cdot \langle \eta, l' \rangle, \mu_2 \cdot \langle \eta, l' \rangle\} - \max\{\nu_1 \cdot \langle \eta, l' \rangle, \nu_2 \cdot \langle \eta, l' \rangle\}. \end{aligned}$$

Both shifted line segments are spanned by the vector l', hence the geometric difference lies also in this span which is demonstrated by setting $\eta = \pm l$ in the above inequality:

$$\langle l, u \rangle \le 0 - 0 = 0, \quad \langle -l, u \rangle \le 0 - 0 = 0$$

Hence, $\langle l, u \rangle = 0$. Let us study the scalar product in the orthogonal directions l' and $-l'$.

$$\langle l', u \rangle \le \max\{\mu_1, \mu_2\} - \max\{\nu_1, \nu_2\} = \mu_2 - \nu_2 \,, \tag{42}$$

$$\langle -l', u \rangle \le \max\{-\mu_1, -\mu_2\} - \max\{-\nu_1, -\nu_2\} = -\mu_1 + \nu_1 . \tag{43}$$

Assume that $\nu_2 - \nu_1 > \mu_2 - \mu_1$ and that $u \in \mathbb{R}^n$ exists with $u \in (\partial g(l) - g(l)l) \stackrel{*}{-} (\partial h(l) - h(l)l)$. Then, (42) and (43) yield the contradiction

$$\mu_1 - \nu_1 \le \langle l', u \rangle \le \mu_2 - \nu_2 \,, \quad \text{i.e.} \quad \nu_2 - \nu_1 \le \mu_2 - \mu_1 \,.$$

Now assume that

$$\nu_2 - \nu_1 \le \mu_2 - \mu_1 \,. \tag{44}$$

We will show that

$$M_1 = M_2$$

holds for

$$M_1 := (\text{co}\{\mu_1, \mu_2\} \cdot l') \stackrel{*}{-} (\text{co}\{\nu_1, \nu_2\} \cdot l') \,, \quad M_2 := \text{co}\{\mu_1 - \nu_1, \mu_2 - \nu_2\} \cdot l' \,.$$

"$\subset$": Let $\eta \in \mathbb{R}^n$. Using $\langle l, u \rangle = 0$ and the orthonormal basis $\{l, l'\}$, we get $\eta = \langle \eta, l \rangle \cdot l + \langle \eta, l' \rangle \cdot l'$ and

$$\begin{aligned}
\delta^*(\eta, M_1) &= \max_{u \in M_1} \langle \eta, u \rangle = \max_{u \in M_1} \left(\langle \eta, l \rangle \cdot \langle l, u \rangle + \langle \eta, l' \rangle \cdot \langle l', u \rangle \right) \\
&= \langle \eta, l' \rangle \cdot \delta^*(l', M_1) \le \langle \eta, l' \rangle \cdot (\mu_2 - \nu_2) \quad \text{(by (42))} \\
&= \langle \eta, l' \rangle \cdot \max_{\alpha \in [\mu_1 - \nu_1, \mu_2 - \nu_2]} \langle l', \alpha \cdot l' \rangle \\
&= \max_{\alpha \in [\mu_1 - \nu_1, \mu_2 - \nu_2]} \left(\langle \eta, l \rangle \cdot \langle l, \alpha \cdot l' \rangle + \langle \eta, l' \rangle \cdot \langle l', \alpha \cdot l' \rangle \right) \\
&= \max_{\alpha \in [\mu_1 - \nu_1, \mu_2 - \nu_2]} \langle \eta, \alpha \cdot l' \rangle = \max_{u \in M_2} \langle \eta, u \rangle = \delta^*(\eta, M_2),
\end{aligned}$$

which shows that $M_1 \subset M_2$.
"$\supset$": Let us first show that $(\mu_1 - \nu_1)l' \in M_1$. Since (44) and $\mu_1 - \nu_1 + \nu_2 \le \mu_2$ hold,

$$\begin{aligned}
(\mu_1 - \nu_1)l' + \text{co}\{\nu_1, \nu_2\} \cdot l' &= (\mu_1 - \nu_1)l' + \text{co}\{\nu_1 l', \nu_2 l'\} \\
&= \text{co}\{(\mu_1 - \nu_1)l' + \nu_1 l', (\mu_1 - \nu_1)l' + \nu_2 l'\} \\
&= \text{co}\{\mu_1 l', (\mu_1 - \nu_1 + \nu_2)l'\} = \text{co}\{\mu_1, (\mu_1 - \nu_1 + \nu_2)\} \cdot l' \\
&\subset \text{co}\{\mu_1, \mu_2\} \cdot l' \,.
\end{aligned}$$

Hence, the first endpoint of the line segment M_2 lies in M_1:

$$(\mu_1 - \nu_1)l' \in \big(\mathrm{co}\{\mu_1, \mu_2\} \cdot l'\big) \stackrel{*}{-} \big(\mathrm{co}\{\nu_1, \nu_2\} \cdot l'\big) = M_1 .$$

Now, we proceed similarly with the second endpoint $(\mu_2 - \nu_2)l'$. Since (44) and $\mu_2 - \nu_2 + \nu_1 \geq \mu_1$ is valid,

$$\begin{aligned}
(\mu_2 - \nu_2)l' + \mathrm{co}\{\nu_1, \nu_2\} \cdot l' &= (\mu_2 - \nu_2)l' + \mathrm{co}\{\nu_1 l', \nu_2 l'\} \\
&= \mathrm{co}\{(\mu_2 - \nu_2)l' + \nu_1 l', (\mu_2 - \nu_2)l' + \nu_2 l'\} \\
&= \mathrm{co}\{(\mu_2 - \nu_2 + \nu_1)l', \mu_2 l'\} = \mathrm{co}\{(\mu_2 - \nu_2 + \nu_1), \mu_2\} \cdot l' \\
&\subset \mathrm{co}\{\mu_1, \mu_2\} \cdot l' .
\end{aligned}$$

An immediate consequence is that the second endpoint of M_2 also lies in M_1:

$$(\mu_2 - \nu_2)l' \in \big(\mathrm{co}\{\mu_1, \mu_2\} \cdot l'\big) \stackrel{*}{-} \big(\mathrm{co}\{\nu_1, \nu_2\} \cdot l'\big) = M_1 .$$

Since M_1 is convex, it follows that

$$M_2 = \mathrm{co}\{(\mu_1 - \nu_1)l', (\mu_2 - \nu_2)l'\} \subset M_1 .$$

This equality for both sets is used to reformulate the geometric difference:

$$\begin{aligned}
\partial g(l) \stackrel{*}{-} \partial h(l) &= (\partial g(l) - g(l)l) \stackrel{*}{-} (\partial h(l) - h(l)l) + f(l)l \\
&= (\mathrm{co}\{\mu_1, \mu_2\} \cdot l') \stackrel{*}{-} (\mathrm{co}\{\nu_1, \nu_2\} \cdot l') + f(l)l \\
&= \mathrm{co}\{\mu_1 - \nu_1, \mu_2 - \nu_2\} \cdot l' + f(l)l \\
&= \mathrm{co}\{(\mu_1 - \nu_1)l' + f(l)l, (\mu_2 - \nu_2)l' + f(l)l\}.
\end{aligned}$$

Let us now calculate both endpoints of the line segment using (41):

$$\begin{aligned}
(\mu_1 - \nu_1)l' + f(l)l &= \big(\langle d_g(l;-l'), l'\rangle - \langle d_h(l;-l'), l'\rangle\big) \cdot l' + g(l)l - h(l)l \\
&= \big(\langle d_g(l;-l'), l'\rangle \cdot l' + \langle d_g(l;-l'), l\rangle \cdot l\big) \\
&\quad - \big(\langle d_h(l;-l'), l'\rangle \cdot l' + \langle d_h(l;-l'), l\rangle \cdot l\big) \\
&= d_g(l;-l') - d_h(l;-l'), \\
(\mu_2 - \nu_2)l' + f(l)l &= \big(\langle d_g(l;l'), l'\rangle - \langle d_h(l;l'), l'\rangle\big) \cdot l' + g(l)l - h(l)l \\
&= \big(\langle d_g(l;l'), l'\rangle \cdot l' + \langle d_g(l;l'), l\rangle \cdot l\big) \\
&\quad - \big(\langle d_h(l;l'), l'\rangle \cdot l' + \langle d_h(l;l'), l\rangle \cdot l\big) \\
&= d_g(l;l') - d_h(l;l').
\end{aligned}$$

This finally shows that

$$\partial g(l) \stackrel{*}{-} \partial h(l) = \mathrm{co}\{d_g(l;-l') - d_h(l;-l'), d_g(l;l') - d_h(l;l')\} .$$

□

The next main theorem shows that the Mordukhovich subdifferential of f at 0 can be represented via visualization parts from (18)–(19) of the directed subdifferential.

Theorem 3.13. *Let $f = g - h$, where $g, h : \mathbb{R}^2 \to \mathbb{R}$ are sublinear functions, and let $\overrightarrow{A} = \overrightarrow{\partial} f(0)$ be the directed subdifferential of f at* 0. *Then,*

$$\partial_{\mathrm{M}} f(0) = P_2(\overrightarrow{A}) \cup \bigcup_{l \in \mathscr{S}_1} Q_{2,l}\left(P_1(\overrightarrow{A_1}(l)) \cup \mathrm{bd}\, N_1(\overrightarrow{A_1}(l))\right) , \tag{45}$$

where bd *denotes the boundary of a set in* $\mathbb{R}$, *and* $Q_{2,l}(y) = R_{2,l}^{\top} \pi_{1,2}^{\top}(y) + a_2(l) l$ *is the reprojection as in [3].*

Proof. First of all, observe that by Proposition 3.1

$$\partial_{\mathrm{M}} f(0) = \partial_{\mathrm{F}} f(0) \cup \bigcup_{l \in \mathscr{S}_1} \left(\partial_{\mathrm{F}} f(l) \cup \operatorname*{Lim\,sup}_{t \downarrow 0} \partial_{\mathrm{F}} f(l - tl') \cup \operatorname*{Lim\,sup}_{t \downarrow 0} \partial_{\mathrm{F}} f(l + tl') \right) , \tag{46}$$

where $l' = R_{2,l}^{\top} e^1$. The proof consists of three parts:
Step 1: We will show that the positive part coincides with the Fréchet subdifferential at $x = 0$:

$$\partial_{\mathrm{F}} f(0) = P_2(\overrightarrow{A}). \tag{47}$$

Step 2: We will conclude that the reprojected positive part is the second term in (46):

$$\partial_{\mathrm{F}} f(l) = \begin{cases} Q_{2,l} P_1(\overrightarrow{A_1}(l)), & \text{if } P_1(\overrightarrow{A_1}(l)) \neq \emptyset\, , \\ \emptyset, & \text{if } P_1(\overrightarrow{A_1}(l)) = \emptyset. \end{cases} \tag{48}$$

Step 3: We will prove the following equality for the reprojected boundary points:

$$\operatorname*{Lim\,sup}_{t \downarrow 0} \partial_{\mathrm{F}} f(l - tl') \cup \operatorname*{Lim\,sup}_{t \downarrow 0} \partial_{\mathrm{F}} f(l + tl') = Q_{2,l}(\mathrm{bd}\, P_1(\overrightarrow{A_1}(l)) \cup \mathrm{bd}\, N_1(\overrightarrow{A_1}(l))). \tag{49}$$

It is not difficult to see that Steps 1–3 together with (46) yield (45).

Step 1: For the Fréchet subdifferential, Lemma 2.2 yields

$$\partial_{\mathrm{F}} f(0) = \{ v \,|\, \langle v, l \rangle \leq f(l) \quad \forall l \in \mathscr{S}_1 \} \, . \tag{50}$$

This equation can be compared with the definition of the positive part of the directed set:

$$P_2(\overrightarrow{A}) = \{ v \in \mathbb{R}^2 \,|\, \langle v, l \rangle \leq a_2(l) \quad \forall l \in \mathscr{S}_1 \}. \tag{51}$$

Since $a_2(l) = g(l) - h(l) = f(l)$, from (50) and (51) we conclude (47).

Step 2: By Lemma 3.12 for all $l \in \mathscr{S}_1$ we have

$$\partial_{\mathrm{F}} f(l) = \begin{cases} \mathrm{co}\{ d_g(l; -l') - d_h(l; -l'), & \\ \quad d_g(l; l') - d_h(l; l') \}, & \text{if case 1 of Lemma 3.12 holds,} \\ \emptyset\, , & \text{if the opposite inequality holds,} \end{cases} \tag{52}$$

where the notation (5) is again used.

At the same time, Lemma 3.10 yields for every $l \in \mathscr{S}_1$

$$P_1(\overrightarrow{A_1}(l)) = P_1(\overrightarrow{[\langle d_g(l;-l') - d_h(l;-l'),l'\rangle, \langle d_g(l;l') - d_h(l;l'),l'\rangle]})$$
$$= \begin{cases} [\langle d_g(l;-l') - d_h(l;-l'),l'\rangle, \\ \langle d_g(l;l') - d_h(l;l'),l'\rangle]\,, & \text{if case 1 of Lemma 3.12 holds,} \\ \emptyset\,, & \text{if case 2 of Lemma 3.12 holds.} \end{cases} \tag{53}$$

Observe that

$$\begin{aligned} &Q_{2,l}(\langle d_g(l;\pm l') - d_h(l;\pm l'),l'\rangle) \\ &= R_{2,l}^\top \pi_{1,2}^\top(\langle d_g(l;\pm l') - d_h(l;\pm l'),l'\rangle) + a_2(l)l \\ &= R_{2,l}^\top(\langle d_g(l;\pm l') - d_h(l;\pm l'),l'\rangle)e^1 + f(l)l \\ &= \langle d_g(l;\pm l') - d_h(l;\pm l'),l'\rangle \cdot l' + (g(l) - h(l))l \\ &= \langle d_g(l;\pm l'),l'\rangle \cdot l' + g(l)l - \langle d_h(l;\pm l'),l'\rangle \cdot l' - h(l)l \\ &= \langle d_g(l;\pm l'),l'\rangle \cdot l' + \langle d_g(l;\pm l'),l\rangle \cdot l \\ &\quad - \langle d_h(l;\pm l'),l'\rangle \cdot l' - \langle d_h(l;\pm l'),l\rangle \cdot l \\ &= d_g(l;\pm l') - d_h(l;\pm l')\,, \end{aligned} \tag{54}$$

and hence

$$\begin{aligned} &Q_{2,l}\{[\langle d_g(l;-l') - d_h(l;-l'),l'\rangle, \langle d_g(l;l') - d_h(l;l'),l'\rangle]\} \\ &= \mathrm{co}\{d_g(l;-l') - d_h(l;-l'), d_g(l;l') - d_h(l;l')\}\,. \end{aligned} \tag{55}$$

Equation (48) follows from (52), (53), and (55).

Step 3: By Lemma 3.6

$$\operatorname*{Lim\,sup}_{t \downarrow 0} \partial_{\mathrm{F}} f(l \pm t l') = \{d_g(l;\pm l') - d_h(l;\pm l')\}\,, \tag{56}$$

since $l \perp l'$. An immediate consequence of (53) and (54) is

$$Q_{2,l}(\mathrm{bd}\,P_1(\overrightarrow{A_1}(l))) = \begin{cases} \{d_g(l;-l') - d_h(l;-l'), \\ \ d_g(l;l') - d_h(l;l')\}, & \text{if } P_1(\overrightarrow{A_1}(l)) \neq \emptyset\,, \\ \emptyset, & \text{if } P_1(\overrightarrow{A_1}(l)) = \emptyset\,. \end{cases} \tag{57}$$

Since $N_1(\overrightarrow{A_1}(l)) = \ominus P_1(-\overrightarrow{A_1}(l))$, the expression for $Q_{2,l}(\mathrm{bd}\,N_1(\overrightarrow{A_1}(l)))$ can be obtained analogously:

$$Q_{2,l}(\mathrm{bd}\,N_1(\overrightarrow{A_1}(l))) = \begin{cases} \{d_g(l;-l') - d_h(l;-l'), \\ \ d_g(l;l') - d_h(l;l')\}\,, & \text{if } N_1(\overrightarrow{A_1}(l)) \neq \emptyset\,, \\ \emptyset\,, & \text{if } N_1(\overrightarrow{A_1}(l)) = \emptyset. \end{cases} \tag{58}$$

There are three possible cases (see [3, Proposition 3.4]): either one of the sets $P_1(\overrightarrow{A_1}(l)))$ or $N_1(\overrightarrow{A_1}(l)))$ is empty or both are singletons and $P_1(\overrightarrow{A_1}(l))) = N_1(\overrightarrow{A_1}(l)))$. Together with (57) and (58), this yields

$$Q_{2,l}(\operatorname{bd} P_1(\overrightarrow{A_1}(l)) \cup \operatorname{bd} N_1(\overrightarrow{A_1}(l))) = \{d_g(l;-l') - d_h(l;-l'), d_g(l;l') - d_h(l;l')\}. \tag{59}$$

Now, (56) and (59) yield (49). □

The Mordukhovich superdifferential and symmetric subdifferential of f at 0 is represented via the directed subdifferential in the following theorem. Besides isolated points from the reprojected lower dimensional positive part of the directed subdifferential, the Mordukhovich superdifferential forms the negative two- and one-dimensional part in the visualization of the directed subdifferential. The positive parts are reflected by the Mordukhovich subdifferential (see Theorem 3.13) so that the Mordukhovich symmetric subdifferential form the complete visualization of the directed subdifferential for DS functions.

Theorem 3.14. *Let $f = g - h$, where $g, h : \mathbb{R}^2 \to \mathbb{R}$ are sublinear functions, and let $\overrightarrow{A} = \overrightarrow{\partial} f(0)$ be the directed subdifferential of f at 0. Then,*

$$\partial_M^+ f(0) = N_2(\overrightarrow{A}) \cup \bigcup_{l \in \mathscr{S}_1} Q_{2,l}\left(N_1(\overrightarrow{A_1}(l)) \cup \operatorname{bd} P_1(\overrightarrow{A_1}(l))\right), \tag{60}$$

$$\partial_M^0 f(0) = V_2(\overrightarrow{\partial} f(0)). \tag{61}$$

Proof. Apply Theorem 3.13 to $-f = h - g$ and use [3, Proposition 3.8]:

$$\overrightarrow{\partial}(-f)(0) = -\overrightarrow{\partial} f(0), \qquad \ominus P_2(-\overrightarrow{A}) = N_2(\overrightarrow{A}),$$
$$\ominus P_1(-\overrightarrow{A_1}(l)) = N_1(\overrightarrow{A_1}(l)), \quad \ominus N_1(-\overrightarrow{A_1}(l)) = P_1(\overrightarrow{A_1}(l)).$$

This, together with (15), immediately yields (60).
Since

$$V_1(\overrightarrow{A_1}(l)) = P_1(\overrightarrow{A_1}(l)) \cup N_1(\overrightarrow{A_1}(l)), \quad M_2(\overrightarrow{A}) \subset Q_{2,l} V_1(\overrightarrow{A_1}(l))$$

and (18) hold, the second equation (61) follows easily. □

Applying the previous Theorems 3.11, 3.13, and 3.14 to the directional derivative generalizes these results to the class of general DC functions (which are not necessarily positively homogeneous).

As a starting point, we will demonstrate that the directed subdifferential of the function at x coincides with the one of its directional derivative evaluated at direction $l = 0$.

Proposition 3.15. *Let $f = g - h$ with $g, h : \mathbb{R}^n \to \mathbb{R}$ be convex functions. Then,*

$$\overrightarrow{\partial}[f'(x;\cdot)](0) = \overrightarrow{\partial} f(x). \tag{62}$$

Proof. By Lemma 3.3, the convex subdifferential of $g'(x;\dot{)}$ in 0 coincides with the one of $g(x)$:

$$\partial[g'(x;\cdot)](0) = Y(0,\partial g(x)) = \partial g(x).$$

The same is true for the convex function h such that

$$\begin{aligned}\overrightarrow{\partial}[f'(x;\cdot)](0) &= J_n(\partial[g'(x;\cdot)](0)) - J_n(\partial[h'(x;\cdot)](0)) \\ &= J_n(\partial g(x)) - J_n(\partial h(x)) = \overrightarrow{\partial} f(x)\,.\end{aligned}$$

□

Since the Mordukhovich subdifferential of the directional derivative may differ from the one for the function itself (see Example 4.3) in contrary to the directed subdifferential, the following results for the Mordukhovich subdifferentials have to be formulated with the directional derivative. The next theorem yields the connection between outer limits of Fréchet subdifferentials and the directed subdifferential.

Theorem 3.16. *Let $g,h : \mathbb{R}^2 \to \mathbb{R}$ be convex functions, and let $f = g - h$. Then the directed subdifferential $\overrightarrow{A} = (\overrightarrow{A_1}(l), a_2(l))_{l \in \mathscr{S}_1}$ of f at x can be constructed via limits of Fréchet normals as follows: for every $l \in \mathscr{S}_1$ let*

$$f_2(l) := f'(x;l)\,,\;\; \overrightarrow{F_1}(l) := \overrightarrow{\left[\pi_{1,2}R_{2,l}\left(D^-(l) - f'(x;l)l\right), \pi_{1,2}R_{2,l}\left(D^+(l) - f'(x;l)l\right)\right]}\,, \tag{63}$$

where

$$D^-(l) := \operatorname*{Lim\,sup}_{t \downarrow 0} \partial_{\mathrm{F}} f'(x;\cdot)(l - tl')\,, \quad D^+(l) := \operatorname*{Lim\,sup}_{t \downarrow 0} \partial_{\mathrm{F}} f'(x;\cdot)(l + tl')\,,$$

and $l' := R_{2,l}^{\top} e^1$.
Then, $\overrightarrow{F} = (\overrightarrow{F_1}(l), f_2(l))_{l \in \mathscr{S}_1}$ coincides with $\overrightarrow{A} = \overrightarrow{\partial} f(x)$.

Proof. Applying [10, Sect. I.3, Proposition 3.1], the directional derivative

$$f'(x;l) = g'(x;l) - h'(x;l)$$

is a DS representation. Hence, Proposition 3.15 and Theorem 3.11 can be applied. □

The next theorem for DC functions, in which we can drop the assumption of positive homogeneity, could be seen as the nonconvex counterpart of the following result for locally Lipschitz and directionally differentiable function in [19, Sect. 3] and [8, (35)]:

$$\partial_{\mathrm{Cl}}[f'(x;\cdot)](0) = \partial_{\mathrm{MP}} f(x)\,,$$

where $\partial_{\mathrm{MP}} f(x)$ is the Michel–Penot subdifferential of f in x (see [8, 21]). In what follows the Mordukhovich symmetric subdifferential for the directional derivative at x in direction 0 coincides with the Rubinov subdifferential at x, that is, its visualized directed subdifferential.

Theorem 3.17. *Let $f = g - h$, where $g, h : \mathbb{R}^2 \to \mathbb{R}$ are convex functions, and let $\overrightarrow{A} = \overrightarrow{\partial} f(x)$ be the directed subdifferential of f at x. Then,*

$$\partial_{\mathrm{M}}[f'(x;\cdot)](0) = P_2(\overrightarrow{A}) \cup \bigcup_{l \in \mathscr{S}_1} Q_{2,l}\left(P_1(\overrightarrow{A_1}(l)) \cup \mathrm{bd}\, N_1(\overrightarrow{A_1}(l))\right), \tag{64}$$

$$\partial_{\mathrm{M}}^{+}[f'(x;\cdot)](0) = N_2(\overrightarrow{A}) \cup \bigcup_{l \in \mathscr{S}_1} Q_{2,l}\left(N_1(\overrightarrow{A_1}(l)) \cup \mathrm{bd}\, P_1(\overrightarrow{A_1}(l))\right), \tag{65}$$

$$\partial_{\mathrm{M}}^{0}[f'(x;\cdot)](0) = V_2(\overrightarrow{\partial} f(x)). \tag{66}$$

Proof. As in the proof of Theorem 3.16, the equality (62) of the directed subdifferential of $f'(x;\cdot)$ in 0 and the one of $f(\cdot)$ in x holds. The claimed equalities are proved by applying Theorems 3.13 and 3.14. □

Remark 3.18. All the lemmas starting from Lemma 3.3 could be adapted to the convex (instead of sublinear) situation. For this purpose, the function must be replaced by its directional derivative, which is sublinear with respect to its second argument. For example, Lemma 3.5 reads for h being only convex:

$$\operatorname*{Lim\,sup}_{t \downarrow 0} \partial[h'(x;\cdot)](l + tl') = Y(l', Y(l, \partial h(x))).$$

4 Examples

For each of the presented examples, we will first calculate theoretically the Mordukhovich subdifferential and superdifferential. Their union, the symmetric subdifferential is compared visually with the Rubinov subdifferential in [4].

We will frequently use Lemma 2.3 for evaluating the Fréchet subdifferential, which is a basic tool for calculating the Mordukhovich subdifferential with (13). Analogously, we proceed with the Fréchet superdifferential and (14) in the same way to evaluate the Mordukhovich superdifferential.

The first example is governed by a parameter r by which three different cases could be studied: the Mordukhovich subdifferential has nonempty interior ($r = 0.5$), the Mordukhovich superdifferential has nonempty interior ($r = 2.0$), and both have empty interior ($r = 1.25$). This corresponds to nonemptiness of the positive part resp. of the negative part as well as the mere presence of the mixed-type part in the directed subdifferential.

Example 4.1 ([4, Example 5.7]). Let $f = g - h$, where

$$g(x) = |x_1| + |x_2|, \quad h(x) = r\sqrt{x_1^2 + x_2^2} = r\|x\|, \; r > 0.$$

To evaluate the Mordukhovich lower/upper/symmetric subdifferential of f at zero directly, we first need to calculate the Fréchet subdifferentials of f at zero and in its neighborhood.

A. The Fréchet subdifferential at 0**.** Observe that f can be represented as follows:

$$f(x) = g(x) - r\sqrt{x_1^2 + x_2^2} = g(x) + \min_{\|w\|=r} \langle w, x\rangle = \min_{\|w\|=r} (g(x) + \langle w, x\rangle) .$$

Let

$$\varphi_w(x) := \langle w, x\rangle + g(x) ,$$

then

$$f(x) = \min_{\|w\|=r} \varphi_w(x) .$$

Since $f'(0;l) = f(l)$, the formula (9) for the Fréchet subdifferential holds

$$\partial_{\mathrm{F}} f(0) = \bigcap_{\|w\|=r} (\partial g(0) + w). \tag{67}$$

It is not difficult to see that

$$\partial g(0) = \mathrm{co}\{(1,1),(-1,1),(1,-1),(-1,-1)\} = [-1,1]^2 .$$

We are going to show that

$$\partial_{\mathrm{F}} f(0) = \begin{cases} [-1+r, 1-r]^2 , & r \le 1 , \\ \emptyset , & r > 1 . \end{cases} \tag{68}$$

Let $u \in \partial_{\mathrm{F}} f(0)$. For every w, $\|w\| = r$, there exists $v \in [-1,1]^2$ by (67) such that the coordinates satisfy

$$u_i = v_i + w_i, \quad i = 1,2 .$$

This yields $-1+r \le u_i \le 1-r$, and hence

$$\partial_{\mathrm{F}} f(0) \subset [-1+r, 1-r]^2 , \qquad \text{if } r \le 1 , \tag{69}$$

and

$$\partial_{\mathrm{F}} f(0) = \emptyset , \qquad \text{if } r > 1 . \tag{70}$$

To show the inclusion opposite to (69), consider an arbitrary u such that $-1+r \le u_i \le 1-r$. For every w, $\|w\| = r$, we set $v := u - w$. Then $v \in [-1,1]^2$ is valid as well as

$$\partial_{\mathrm{F}} f(0) \supset [-1+r, 1-r]^2 . \tag{71}$$

Now, (68) follows from (69)–(71).

B. The Fréchet superdifferential at 0**.** Observe that

$$\begin{aligned} f(x) &= \max_{i=1,\ldots,4} \langle v^i, x\rangle - \max_{\|w\|=r} \langle w, x\rangle \\ &= \max_{i=1,\ldots,4} \left\{ \langle v^i, x\rangle - \max_{\|w\|=r} \langle w, x\rangle \right\} = - \min_{i=1,\ldots,4} \left\{ \max_{\|w\|=r} \langle w, x\rangle - \langle v^i, x\rangle \right\} , \end{aligned}$$

where

$$v^1 = (1,1), \quad v^2 = (1,-1), \quad v^3 = (-1,1), \quad v^4 = (-1,-1).$$

Let

$$\varphi_i(x) = \max_{\|w\|=r} \langle w, x \rangle - \langle v^i, x \rangle .$$

It is not difficult to observe that

$$\partial \varphi_i(x) = \mathbb{B}_r(0) - v^i = \mathbb{B}_r(-v^i),$$

where $\mathbb{B}_r(m) = \{x \mid \|x-m\| = r\}$. Using (10), the Fréchet superdifferential can be calculated as

$$\partial_{\mathrm{F}}^+ f(x) = \ominus \bigcap_{i=1}^{4} \mathbb{B}_r(-v^i) = \bigcap_{i=1}^{4} \mathbb{B}_r(v^i). \tag{72}$$

C. The Fréchet sub- and superdifferentials around 0**.** For every $x \neq 0$ the function h is smooth, hence

$$\partial_{\mathrm{F}} f(x) = \partial f(x) = \partial g(x) - h'(x) = \partial g(x) - r\frac{x}{\|x\|} \quad \forall x \neq 0.$$

For the Fréchet superdifferential in all points $x \neq 0$ we have

$$\partial_{\mathrm{F}}^+ f(x) = \begin{cases} g'(x) - r\frac{x}{\|x\|}, & \text{if } g \text{ is differentiable at } x, \\ \emptyset, & \text{otherwise, since } g \text{ is not Fréchet superdifferentiable} \\ & \text{due to [18, Proposition 1.3].} \end{cases}$$

Observe that for $x \neq 0$, the subdifferential of g is given by

$$\partial g(x) = \begin{cases} \{(\mathrm{sgn}(x_1), \mathrm{sgn}(x_2))\}, & x_1 \neq 0, x_2 \neq 0, \\ \mathrm{co}\{(1,1),(1,-1)\}, & x_1 > 0, x_2 = 0, \\ \mathrm{co}\{(1,1),(-1,1)\}, & x_1 = 0, x_2 > 0, \\ \mathrm{co}\{(-1,-1),(-1,1)\}, & x_1 < 0, x_2 = 0, \\ \mathrm{co}\{(-1,-1),(1,-1)\}, & x_1 = 0, x_2 < 0. \end{cases}$$

Therefore,

$$\partial_{\mathrm{F}} f(x) = \begin{cases} \{(\mathrm{sgn}(x_1), \mathrm{sgn}(x_2)) - r\frac{x}{\|x\|}\}, & x_1 \neq 0, x_2 \neq 0, \\ \mathrm{co}\{(1,1),(1,-1)\} - \{r\frac{x}{\|x\|}\}, & x_1 > 0, x_2 = 0, \\ \mathrm{co}\{(1,1),(-1,1)\} - \{r\frac{x}{\|x\|}\}, & x_1 = 0, x_2 > 0, \\ \mathrm{co}\{(-1,-1),(-1,1)\} - \{r\frac{x}{\|x\|}\}, & x_1 < 0, x_2 = 0, \\ \mathrm{co}\{(-1,-1),(1,-1)\} - \{r\frac{x}{\|x\|}\}, & x_1 = 0, x_2 < 0, \end{cases} \tag{73}$$

and

$$\partial_{\mathrm{F}}^{+} f(x)=\begin{cases}\{(\operatorname{sgn}(x_1),\operatorname{sgn}(x_2))-r\frac{x}{\|x\|}\}\,, & x_1\neq 0, x_2\neq 0,\\ \emptyset\,, & x_1x_2=0, x\neq 0.\end{cases} \tag{74}$$

It is not difficult to observe that for every $l\in S_1$ with $l_1l_2\neq 0$, we have

$$\operatorname*{Lim\,sup}_{t\downarrow 0}\partial_{\mathrm{F}} f(l+tl')=\partial_{\mathrm{F}} f(l)\,,\quad \operatorname*{Lim\,sup}_{t\downarrow 0}\partial_{\mathrm{F}}^{+} f(l+tl')=\partial_{\mathrm{F}}^{+} f(l)\,, \tag{75}$$

by applying (73) and (74). For $l=(1,0)$ and $l'=(0,1)$

$$\begin{aligned}\operatorname*{Lim\,sup}_{t\downarrow 0}\partial_{\mathrm{F}} f(l+tl') &= \operatorname*{Lim\,sup}_{t\downarrow 0}\partial_{\mathrm{F}}^{+} f(l+tl')=\operatorname*{Lim\,sup}_{t\downarrow 0}\left\{(1,1)-r\frac{l+tl'}{\|l+tl'\|}\right\}\\ &= \operatorname*{Lim\,sup}_{t\downarrow 0}\left\{\left(1-\frac{r}{\|l+tl'\|},1-\frac{rt}{\|l+tl'\|}\right)\right\}\\ &= \{(1-r,1)\}\,.\end{aligned}$$

The corresponding outer limits for the remaining directions can be evaluated analogously. We have

$$\begin{aligned}&\operatorname*{Lim\,sup}_{t\downarrow 0}\partial_{\mathrm{F}} f(l+tl')\\ &=\operatorname*{Lim\,sup}_{t\downarrow 0}\partial_{\mathrm{F}}^{+} f(l+tl')=\begin{cases}\{(1-r,1)\}\,, & l=(1,0),\ l'=(0,1),\\ \{(1-r,-1)\}\,, & l=(1,0),\ l'=(0,1),\\ \{(1,1-r)\}\,, & l=(0,1),\ l'=(1,0),\\ \{(-1,1-r)\}\,, & l=(0,1),\ l'=(-1,0),\\ \{(-1+r,1)\}\,, & l=(-1,0),\ l'=(1,0),\\ \{(-1+r,-1)\}\,, & l=(-1,0),\ l'=(-1,0),\\ \{(1,-1+r)\}\,, & l=(0,-1),\ l'=(1,0),\\ \{(-1,-1+r)\}\,, & l=(0,-1),\ l'=(-1,0).\end{cases}\end{aligned} \tag{76}$$

D. The Mordukhovich subdifferentials at 0. To finish the evaluation of the Mordukhovich subdifferential, we use Proposition 3.1.

From (68), (73), (75), and (76), the Mordukhovich subdifferential is given by

$$\begin{aligned}\partial_{\mathrm{M}} f(0) &= \partial_{\mathrm{F}} f(0)\cup\operatorname*{Lim\,sup}_{\substack{x\to 0,\\ x\neq 0}}\partial_{\mathrm{F}} f(x)\\ &=\{u\,|\,-1+r\leq u_i\leq 1-r, i=1,2\}\\ &\quad\cup[\{(1,1)\}+\{w\,|\,\|w\|=r,\ w_1\leq 0,\ w_2\leq 0\}]\\ &\quad\cup[\{(-1,1)\}+\{w\,|\,\|w\|=r,\ w_1\geq 0,\ w_2\leq 0\}]\\ &\quad\cup[\{(1,-1)\}+\{w\,|\,\|w\|=r,\ w_1\leq 0,\ w_2\geq 0\}]\\ &\quad\cup[\{(-1,-1)\}+\{w\,|\,\|w\|=r,\ w_1\geq 0,\ w_2\geq 0\}]\\ &\quad\cup\operatorname{co}\{(1-r,1),(1-r,-1)\}\end{aligned}$$

$$\cup \mathrm{co}\{(-1,1-r),(1,1-r)\}$$
$$\cup \mathrm{co}\{(-1+r,-1),(-1+r,1)\}$$
$$\cup \mathrm{co}\{(-1,-1+r),(1,-1+r)\}\,.$$

Analogously, from (72) and (74)–(76)

$$\begin{aligned}\partial_{\mathrm{M}}^{+} f(0) &= \partial_{\mathrm{F}}^{+} f(0) \cup \operatorname*{Lim\,sup}_{\substack{x\to 0,\\ x\neq 0}} \partial_{\mathrm{F}}^{+} f(x) \\ &= \bigcap_{\mu=1}^{4} \mathbb{B}_r(v^{\mu}) \\ &\quad \cup [\{(1,1)\} + \{w \mid \|w\| = r,\ w_1 \le 0,\ w_2 \le 0\}] \\ &\quad \cup [\{(-1,1)\} + \{w \mid \|w\| = r,\ w_1 \ge 0,\ w_2 \le 0\}] \\ &\quad \cup [\{(1,-1)\} + \{w \mid \|w\| = r,\ w_1 \le 0,\ w_2 \ge 0\}] \\ &\quad \cup [\{(-1,-1)\} + \{w \mid \|w\| = r,\ w_1 \ge 0,\ w_2 \ge 0\}]\,.\end{aligned}$$

The Mordukhovich subdifferentials of f at 0 for the values of $r = 0.5$, 1.25, and 2.0 are plotted in Figs. 1–3.

The corresponding series for the visualization of the directed subdifferentials with the same values of r are plotted in Fig. 4, see also [4, Example 5.7] for further explanations. The plots coincide with the pictures of the Mordukhovich symmetric subdifferentials. Since the subdifferentials of the convex functions g and h are known, the Rubinov subdifferential could be easily calculated as the visualization of the difference of these embedded convex sets.

The arrows in Fig. 4 indicate outer normals to the directed "supporting faces". They also form the parametrizing directions in (16) for the directed subdifferential. The positive part in the left picture of Fig. 4 is a convex set. It is colored in gray and only outer normals are attached to its boundary. The other nonconvex part belongs to the mixed-type part. Similarly for the right picture in Fig. 4. The gray convex subset is the negative part and has only inner normals attached to its boundary. The positive and negative part in the middle picture are empty and the Rubinov subdifferential consists only of the mixed-type part. Note that the unique "supporting points" belong both to the Mordukhovich subdifferential and superdifferential due to Theorems 3.13 and 3.14, since for such a point the lower dimensional positive and negative parts coincide with the point itself.

Example 4.2 ([23, Example 2.49]). Let

$$f(x_1,x_2) := \big||x_1| + x_2\big|\,.$$

Straightforward computation of the Mordukhovich subdifferentials of f (see [23, Example 2.49]) gives

$$\partial_{\mathrm{M}} f(0,0) = \mathrm{co}\{(0,0),(1,1),(-1,1)\} \cup \mathrm{co}\{(0,0),(-1,-1)\} \cup \mathrm{co}\{(0,0),(1,-1)\}\,,$$
$$\partial_{\mathrm{M}}^{+} f(0,0) = \mathrm{co}\{(1,-1),(-1,-1)\} \cup \{(1,1),(-1,1)\}\,,$$

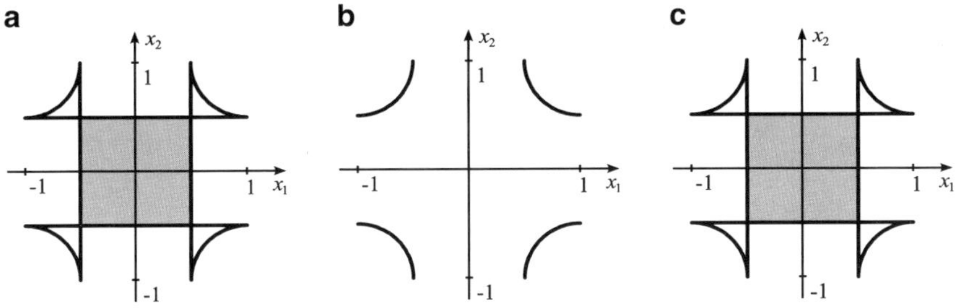

Fig. 1 Mordukhovich subdifferentials of f if $r = 0.5$: (**a**) $\partial_{\mathrm{M}} f(0)$; (**b**) $\partial_{\mathrm{M}}^{+} f(0)$; (**c**) $\partial_{\mathrm{M}}^{0} f(0)$

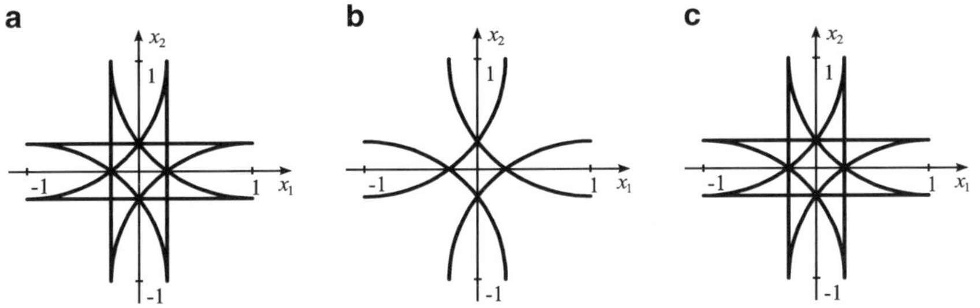

Fig. 2 Mordukhovich subdifferentials of f if $r = 1.25$: (**a**) $\partial_{\mathrm{M}} f(0)$; (**b**) $\partial_{\mathrm{M}}^{+} f(0)$; (**c**) $\partial_{\mathrm{M}}^{0} f(0)$

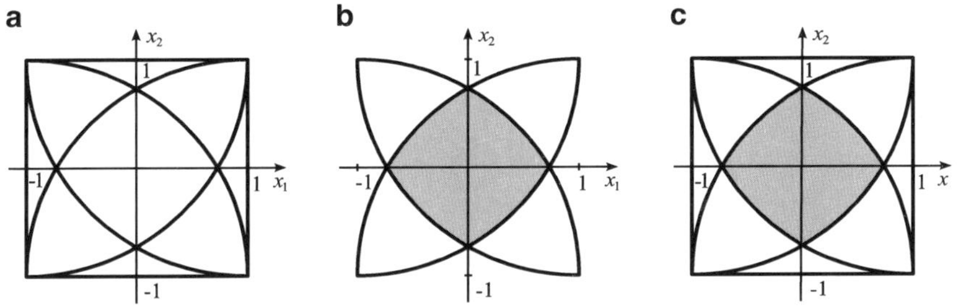

Fig. 3 Mordukhovich subdifferentials of f if $r = 2$: (**a**) $\partial_{\mathrm{M}} f(0)$; (**b**) $\partial_{\mathrm{M}}^{+} f(0)$; (**c**) $\partial_{\mathrm{M}}^{0} f(0)$

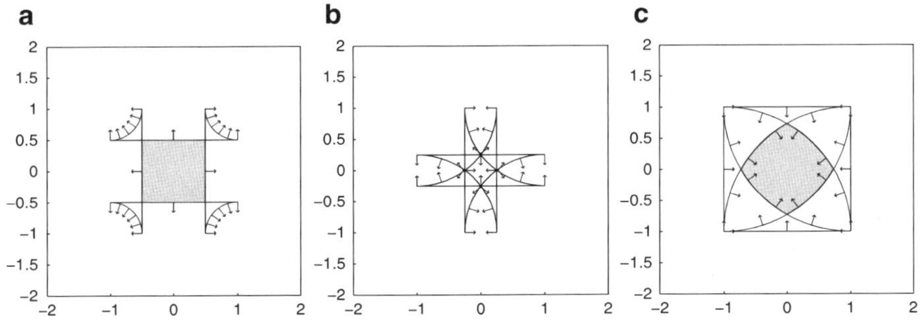

Fig. 4 Visualization of directed subdifferential for Example 4.1 for (**a**) $r = 0.5$; (**b**) $r = 1.25$; (**c**) $r = 2.0$

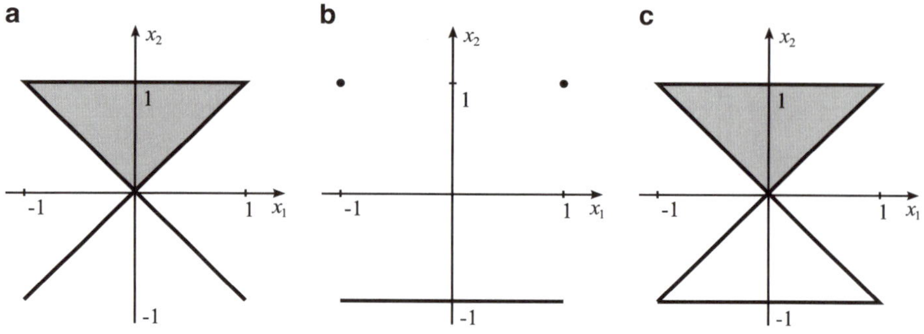

Fig. 5 Mordukhovich subdifferentials for Example 4.2: (**a**) $\partial_{\mathrm{M}} f(0)$; (**b**) $\partial_{\mathrm{M}}^{+} f(0)$; (**c**) $\partial_{\mathrm{M}}^{0} f(0)$

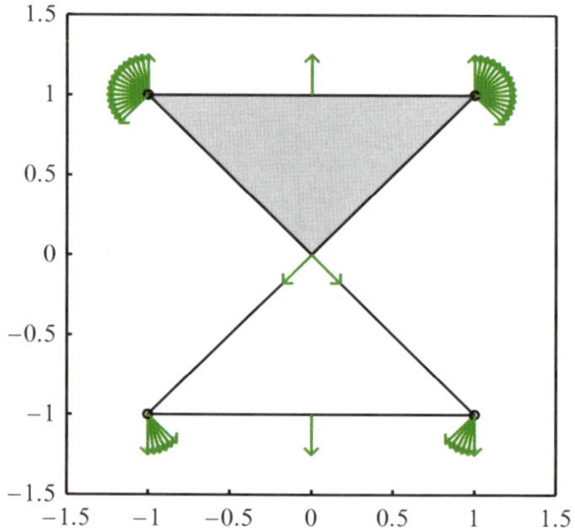

Fig. 6 Visualization of directed subdifferential for Example 4.2

and

$$\partial_{\mathrm{M}}^{0} f(0,0) = \partial_{\mathrm{M}} f(0,0) \cup \operatorname{co}\{(1,-1),(-1,-1)\} \,.$$

Figures 5–6 show the comparison between the Mordukhovich lower/upper/symmetric subdifferential with the Rubinov subdifferential. The calculation of the latter is based on one DC representation of f, for example

$$f(x) = \max\{2x_1 + 2x_2, -2x_1 + 2x_2, 0\} - \max\{x_1 + x_2, -x_1 + x_2\} \,.$$

As in Example 4.1, one can see that the four unique directed "supporting points" $(\pm 1, \pm 1)$ (see Fig. 5) are present both in the Mordukhovich subdifferential and superdifferential.

The only segment $\operatorname{co}\{(-1,-1),(1,-1)\}$ in the Mordukhovich superdifferential may be recognized from the Rubinov subdifferential in Fig. 6 as coming from a negative part of a directed interval, since there are outer normals attached to its ends (see Fig. 6) where the projections are pointing inside the interval, contrary to all the segments in the Mordukhovich subdifferential.

Also the Rubinov subdifferential (the visualization of the directed one in Fig. 6) coincides with the Mordukhovich symmetric subdifferential, according to Theorem 3.14.

The last example shows the difference between Theorems 3.13/3.14 and 3.17. In this example, the function f is DC, but not positive homogeneous. So we cannot expect that we have equality between the Mordukhovich symmetric subdifferential and the Rubinov one (the visualization of the directed subdifferential) as in Theorem 3.14.

Example 4.3 ([10, Sect. III.4, Example 4.2] and [4, Example 4.7]). Let $f = g - h$, where

$$g(x) = \max\{2x_2, x_1^2 + x_2\}\,, \quad h(x) = \max\{0, x_1^2 + x_2\}\,.$$

Together with

$$\varphi_1(x) = \max\{2x_2, x_1^2 + x_2\}\,, \quad \varphi_2(x) = \max\{0, x_2 - x_1^2\},$$

it follows that

$$\begin{aligned} f(x) &= \max\{2x_2, x_1^2 + x_2\} + \min\{0, -x_1^2 - x_2\} \\ &= \min\{\max\{2x_2, x_1^2 + x_2\}, \max\{0, x_2 - x_1^2\}\} \\ &= \min\{\varphi_1(x), \varphi_2(x)\}\,. \end{aligned}$$

We have

$$\partial\varphi_1(x) = \begin{cases} \{(0,2)\}\,, & \text{if } x_2 > x_1^2\,, \\ \{(2x_1,1)\}\,, & \text{if } x_2 < x_1^2\,, \\ \operatorname{co}\{(0,2),(2x_1,1)\}\,, & \text{if } x_2 = x_1^2\,, \end{cases}$$

$$\partial\varphi_2(x) = \begin{cases} \{(-2x_1,1)\}\,, & \text{if } x_2 > x_1^2\,, \\ \{(0,0)\}\,, & \text{if } x_2 < x_1^2\,, \\ \operatorname{co}\{(0,0),(-2x_1,1)\}\,, & \text{if } x_2 = x_1^2\,. \end{cases}$$

It is not difficult to observe that the set of active indices of f in x, that is

$$I_f(x) = \{i \in \{1,2\} \mid f(x) = \varphi_i(x)\} = \begin{cases} \{1\}\,, & \text{if } x_2 < -x_1^2\,, \\ \{2\}\,, & \text{if } x_2 > -x_1^2\,, \\ \{1,2\}\,, & \text{if } x_2 = -x_1^2\,. \end{cases}$$

From Lemma 2.3 follows that

$$\partial_F f(x) = \begin{cases} \partial\varphi_1(x)\,, & \text{if } x_2 < -x_1^2\,, \\ \partial\varphi_2(x)\,, & \text{if } x_2 > -x_1^2\,, \\ \partial\varphi_1(x) \cap \partial\varphi_2(x)\,, & \text{if } x_2 = -x_1^2\,, \end{cases}$$

$$= \begin{cases} \{(0,0)\}\,, & \text{if } -x_1^2 < x_2 < x_1^2\,, \\ \{(-2x_1,1)\}\,, & \text{if } x_2 > x_1^2\,, \\ \{(2x_1,1)\}\,, & \text{if } x_2 < -x_1^2\,, \\ \operatorname{co}\{(0,0),(-2x_1,1)\}\,, & \text{if } x_2 = x_1^2\,,\, x_1, x_2 \neq 0\,, \\ \emptyset\,, & \text{if } x_2 = -x_1^2\,,\, x_1, x_2 \neq 0\,, \\ \{(0,1)\}\,, & \text{if } x_1 = x_2 = 0\,. \end{cases}$$

The evaluation of the outer limit in (13) is straight forward:

$$\partial_{\mathrm{M}} f(0) = \operatorname{Lim\,sup}_{x \to 0} \partial_{\mathrm{F}} f(x) = \operatorname{co}\{(0,0),(0,1)\}\,.$$

Since f is Fréchet differentiable, the Rubinov subdifferential yields just the gradient (see [4]):

$$V_2(\overrightarrow{\partial} f(0)) = \{(0,1)\},$$

which is a strict subset of the Mordukhovich subdifferential, see Fig. 7.

Let us try to apply Theorem 3.17. The formula for the directional derivatives of a DC function is proved in [10, Sect. I.3, Proposition 3.1]:

$$f'(x;l) = g'(x;l) - h'(x;l)$$

Since the directional derivatives of g and h involve a maximum, we set

$$\begin{array}{ll} g_1(x) = 2x_2\,, & g_2(x) = x_1^2 + x_2\,, \\ h_1(x) = 0\,, & h_2(x) = g_2(x), \end{array}$$

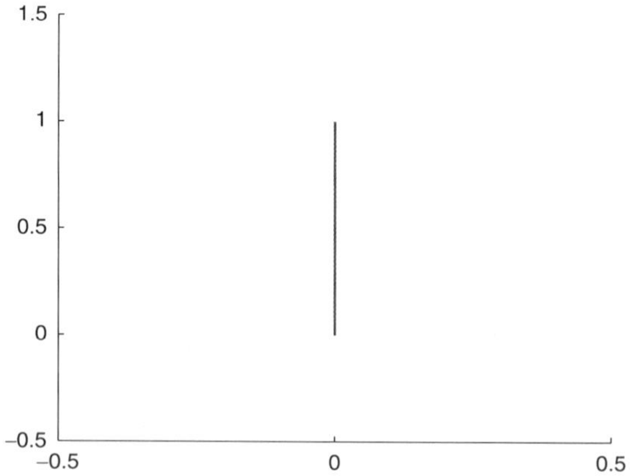

Fig. 7 Mordukhovich and Clarke subdifferential for Example 4.3

and apply [10, Sect. I.3, Proposition 3.1]:

$$
\begin{aligned}
g'(x;l) &= \max_{i \in I_g(x)} g_i'(x;l)\ , \quad I_g(x) = \{i \in \{1,2\} \,|\, g(x) = g_i(x)\}\ ,\\
h'(x;l) &= \max_{i \in I_h(x)} h_i'(x;l)\ , \quad I_h(x) = \{i \in \{1,2\} \,|\, h(x) = h_i(x)\}\ .
\end{aligned}
$$

Now,

$$
g'(x;l) = \begin{cases} g_1'(x;l) = \nabla g_1(x) l = (0,2)\cdot \binom{l_1}{l_2} = 2l_2\ , & \text{if } x_2 > x_1^2\ ,\\ g_2'(x;l) = \nabla g_2(x) l = (2x_1,1)\cdot \binom{l_1}{l_2} = 2x_1 l_1 + l_2\ , & \text{if } x_2 < x_1^2\ ,\\ \max\{g_1'(x;l), g_2'(x;l)\} = \max\{2l_2, 2x_1 l_1 + l_2\}\ , & \text{if } x_2 = x_1^2\ , \end{cases}
$$

$$
h'(x;l) = \begin{cases} h_1'(x;l) = \nabla h_1(x) l = (0,0)\cdot \binom{l_1}{l_2} = 0\ , & \text{if } x_2 < -x_1^2\ ,\\ h_2'(x;l) = \nabla h_2(x) l = (2x_1,1)\cdot \binom{l_1}{l_2} = 2x_1 l_1 + l_2\ , & \text{if } x_2 > -x_1^2\ ,\\ \max\{h_1'(x;l), h_2'(x;l)\} = \max\{0, 2x_1 l_1 + l_2\}\ , & \text{if } x_2 = -x_1^2\ . \end{cases}
$$

Since we fix $x = 0$, we have $x_2 = -x_1^2$ and $x_2 = x_1^2$ and hence,

$$
\begin{aligned}
f'(0;l) &= \max\{2l_2, 2\cdot 0\cdot l_1 + l_2\} - \max\{0, 2\cdot 0 \cdot l_1 + l_2\} = \max\{2l_2, l_2\} - \max\{0, l_2\}\\
&= l_2 + \max\{l_2, 0\} - \max\{0, l_2\} = l_2.
\end{aligned}
$$

The function $f'(0;\cdot)$ is continuously differentiable with respect to l, hence strict differentiable by [7, Corollary to Proposition 2.2.1]. One can apply [7, Proposition 2.2.4] to show

$$
\partial_{\mathrm{MP}} f(0) = \partial_{\mathrm{Cl}}[f'(0;\cdot)](0) = \{\nabla_l f'(0;\cdot)(0)\} = \{(0,1)\},
$$

which coincides with the Rubinov subdifferential.
A similar reasoning shows that the Fréchet subdifferential and superdifferential of the directional derivative coincides with the gradient of $f'(0;\cdot)$ with respect to l in any direction η by [18, Proposition 1.3]. Hence, the Mordukhovich subdifferential and the Mordukhovich superdifferential also equal to the point $(0,1)$ due to (14).

5 Conclusions

As we have shown in this chapter, the connection between the Mordukhovich subdifferential/superdifferential and the Rubinov subdifferential may provide substantial information related to their computing and their applications. This relation will be investigated and explored in more detail in our further research. Especially, we are currently working on the extension of our results from the class of DC functions to quasidifferentiable functions and on their application to quasidifferential calculus. Another focus of future research will be the case of dimension higher than two.

Acknowledgements The authors thank the referees for their valuable remarks. They also express their gratitude to Boris Mordukhovich for encouraging them to explore the connections between the Rubinov subdifferential and his subdifferential as well as to Vladimir Demyanov, who initiated the research to find connections between the quasidifferential and the Mordukhovich subdifferential. This work was partially supported by The Hermann Minkowski Center for Geometry at Tel Aviv University.

References

1. Amahroq, T., Penot, J.P., Syam, A.: On the subdifferentiability of the difference of two functions and local minimization. Set-Valued Anal. **16**(4), 413–427 (2008)
2. Baier, R., Farkhi, E.: Differences of convex compact sets in the space of directed sets. Part I: The space of directed sets. Set-Valued Anal. **9**(3), 217–245 (2001)
3. Baier, R., Farkhi, E.: Differences of convex compact sets in the space of directed sets. Part II: Visualization of directed sets. Set-Valued Anal. **9**(3), 247–272 (2001)
4. Baier, R., Farkhi, E.: The directed subdifferential of DC functions. In: A. Leizarowitz, B.S. Mordukhovich, I. Shafrir, A.J. Zaslavski (eds.) Nonlinear Analysis and Optimization II: Optimization. A Conference in Celebration of Alex Ioffe's 70th and Simeon Reich's 60th Birthdays, June 18–24, 2008, Haifa, Israel, *AMS Contemporary Mathematics*, vol. 513, pp. 27–43. AMS and Bar-Ilan University (2010)
5. Bazaraa, M.S., Goode, J.J., Nashed, M.Z.: On the cones of tangents with applications to mathematical programming. J. Optim. Theory Appl. **13**, 389–426 (1974)
6. Borwein, J.M., Strójwas, H.M.: Proximal analysis and boundaries of closed sets in Banach space. I. Theory. Canad. J. Math. **38**(2), 431–452 (1986)
7. Clarke, F.H.: Optimization and nonsmooth analysis, *Classics in Applied Mathematics*, vol. 5. SIAM, Philadelphia, PA (1990). First edition published in Wiley, New York, 1983
8. Demyanov, V.F., Jeyakumar, V.: Hunting for a smaller convex subdifferential. J. Global Optim. **10**(3), 305–326 (1997)
9. Demyanov, V.F., Roshchina, V.A.: Exhausters, optimality conditions and related problems. J. Global Optim. **40**(1-3), 71–85 (2008)
10. Demyanov, V.F., Rubinov, A.M.: Constructive nonsmooth analysis, *Approximation and Optimization*, vol. 7. Peter Lang, Frankfurt/Main (1995). Russian original "Foundations of nonsmooth analysis, and quasidifferential calculus" published in Nauka, Moscow, 1990
11. Dinh, N., Mordukhovich, B.S., Nghia, T.T.A.: Qualification and optimality conditions for DC programs with infinite constraints. Acta Math. Vietnam. **34**(1), 125–155 (2009)
12. Eberhard, A., Nyblom, M.: Jets, generalised convexity, proximal normality and differences of functions. Nonlinear Anal. **34**(3), 319–360 (1998)
13. Hadwiger, H.: Minkowskische Addition und Subtraktion beliebiger Punktmengen und die Theoreme von Erhard Schmidt. Mathematische Zeitschrift **53**(3), 210–218 (1950)
14. Hiriart-Urruty, J.B.: Generalized differentiability, duality and optimization for problems dealing with differences of convex functions. In: J. Ponstein (ed.) Convexity and Duality in Optimization. Proceedings of the Symposium on Convexity and Duality in Optimization Held at the University of Groningen, The Netherlands June 22, 1984, *Lecture Notes in Econom. and Math. Systems*, vol. 256, pp. 37–70. Springer, Berlin (1985)
15. Hiriart-Urruty, J.B., Lemaréchal, C.: Convex Analysis and Minimization Algorithms I. Fundamentals, *Grundlehren der mathematischen Wissenschaften*, vol. 305. Springer, Berlin (1993)
16. Hörmander, P.L.: Sur la fonction d'appui des ensembles convexes dans un espace localement convexe. Ark. Mat. **3**(12), 181–186 (1954)
17. Ioffe, A.D.: Sous-différentielles approchées de fonctions numériques. C. R. Acad. Sci. Paris Sér. I Math. **292**(14), 675–678 (1981)
18. Kruger, A.Y.: On Fréchet subdifferentials. J. Math. Sci. (N.Y.) **116**(3), 3325–3358 (2003). Optimization and related topics, Vol. 3

19. Kuntz, L., Scholtes, S.: Constraint qualifications in quasidifferentiable optimization. Math. Program. Ser. A **60**(3), 339–347 (1993)
20. Martínez-Legaz, J.E., Seeger, A.: A formula on the approximate subdifferential of the difference of convex functions. Bull. Austral. Math. Soc. **45**(1), 37–41 (1992)
21. Michel, P., Penot, J.P.: Calcul sous-différentiel pour des fonctions lipschitziennes et non lipschitziennes. C. R. Acad. Sci. Paris Sér. I Math. **298**(12), 269–272 (1984)
22. Mordukhovich, B.S.: Generalized differential calculus for nonsmooth and set-valued mappings. J. Math. Anal. Appl. **183**(1), 250–288 (1994)
23. Mordukhovich, B.S.: Variational analysis and generalized differentiation. I Basic theory, *Grundlehren der Mathematischen Wissenschaften [Fundamental Principles of Mathematical Sciences]*, vol. 330. Springer, Berlin (2006)
24. Mordukhovich, B.S.: Variational analysis and generalized differentiation. II Applications, *Grundlehren der Mathematischen Wissenschaften [Fundamental Principles of Mathematical Sciences]*, vol. 331. Springer, Berlin (2006)
25. Mordukhovich, B.S., Shao, Y.H.: Nonsmooth sequential analysis in Asplund spaces. Trans. Amer. Math. Soc. **348**(4), 1235–1280 (1996)
26. Penot, J.P.: Sous-différentiels de fonctions numériques non convexes. C. R. Acad. Sci. Paris Sér. I Math. **278**, 1553–1555 (1974)
27. Penot, J.P.: Calcul sous-différentiel et optimisation. J. Funct. Anal. **27**(2), 248–276 (1978)
28. Pontryagin, L.S.: Linear differential games. II. Sov. Math., Dokl. **8**(4), 910–912 (1967)
29. Rockafellar, R.T.: Convex Analysis, *Princeton Mathematical Series*, vol. 28, 2nd edn. Princeton University Press, Princeton, NJ (1972). First published in 1970
30. Rockafellar, R.T., Wets, R.J.B.: Variational analysis, *Grundlehren der Mathematischen Wissenschaften [Fundamental Principles of Mathematical Sciences]*, vol. 317. Springer, Berlin (1998)

Future Challenges for Variational Analysis*

Jonathan M. Borwein

Abstract Modern nonsmooth analysis is now roughly 35 years old. In this chapter, I shall attempt to analyse (briefly): where the subject stands today, where it should be going, and what it will take to get there? In summary, the conclusion is that the first-order theory is rather impressive, as are many applications. The second-order theory is by comparison somewhat underdeveloped and wanting of further advance.

> It is not knowledge, but the act of learning, not possession but the act of getting there, which grants the greatest enjoyment. When I have clarified and exhausted a subject, then I turn away from it, in order to go into darkness again; the never-satisfied man is so strange if he has completed a structure, then it is not in order to dwell in it peacefully,but in order to begin another. I imagine the world conqueror must feel thus, who, after one kingdom is scarcely conquered, stretches out his arms for others. – Carl Friedrich Gauss (1777–1855).[1]

1 Preliminaries and Precursors

I intend to first discuss *First-Order Theory*, and then *Higher-Order Theory* – mainly second-order – and only mention passingly higher-order theory, which really devolves to second-order theory. I will finish by touching on *Applications of Variational Analysis* or VA both inside and outside Mathematics, mentioning both successes and limitations or failures. Each topic leads to open questions even in the convex case, which I will refer to as CA. Some issues are technical and

Jonathan M. Borwein
CARMA, University of Newcastle, Newcastle, NSW, Australia
e-mail: jonathan.borwein@newcastle.edu.au

* This chapter is dedicated to Boris Mordukhovich on the occasion of his 60th birthday. It is based on a talk presented at the *International Symposium on Variational Analysis and Optimization* (ISVAO), Department of Applied Mathematics, Sun Yat-Sen University, 28–30 November 2008.

[1] From an 1808 letter to his friend Farkas Bolyai (the father of Janos Bolyai).

R.S. Burachik and J.-C. Yao (eds.), *Variational Analysis and Generalized Differentiation in Optimization and Control*, Springer Optimization and Its Applications 47,
DOI 10.1007/978-1-4419-0437-9_4,

specialized, others are some broader and more general. In nearly every case, Boris Mordukhovich has made prominent or seminal contributions; many of which are elaborated in [24] and [25].

To work fruitfully in VA, it is really important to understand both CA and *smooth analysis* (SA); they are the motivating foundations and very often provide the key technical tools and insights. For example, Fig. 1 shows how an essentially strictly convex [8, 11] function defined on the orthant can fail to be strictly convex.

$$(x,y) \mapsto \max\left\{(x-2)^2+y^2-1, -(xy)^{1/4}\right\}.$$

Understanding this sort of boundary behaviour is clearly prerequisite to more delicate variational analysis of lower semicontinuous functions as are studied in [8, 13, 24, 28].

In this note, our terminology is for the most part consistent with those references and since I wish to discuss patterns, not proofs, I will not worry too much about exact conditions. That said, f will at least be a proper and lower semicontinuous extended-real valued function on a Banach space X.

Fig. 1 A function that is essentially strictly but not strictly convex with nonconvex subgradient domain

Let us first recall the two main starting points:

1.1 A Descriptive Approach

By 1968 Pshenichnii, as described in his book [27], had started a study of the large class of *quasi-differentiable* locally Lipschitz functions for which

$$f'(x;h) := \limsup_{t\to 0^+} \frac{f(x+th)-f(x)}{t}$$

is *required* to exist and be convex as a function of h. We define $\partial' f(x) := \partial_2 f'(x;0)$, where we take the classical convex subdifferential with respect to the second variable.

1.2 A Prescriptive Approach

By contrast, Clarke in his 1972 thesis (described in his now classic book [15]) considered *all* locally Lipschitz functions for which

$$f^\circ(x;h) := \limsup_{t \to 0^+ \; y \to x} \frac{f(y+th) - f(y)}{t}$$

is *constructed* to be convex. In convex terms, we may now define a generalized subdifferential by $\partial^o f(x) := \partial_2 f^0(x;0)$. (Here the later is again the convex subdifferential with respect to the h variable.)

Both ideas capture the smooth and the convex case, both are closed under $+$ and $\vee$, and both satisfy a reasonable calculus; so we are off to the races. Of course, we now wish to do as well as we can with more general lsc functions.

2 First-Order Theory of Variational Analysis

The key players are as I shall outline below. We start with:

1. *The (Fréchet) subgradient*: $\partial_{\mathrm{F}} f(x)$, which denotes a one-sided lower Fréchet subgradient (i.e., the appropriate limit is taken uniformly on bounded sets) and which can (for some purposes) be replaced by a Gâteaux (uniform on finite sets), Hadamard (uniform on norm-compact sets), or weak Hadamard (uniform on weakly compact sets) object. These are denoted by $\partial_{\mathrm{G}} f(x), \partial_{\mathrm{H}} f(x)$, and $\partial_{\mathrm{WH}} f(x)$ respectively.
   That is $\phi \in \partial_{\mathrm{F}} f(x)$, exactly when
   $$\phi(h) \leq \liminf_{t \to 0^+ \; \|h\|=1} \frac{f(x+th) - f(x)}{t}.$$
   A formally smaller and more precise object is a derivative bundle of F, G, H, or WH-smooth (local) minorants:
2. *The viscosity subgradient*:
   $$\partial^v_{\mathrm{F}} f(x) := \{\phi \colon \phi = \nabla_{\mathrm{F}} g(x), f(y) - g(y) \geq f(x) - g(x) \text{ for } y \text{ near } x\}$$
   as illustrated in Fig. 2. By its very definition $0 \in \partial^v_{\mathrm{F}} f(x)$ when x is a local minimizer of f. In nice spaces, say those with Fréchet-smooth renorms as have

reflexive spaces, these two subgradient notions coincide [13]. In this case, we have access to a good generalization of the sum rule from convex calculus [11]:

3. *(Fuzzy) sum rule*: For each $\varepsilon > 0$

$$\partial_F(f+g)(x) \subseteq \partial_F f(x_1) + \partial_F g(x_2) + \varepsilon B_{X^*}$$

for points x_1, x_2 each within ε of x. In Euclidean space and even in Banach space – under quite stringent compactness conditions except in the Lipschitz case – with the addition of *asymptotic subgradients*, one can pass to the limit and recapture *approximate* subdifferentials [13, 24, 25, 28].

For now, we let ∂f denote any of a number of subgradients and have the appropriate tools to define a workable normal cone.

4. *Normal cones*: We define

$$N_{\mathrm{epi} f} := \partial \iota_{\mathrm{epi} f}.$$

Here ι_C denotes the convex *indicator function* of a set C.

Key to establishing the fuzzy sum rule and its many equivalences [13, 24] are:

5. *Smooth variational principles (SVP)* which establish the existence of many points, x, and locally smooth (with respect to an appropriate topology) minorants g such that

$$f(y) - g(y) \geq f(x) - g(x)$$

for y near x.

We can now establish the existence and structure of:

6. *Limiting subdifferentials* such as

$$\partial^a f(x) := \limsup_{y \to_f x} \partial_F f(x),$$

for appropriate topological limits superior, and of:

7. *Coderivatives of multifunctions*: As in [24], one may write

$$D^* \Omega(x,y)(y^*) = \left\{x^* : (x^*, -y^*) \in N_{\mathrm{gph}(\Omega)}(x,y)\right\}.$$

The fuzzy sum rule and its related calculus also leads to fine results about the notion of:

8. *Metric regularity*:

Indeed, we can provide very practicable conditions on a multifunction Ω, see [12, 13, 17, 24], so that locally around $y_0 \in \Omega(x_0)$ one has

$$Kd(\Omega(x), y) \geq d(x, \Omega^{-1}(y)). \tag{1}$$

Estimate (1) allows one to show many things easily. For example, it allows one straight forwardly to produce C^k-implicit function theorems under second-order sufficiency conditions [3, 13]. Estimate (1) is also really useful in the very concrete setting of alternating projections on two closed convex sets C and D, where one uses $\Omega(x) := x - D$ for $x \in C$ and $\Omega(x) := \emptyset$ otherwise [13].

The very recent book by Dontchev and Rockafellar [17] gives a comprehensive treatment of implicit function theory for Euclidean multifunctions (and much more).

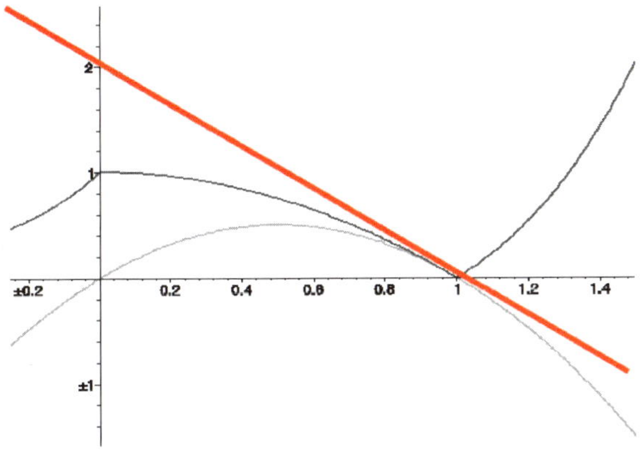

Fig. 2 A function and its smooth minorant and a viscosity subdifferential (in *red*)

2.1 Achievements and Limitations

Variational principles meshed with viscosity subdifferentials provide a fine first-order theory. Sadly, $\partial^a f(x)$ is inapplicable outside of *Asplund space* (such as reflexive space or spaces with separable duals) and extensions using $\partial_{\mathrm{H}} f$ are limited and technically complicated. Correspondingly, the coderivative is very beautiful theoretically but is hard to compute even for "nice" functions. Moreover, the compactness restrictions (e.g., *sequential normal compactness* as described in [24]) are fundamental, not technical. Better results rely on restricting classes of functions (and spaces) such as considering, *prox-regular* [28], *lower* C^2 [28], or *essentially smooth* functions [13].

Moreover, the limits of a prescriptive approach are highlighted by the fact that one can prove results showing that in all (separable) Banach spaces X a *generic* nonexpansive function has no information in its generalized derivative:

$$\partial^a f(x) = \partial^o f(x) \equiv B_{X^*}$$

for all points $x \in X$ [10, 13]. Similarly, one can show that nonconvex equilibrium results will frequently contain little or no nontrivial information [13].

3 Higher-Order Theory of Variational Analysis

Recall that for closed proper *convex functions* the *difference quotient* of f is given by

$$\Delta_t f(x) : h \mapsto \frac{f(x+th)-f(x)}{t};$$

and the *second-order difference quotient* of f by

$$\Delta_t^2 f(x) : h \mapsto \frac{f(x+th)-f(x)-t\langle \nabla f(x), h\rangle}{\frac{1}{2}t^2}.$$

Analogously let

$$\Delta_t[\partial f](x) : h \mapsto \frac{\partial f(x+th)-\nabla f(x)}{t}.$$

For any $t > 0$, $\Delta_t f(x)$ is closed, proper, convex and nonnegative [11, 28]. Quite beautifully, as Rockafellar [11,28] discovered,

$$\partial\left[\frac{1}{2}\Delta_t^2 f(x)\right] = \Delta_t[\partial f](x).$$

Hence, we reconnect the two most natural ways of building a second-order convex approximation.

This relates to a wonderful result [1, 11]:

Theorem 1 ([1]). *In Euclidean space a real-valued continuous convex function admits a second-order Taylor expansion at almost all points (with respect to Lebesgue measure).*

My favourite proof is a specialization of Mignot's 1976 extension of Alexandrov's theorem for monotone operators [11,28]. The theorem relies on many happy coincidences in Euclidean space. This convex result is quite subtle and so the paucity of definitive nonconvex results is no surprise.

3.1 The State of Higher-Order Theory

Various lovely patterns and fine theorems are available in Euclidean space [11, 24, 28] but no definitive corpus of results exists, nor even canonical definitions, outside of the convex case. There is interesting work by Jeyakumar and Luc [21], by Dutta, and others, much of which is surveyed in [18].

Starting with Clarke, many have noted that

$$\partial^2 f(x) := \partial \nabla_G f(x)$$

is a fine object when the function f is Lipschitz smooth in a separable Banach space – so that the Banach space version of Rademacher's Theorem [11] applies.

More interesting are the quite fundamental results by Ioffe and Penot [20] on limiting 2-subjets and 2-coderivatives in Euclidean space, with a more refined calculus of "efficient" sub-Hessians given by Eberhard and Wenczel [19]. Ioffe and Penot [20] exploit Alexandrov-like theory, again starting with the subtle analysis in [16], to carefully study a *subjet* of a reasonable function f at x, the subjet $\partial_-^2 f(x)$ being defined as the collection of second-order expansions of all C^2 local minorants g with $g(x) = f(x)$. The (nonempty) *limiting 2-subjet* is then defined by

$$\overline{\partial}^2 f(x) := \limsup_{y \to_f x} \partial_-^2 f(x).$$

Various distinguished subsets and limits are also considered in their paper. They provide a calculus, based on a sum rule for limiting 2-subjets (that holds for all lower-C^2 functions and so for all continuous convex functions) making note of both the similarities and differences from the first-order theory. As noted, interesting refinements have been given by Eberhard and Wenczel in [19].

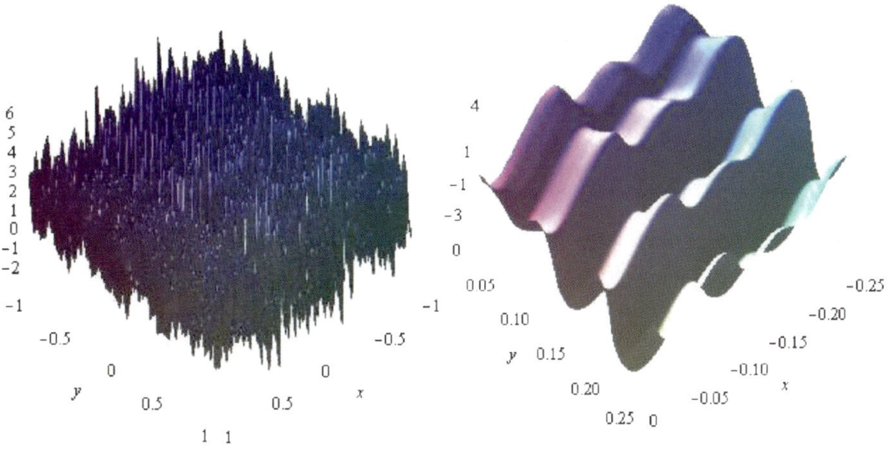

Fig. 3 Nick Trefethen's digit-challenge function (2)

There is little "deep" work in infinite dimensions, that is, when reasonably obvious extensions fail even in Hilbert space. Outside separable Hilbert space, general positive results are not to be expected [11]. So it seems clear to me that research should focus on structured classes of functions for which more can be obtained; such as *integral functionals* as in Moussaoui and Seeger [26], *semi-smooth* and *prox-regular functions* [8], or *composite convex functions* [28].

4 Some Reflections on Applications of Variational Analysis

The tools of variational analysis are now an established part of pure nonlinear and functional analysis. This is a major accomplishment.

There are also more concrete successes:

- There is a convergence theory for "pattern search" derivative-free optimization algorithms (see [23] for an up to date accounting of such methods) based on the Clarke subdifferential.
- Eigenvalue and singular value optimization theory has been beautifully developed [8], thanks largely to Adrian Lewis. There is a quite delicate second-order theory due to Lewis and Sendov [22]. There are even some results for Hilbert–Schmidt operators [11, 13].
- We can also handle a wide variety of differential inclusions and optimal control problems well [25].
- There is a fine approximate Maximum Principle and a good accounting of Hamilton–Jacobi equations [13, 24, 25].
- Nonconvex mathematical economics and *Mathematical Programs with Equilibrium Constraints* (MPECS) are much better understood than before [24, 25].
- Exact penalty and universal barrier methods are well developed, especially in finite dimensions [11].
- Counting convex optimization – as we certainly should – we have many more successes [14].

That said, there has been only limited numerical success even in the convex case – excluding somewhat spectral optimization, semidefinite programming code, and bundle methods.

For example, consider the following two-variable well-structured very smooth function taken from [4], in which only the first two rather innocuous terms couple the variables

$$\begin{aligned}(x,y) \mapsto &+ (x^2+y^2)/4 - \sin(10(x+y)) + \exp(\sin(50x)) \\ &+ \sin(\sin(80y)) + \sin(70\sin x) + \sin(60e^y). \end{aligned} \tag{2}$$

This function is quite hard to minimize. Actually, the global minimum occurs at $(x^*,y^*) \approx (-0.024627\ldots,0.211789\ldots)$ with minimal value of $\approx -3.30687\ldots$.

The pictures in Fig. 3, plotted using 10^6 grid points on $[0,1]\times[0,1]$ and also – after "zooming in" – on $[-00.25,0]\times[0,0.25]$, shows that we really cannot robustly distinguish the function from a nonsmooth function. Hence, it makes little sense to look at practicable nonsmooth algorithms without specifying a realistic subclass of functions on which they should operate.

Perhaps we should look more towards projects like Robert Vanderbei's SDP/Convex package *LOQO/LOCO*[2] and Janos Pinter's Global Optimization

[2] http://www.princeton.edu/~rvdb/.

LGO[3] package, while working with composite convex functions and smoothing techniques, and adopting the "disciplined convex programming"[4] approach advocated by Steve Boyd.

5 Open Questions and Concluding Remarks

I pose six problems below, which should either have variational solutions or instructive counterexamples. Details can be found in the specified references.

5.1 Alexandrov Theorem in Infinite Dimensions

For me, the most intriguing open question about convex functions is:

> Does every continuous convex function on separable Hilbert space admit a second order Gâteaux expansion at at least one point (or perhaps on a dense set of points)? [7,9,13]

This fails in nonseparable Hilbert space and in every separable $\ell_p(\mathbf{N}), 1 \le p < \infty, p \neq 2$. It also fails in the Fréchet sense even in $\ell_2(\mathbf{N})$.

The following example from [9] provides a continuous convex function d on any nonseparable Hilbert space which is nowhere second-order differentiable: *Let A be uncountable and let C the positive cone of $\ell_2(A)$. Denote by d the distance function to C and let $P := \nabla d$. Then d is nowhere second-order differentiable and P is nowhere differentiable (in the sense of Mignot [28]).*

Proof. Clearly, $P(a) = a^+$ for all $a \in \ell_2(A)$, where $a^+ = (a^+_\alpha)_{\alpha \in A}$ and $a^+_\alpha = \max\{0, a_\alpha\}$. Pick $x \in \ell_2(A)$ and $\alpha \in A$, then P is differentiable in the direction e_α if and only if $x_\alpha \neq 0$. Here e_α stands for an element of the canonical basis. Since each $x \in \ell_2(A)$ has only countably many nonzero coordinates, d is nowhere second-order differentiable. Likewise the maximal monotone operator P is nowhere differentiable. □

So I suggest to look for a counterexample. I might add that, despite the wonderful results of Preiss, see [11], and others on differentiability of Lipschitz functions, it is still also unknown whether two arbitrary real-valued Lipschitz functions on a separable Hilbert space must share a point of Fréchet differentiability.

---

[3] http://myweb.dal.ca/jdpinter/index.html.

[4] http://www.stanford.edu/~boyd/cvx/.

5.2 Subjets in Hilbert Space

I turn to a question about nonsmooth second-order behaviour:

> Are there sizeable classes of functions for which *subjets* or other useful second order expansions can be built in separable Hilbert space? [11, 19, 20]

I have no precise idea what "useful" means and even in convex case this is a tough request; if one could handle the convex case then one might be able to use Lasry–Lions regularization or other such tools more generally. A potentially tractable case is that of continuous integral functionals for which positive Alexandrov-like results are known in the convex case [9].

5.3 Chebyshev Sets

The *Chebyshev problem* as posed by Klee (1961) asks:

> Given a nonempty set C in a Hilbert space H such that every point in H has a unique nearest (also called proximal) point in C must C convex? [5, 8, 11]

Such sets are called *Chebyshev sets*. Clearly convex closed sets in Hilbert space are Chebyshev sets. The answer is "yes" in finite dimensions. This is the *Motzkin–Bunt theorem* of which four proofs are given in Euclidean space in [8] and [5]. In [5, 11], a history of the problem, which fails in some incomplete normed spaces, is given.

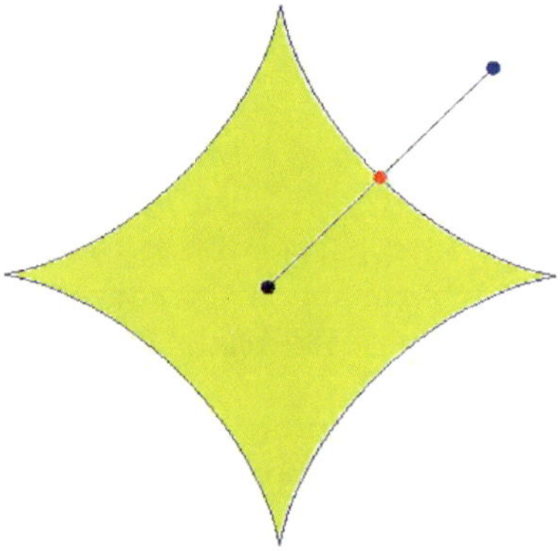

Fig. 4 A proximal point on the boundary of the $(2/3)$-ball

5.4 Proximality

The most striking open question I know regarding proximality is:

> (a) Let C be a closed subset of a Hilbert space H. Fix an arbitrary equivalent renorming of H. Must some (many) points in H have a nearest point in C in the given renorming?

(b) More generally, is it possible that in every reflexive Banach space, the *proximal points* on the boundary of C (see Fig. 4) are dense in the boundary of C? [6, 13]

The answer is "yes" in if the set is bounded or the norm is *Kadec–Klee* and hence if the space is finite dimensional or if it is locally uniformly rotund [6, 11, 13].

So any counterexample must be a wild set in a weird equivalent norm on H.

5.5 Legendre Functions in Reflexive Space

Recall that a convex function is of *Legendre-type* if it is both essentially smooth and essentially strictly convex. In the reflexive setting, the property is preserved under Fenchel conjugacy:

Find a generalization of the notion of a Legendre function for convex functions on a reflexive space that applies when the functions have no points of continuity such as is the case of the (negative) Shannon entropy. [2, 11]

When f has a point of continuity, a quite useful theory is available but it does not apply to entropy functions like $x \mapsto \int_0^1 x(t) \log x(t)\, \mu(\mathrm{d}t)$ or $x \mapsto -\int_0^1 \log x(t)\, \mu(\mathrm{d}t)$, whose domains are subsets of the nonnegative cone when viewed as operators on $L_2(T, \mu)$. More properly to cover these two examples, the theory should really apply to integral functionals on nonreflexive spaces such as $L_1(T, \mu)$.

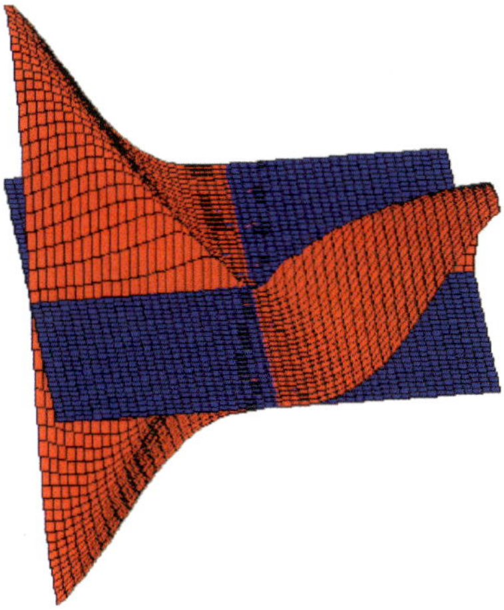

Fig. 5 A function and nonviscosity *subderivative* of 0

5.6 Viscosity Subdifferentials in Hilbert Space

A more technical but fundamental question is:

Is there a real-valued locally Lipschitz function f on $\ell_2(\mathbf{N})$ such that properly

$$\partial^v_{\mathrm{H}} f(x) \subset \partial_{\mathrm{H}} f(x)$$

for some $x \in \ell_2(\mathbf{N})$? [12, 13]

As shown in Fig. 5, the following continuous but non-Lipschitz function

$$(x,y) \mapsto \frac{xy^3}{x^2+y^4}$$

with value zero at the origin has $0 \in \partial_{\mathrm{H}} f(0)$ but $0 \notin \partial^v_{\mathrm{H}} f(0)$ [12, 13].

For a Lipschitz function in Euclidean space the answer is "no" since $\partial_{\mathrm{F}} f = \partial_{\mathrm{H}} f$ in this setting. And as we have noted $\partial_{\mathrm{F}} f = \partial^v_{\mathrm{F}} f$ in reflexive space. A counterexample would be very instructive, while a positive result would allow for many results to be extended from the Fréchet case to the Gateaux case: as $\partial_{\mathrm{G}} f = \partial_{\mathrm{H}} f$ for all locally Lipschitz f.

5.7 Final Comments

My view is that rather than looking for general prescriptive results based on universal constructions, we would do better to spend some real effort, or "brain grease" as Einstein called it,[5] on descriptive results for problems such as the six above. Counterexamples or complete positive solutions would be spectacular, but even somewhat improving best current results will require sharpening the tools of variational analysis in interesting ways. That would also provide great advertising for our field.

References

1. Alexandrov AD (1939) Almost everywhere existence of the second differential of a convex function and some properties of convex surfaces connected with it. Leningrad State University, Annals, Mathematical Series 6:3–35
2. Bauschke HH, Borwein JM, and Combettes PL (2001) Essential smoothness, essential strict convexity, and Legendre functions in Banach spaces. Communications in Contemporary Mathematics 3:615–647
3. Borwein JM (1986) Stability and regular points of inequality systems. Journal of Optimization Theory and Applications 48:9–52

[5] "On quantum theory, I use up more brain grease than on relativity." (Albert Einstein in a letter to Otto Stern in 1911).

4. Borwein JM (2005) The SIAM 100 digits challenge. Extended review in The Mathematical Intelligencer 27:40–48
5. Borwein JM (2007) Proximality and Chebyshev sets. Optimization Letters 1:21–32
6. Borwein JM and Fitzpatrick S (1989) Existence of nearest points in Banach spaces. Canadian Journal of Mathematics 61:702–720
7. Borwein JM and Ioffe A (1996) Proximal analysis in smooth spaces. Set-Valued Analysis 4:1–24
8. Borwein JM and Lewis AS (2005) Convex Analysis and Nonlinear Optimization. Theory and Examples, 2nd Edn. Canadian Mathematical Society – Springer, Berlin
9. Borwein JM and Noll D (1994) Second order differentiability of convex functions in Banach spaces. Transactions of the American Mathematical Society 342:43–82
10. Borwein JM and Sciffer S (2010) An explicit non-expansive function whose subdifferential is the entire dual ball. Contemporary Mathematics 514:99–103
11. Borwein JM and Vanderwerff JM (2010) Convex Functions: Constructions, Characterizations and Counterexamples. Cambridge University Press, Cambridge
12. Borwein JM and Zhu Q (1996) Viscosity solutions and viscosity subderivatives in smooth Banach spaces with applications to metric regularity. SIAM Journal of Optimization 34:1568–1591
13. Borwein JM and Zhu Q (2005) Techniques of Variational Analysis. Canadian Mathematical Society – Springer, Berlin
14. Boyd S and Vandenberghe L (2004) Convex Optimization. Cambridge University Press, Cambridge
15. Clarke FH (1983) Optimization and Nonsmooth Analysis. Canadian Mathematical Society Series of Monographs and Advanced Texts. Wiley, New York
16. Crandall M, Ishii H, and Lions P-L (1992) User's guide to viscosity solutions of second-order partial differential equations. Bulletin of the American Mathematical Society 27:1–67
17. Dontchev AL and Rockafellar RT (2009) Implicit Functions and Solution Mappings: A View from Variational Analysis. Springer Monographs in Mathematics. Springer, New York
18. Dutta J (2005) Generalized derivatives and nonsmooth optimization, a finite dimensional tour. With discussions and a rejoinder by the author. TOP 11:85–314
19. Eberhard AC and Wenczel R (2007) On the calculus of limiting subhessians. Set-Valued Analysis 15:377–424
20. Ioffe AD and Penot J-P (1997) Limiting subhessians, limiting subjets and their calculus. Transactions of the American Mathematical Society 349:789–807
21. Jeyakumar V and Luc DT (2008) Nonsmooth Vector Functions and Continuous Optimization. Springer Optimization and Its Applications, vol. 10. Springer, Berlin
22. Lewis AS and Sendov HS (2005) Nonsmooth analysis of singular values, I & II. Set-Valued Analysis 13:213–264
23. Macklem M (2009) Parallel Continuous Optimization. Doctoral Thesis, Dalhousie University
24. Mordukhovich BS (2006) Variational Analysis & Generalized Differentiation I. Basic Theory. Springer Series Fundamental Principles of Mathematical Sciences, vol. 330. Springer, Berlin
25. Mordukhovich BS (2006) Variational Analysis & Generalized Differentiation II. Applications. Springer Series Fundamental Principles of Mathematical Sciences, vol. 331. Springer, Berlin
26. Moussaoui M and Seeger A (1999) Second-order subgradients of convex integral functionals. Transactions of the American Mathematical Society 351:3687–3711
27. Pshenichnii B (1971) Necessary Conditions for an Extremum. Marcel Dekker, New York
28. Rockafellar RT and Wets RJ-B (1998) Variational Analysis. A Series of Comprehensive Studies in Mathematics, vol. 317. Springer, Berlin

A Deflected Subgradient Method Using a General Augmented Lagrangian Duality with Implications on Penalty Methods

Regina S. Burachik and C. Yalçın Kaya

Abstract We propose a duality scheme for solving constrained nonsmooth and nonconvex optimization problems. Our approach is to use a new variant of the deflected subgradient method for solving the dual problem. Our augmented Lagrangian function induces a primal–dual method with strong duality, that is, with zero duality gap. We prove that our method converges to a dual solution if and only if a dual solution exists. We also prove that all accumulation points of an auxiliary primal sequence are primal solutions. Our results apply, in particular, to classical penalty methods, since the penalty functions associated with these methods can be recovered as a special case of our augmented Lagrangians. Besides the classical augmenting terms given by the ℓ_1- or ℓ_2-norm forms, terms of many other forms can be used in our Lagrangian function. Using a practical selection of the step-size parameters, as well as various choices of the augmenting term, we demonstrate the method on test problems. Our numerical experiments indicate that it is more favourable to use an augmenting term of an exponential form rather than the classical ℓ_1- or ℓ_2-norm forms.

1 Introduction

Subgradient methods were introduced in the middle of 1960s, in the works of Demyanov [12], Polyak [20–22] and Shor [34, 35] for solving nonsmooth and convex unconstrained optimization problems (for a description of these methods, see [3, 4, 18, 19] and the references therein).

When an optimization problem is constrained it is common practice to apply Lagrange relaxation techniques: a Lagrange function, in other words the Lagrangian, is used to construct a duality scheme, and a dual function is maximized in the dual

Regina S. Burachik and C. Yalçın Kaya
Mathematics and Statistics, University of South Australia, Mawson Lakes SA 5095, Australia
e-mail: regina.burachik@unisa.edu.au, yalcin.kaya@unisa.edu.au

R.S. Burachik and J.-C. Yao (eds.), *Variational Analysis and Generalized Differentiation in Optimization and Control*, Springer Optimization and Its Applications 47,
DOI 10.1007/978-1-4419-0437-9_5,

variable space, typically applying the subgradient method, because the dual problem is convex. An alternative approach is to use projections; for example, D'Antonio et al. [1] considers methods for nonsmooth and convex constrained problems combining deflected subgradient directions and projections on the constraint set.

There is an extensive literature on the study of general augmented Lagrangians and the duality schemes they give rise to (see, for example, the works of Rubinov, Yang and their co-workers [26, 29–32, 36]). Strong duality (or zero-duality gap property), which states that the optimal values of the dual and the primal problem coincide, is clearly desirable for a duality scheme. The augmented Lagrangian duality, introduced by Rockafellar and Wets [27, Chap. 11] is known to have this property. With this kind of Lagrangians, one can apply the subgradient method for solving the (convex) dual problem. Gasimov and Ismayilova [13, 14] exploited this idea by considering a deflected subgradient direction for solving the dual problem. A main advantage of their method is the strict improvement of the dual values. This monotonic improvement does not hold in general for the classical subgradient method and its variants. Gasimov and Ismayilova proved convergence of their method to an optimal dual value, under a restricted form of a Polyak-type step-size.

We describe below the current literature on deflected methods.

Burachik et al. [8] further studied the deflected subgradient method given in [13, 14], and established, under a much broader selection of step-size parameters, convergence of the dual iterates to a dual solution, and primal optimality of the accumulation points of an auxiliary primal sequence. Also in [8], a new kind of step-size parameter (differing in form from the classical Polyak-type step-size parameter) was proposed for a numerical implementation of the new method, and comprehensive numerical experiments were given for the first time.

Primal–dual schemes may provide stable rules for updating the parameter in penalty methods. These methods are known to be prone to ill-conditioning because of the uncontrolled growth of the penalty parameter. In order to avoid this ill-conditioning, the papers [15, 24, 25] propose a dynamic update of the penalty parameter for convex and smooth problems. For addressing the nonsmooth and nonconvex case, Burachik and Kaya [10] devised a penalty parameter update rule using a deflected subgradient method and proved that primal convergence is equivalent to the differentiability of the dual function at the dual limit. This equivalence was recently extended in [6] to more general families of augmented Lagrangians.

An inexact version of a deflected subgradient method, introduced in [11], was proved to have the same convergence and existence properties as those given in [8]. Numerical experiments given in [11] illustrate significant computational savings as a result of this inexact scheme.

The deflected methods considered in [8, 11] approach a dual solution, but convergence to a primal solution was only established for an auxiliary convergence. Indeed, the primal sequence may stay far from the primal solution set (see [8, Example 1]). This drawback was recently overcome in [9], where it is proved that, for a specific choice of the step-size, all accumulation points of the primal sequence are primal solutions.

Our aim is to develop variants of the deflected subgradient method, which are likely to be more convenient from the computational point of view. We do this by considering a more general Lagrangian form and prove that the main convergence properties established in [8] are preserved.

Our general Lagrangian is the sum of the objective function plus two additional terms. One of these terms is an augmenting function, and the other is the linear (classical) term. The augmenting term is a penalty expression given by a general function satisfying some mild assumptions. The classical linear term is transformed by means of a general symmetric (scaling) matrix. The constraint functions appearing in the penalty term remain unscaled, though.

For this general kind of Lagrangian, we prove convergence and existence results for the deflected subgradient method. Because the scaling matrix can be any symmetric matrix, the case when the matrix is zero represents the classical penalty function method, with any positive valued continuous function acting as the penalty term. Therefore our results apply, in particular, to the analysis of classical penalty methods, as well as a family of new penalty function methods, because of the generality of the function in the penalizing term.

For a numerical implementation of our method, we propose seven functional forms for the function in the penalty term. Six of these can be collected into two groups: one involving ℓ_1, ℓ_2 and ℓ_∞ norms and the other involving an exponential of these norms. The seventh form is a hybrid form involving the ℓ_2 norm. Although it is possible to consider many other functional forms (e.g., as in [5], albeit in [5] different Lagrangian forms are investigated for differentiable problems), we restrict our attention to these seven forms only, for the purposes of illustrating the use of the method with alternative choices of the function in the penalty term and the scaling matrix. As the scaling matrix we choose the identity and zero matrices, the latter giving rise to penalty functions.

We have chosen 16 test problems from the literature for the numerical experiments. We report the results of the experiments in terms of the number of iterations the deflected subgradient methods takes as well as the number of Lagrange function evaluations.

The chapter is organized as follows. In Sect. 2, we recall the duality properties our Lagrangian scheme. In Sect. 3, we state the deflected subgradient method. In Sects. 4 and 5, we give the convergence and existence results. In Sect. 6, we define practical step-sizes for a numerical implementation, and demonstrate their use on test problems.

2 Duality Framework for Augmented Lagrangians

We consider the nonlinear programming problem:

$$(P) \qquad \textit{minimize } f_0(x) \text{ over all } x \text{ in } X \text{ satisfying } f(x) = 0,$$

where X is a compact subset of $\mathbf{R}^n$, and the functions $f_0 : \mathbf{R}^n \to \mathbf{R}$ and $f : \mathbf{R}^n \to \mathbf{R}^m$ are continuous. Our duality analysis is based on the work [7] and hence we start off by recalling the duality framework presented there. Denote by $d(\cdot,\cdot)$ the metric distance in X and by $X_0 := \{x \in X \,:\, f(x) = 0\}$ the constraint set of problem (P). Let

$$M_P := \inf_{x \in X_0} f_0(x),$$

be the optimal value of the problem (P). Let Λ be a nonempty set (to which dual variables will belong). Consider a function $L : X \times \Lambda \to \mathbf{R}$. Given the set Λ and the function $L : X \times \Lambda \to \mathbf{R}$, define the *dual function* $H : \Lambda \to \mathbf{R} \cup \{-\infty\}$ as $H(\lambda) := \inf_{x \in X} L(x,\lambda)$, so that the *dual problem* for (P) is

$$(D) \qquad \max_{\lambda \in \Lambda} H(\lambda),$$

and hence the *optimal dual value* is $M_D := \sup_{\lambda \in \Lambda} H(\lambda)$. In such a context we consider L as a Lagrange-type function. We say that the *weak duality with respect to* L holds if

$$\inf_{x \in X} L(x,\lambda) \leq M_P,$$

for all $\lambda \in \Lambda$. And we say that the *strong duality with respect to* L (in another terminology, *zero duality gap with respect to* L) holds if $M_P = M_D$, that is

$$\sup_{\lambda \in \Lambda} \inf_{x \in X} L(x,\lambda) = M_P.$$

In the study developed in [7], the conditions for zero duality gap are given through a set of requirements on the Lagrangian and the data of the problem. We list these requirements below:

(A_1) For all $\alpha < M_P$, the level set $\{z \in X \mid f_0(z) \leq \alpha\}$ is compact.
(H_1) $f_0(x) = L(x,\lambda)$ for all $x \in X_0$ and all $\lambda \in \Lambda$.

Assume there exists a subset $\Lambda_0 \subset \Lambda$ such that

$(H_2(\Lambda_0))$ $\forall \alpha < M_P$ and $\forall \delta > 0$

$$\sup_{\lambda \in \Lambda_0} \left[\inf_{\substack{x \in X \\ d(x,X_0) \geq \delta}} L(x,\lambda) \right] > \alpha.$$

$(H_3(\Lambda_0))$

$$f_0(x) \leq L(x,\lambda),$$

for all $x \in X$, $\lambda \in \Lambda_0$.

We quote the following result.

Theorem 1 ([7], Theorem 2). *Assume that there exists $\Lambda_0 \subset \Lambda$ such that (H_1), $(H_2(\Lambda_0))$ and $(H_3(\Lambda_0))$ are satisfied, and suppose also that condition (A_1) holds for f_0, then $M_P = M_D$.*

Definition 1. Consider the set of dual variables given by $\Lambda := \mathbf{R}^m \times \mathbf{R}_+$ and a continuous function $\sigma : \mathbf{R}^m \to \mathbf{R}$ such that $\sigma(z) > 0$ for all $z \in \mathbf{R}^m \setminus \{0\}$ and $\sigma(0) = 0$. Let $A \in \mathbf{R}^{m\times m}$ be a symmetric matrix. Our Lagrangian function $L : \mathbf{R}^n \times \Lambda \to \mathbf{R}$ is defined as

$$L(x,(u,c)) := f_0(x) + c\sigma(f(x)) - \langle Au, f(x)\rangle. \tag{1}$$

From now on, we will always assume that σ and A are as stated in Definition 1.

Remark 1. The Lagrangian in Definition 1 verifies the assumptions of Theorem 1 for the choice $\Lambda_0 := \{0\} \times \mathbf{R}_+$. Indeed, note that our basic assumptions on X and f_0 trivially yield (A_1). It can be checked in a way similar to the one in [7, Example 3.1] that conditions (H_1), $(H_2(\Lambda_0))$ and $(H_3(\Lambda_0))$ also hold. As a consequence, our duality scheme enjoys the zero duality gap property. Moreover, the dual function $H : \mathbf{R}^m \times \mathbf{R}_+ \to \mathbf{R}$ is defined as

$$H(u,c) = \min_{x\in X}\left[f_0(x) + c\sigma(f(x)) - \langle Au, f(x)\rangle\right]. \tag{2}$$

Since H is the minimum of concave and upper-semicontinuous (more precisely, affine) functions of c, u, we conclude that H is concave and upper-semicontinuous. The dual problem of (P) is therefore a convex problem given by

$$(D): \qquad \max_{(u,c)\in\mathbf{R}^m\times\mathbf{R}_+} H(u,c).$$

Since H is finite on $\mathbf{R}^m \times \mathbf{R}_+$, this maximization problem is effectively unconstrained. By concavity, H is continuous everywhere.

Remark 2. For $A = I$ and $\sigma(\cdot) = \|\cdot\|$, where $\|\cdot\|$ is the (Euclidean) ℓ_2-norm, the Lagrangian in Definition 1 becomes the so-called sharp Lagrangian, introduced in [27, Example 11.58] and used in [8, 10, 11]. For $A = 0$, the Lagrangian becomes a penalty function with the penalty term given by $c\sigma(f(x))$.

The solution set of problem (P) is denoted by $S(P)$. We typically denote an element of $S(P)$ by $\bar{x}$. The solution set of problem (D) is denoted by $S(D)$. We typically denote an element in $S(D)$ by $\bar{z} = (\bar{u}, \bar{c})$. For convenience, we introduce the set

$$X(u,c) = \operatorname*{argmin}_{x\in X}\left[f_0(x) + c\sigma(f(x)) - \langle Au, f(x)\rangle\right]. \tag{3}$$

The proof of the next lemma is similar to the one in [11] for the special case of $\sigma(\cdot) = \|\cdot\|$ and $A = I$. We include it here for the reader's convenience.

Lemma 1. *Let H be the dual function defined above. For every $(u,c) \in \mathbf{R}^m \times \mathbf{R}_+$, take $\tilde{x} \in X(u,c)$. Then:*

(a) $(-Af(\tilde{x}), \sigma(f(\tilde{x}))(1+\gamma)) \in \partial_{c\gamma\sigma(f(\tilde{x}))} H(u,c)$ for every $\gamma \geq 0$. In particular, we have that $(-Af(\tilde{x}), \sigma(f(\tilde{x}))) \in \partial H(u,c)$.
(b) If $(u,c) \in S(D)$, then $(u,d) \in S(D)$ for each $d > c$. In this situation, for every $\hat{x} \in X(u,d)$ we must have $f(\hat{x}) = 0$.

Proof. (a) We must prove that for all $(u',c') \in \mathbf{R}^m \times \mathbf{R}_+$ it holds

$$H(u',c') \leq H(u,c) + \langle u' - u, -Af(\tilde{x})\rangle + (c'-c)(1+\gamma)\sigma(f(\tilde{x})).$$

Indeed, by definition of H we have that

$$\begin{aligned} H(u',c') &= \min_{x \in X} L(x,(u',c')) \\ &\leq f_0(\tilde{x}) + c\sigma(f(\tilde{x})) - \langle Au, f(\tilde{x})\rangle \\ &\quad + (c'-c)\sigma(f(\tilde{x})) + \langle u'-u, -Af(\tilde{x})\rangle \\ &\leq H(u,c) + \langle u'-u, -Af(\tilde{x})\rangle + (c'-c)(1+\gamma)\sigma(f(\tilde{x})) - (c'-c)\gamma\sigma(f(\tilde{x})) \\ &\leq H(u,c) + \langle u'-u, -Af(\tilde{x})\rangle + (c'-c)(1+\gamma)\sigma(f(\tilde{x})) + c\gamma\sigma(f(\tilde{x})), \end{aligned}$$

where we used the fact that $c' \geq 0$ is the last inequality. The second statement follows from the first one for $\gamma = 0$. The proof of (a) is complete.
(b) Since $(u,c) \in S(D)$, we must have $H(u,c) \geq H(u,d)$. On the other hand, take $\hat{x} \in X(u,d)$, where $d > c$. By item (a), we have

$$\begin{aligned} H(u,d) \leq H(u,c) &\leq H(u,d) + \langle u-u, -Af(\hat{x})\rangle + (c-d)\sigma(f(\hat{x})) \\ &= H(u,d) + (c-d)\sigma(f(\hat{x})) \\ &\leq H(u,d), \end{aligned}$$

where we also used that $\sigma \geq 0$. Altogether, $H(u,c) = H(u,d)$ and hence (u,d) is also a dual solution. Since $d - c > 0$ we must have $f(\hat{x}) = 0$.

The distance between two given points in the dual space, $w, z \in \mathbf{R}^m \times \mathbf{R}_+$, will be taken as $d(w,z) := \|w - z\|^2$. The following notation will be used throughout the chapter.

$$\begin{aligned} & z_k := (u_k, c_k), \\ & x_k \in X(u_k, c_k), \\ & f_k := f(x_k), \\ & H_k := H(u_k, c_k), \\ & \overline{H} := H(\bar{u}, \bar{c}), \\ & d_k := d(z, z_k). \end{aligned}$$

The theorem below is a trivial modification of the analogous one proved in [13]. It will be used as a stopping criteria in the next section, so we include here its short proof.

Theorem 2. *Let $M_P \geq M_D$ and suppose that $\overline{x} \in X(\overline{u}, \overline{c})$. Then $\overline{x}$ is a solution of (P) and $(\overline{u}, \overline{c})$ is a solution of (D) if and only if $f(\overline{x}) = 0$.*

Proof. It is enough to prove that $\bar{x}$ is a solution, since the converse is trivial. Assume that $f(\bar{x}) = 0$ and $\bar{x} \in X(\bar{u},\bar{c})$, then $L(\bar{x},(\bar{u},\bar{c})) = f_0(\bar{x}) \geq M_P$. On the other hand, $L(\bar{x},(\bar{u},\bar{c})) = \min_{x \in X} L(x,(\bar{u},\bar{c})) = H(\bar{u},\bar{c}) \leq M_D \leq M_P$. Where we used weak duality in the last inequality. Thus we have $f_0(\bar{x}) = M_P$ and hence $\bar{x}$ is a solution of (P).

3 The DSG Algorithm and Motivation

We propose the following *deflected subgradient* (DSG) *algorithm* for solving Problem (P), which is described as follows.

The DSG Algorithm:

Step 0 Choose (u_0, c_0) with $c_0 \geq 0$. Set $k = 1$.
Step k Given (u_k, c_k):

Step k.1 Find the vector

$$x_k \in X(u_k, c_k).$$

If $f_k = 0$, STOP.
Step k.2 Set

$$u_{k+1} := u_k - s_k A(f_k)\ ,$$
$$c_{k+1} := c_k + (s_k + \varepsilon_k)\sigma(f_k)\ ,$$

where $s_k, \varepsilon_k > 0$. Set $k = k+1$ and repeat Step k.

Let $\|\cdot\|$ be the ℓ_2 norm in $\mathbf{R}^m$. We will make the following basic assumption on σ and A.

(L1) $\sigma(z) \geq \|A(z)\|$ for all $z \in \mathbf{R}^m$.

Remark 3. The sharp Lagrangian (i.e., when $\sigma(\cdot) := \|\cdot\|$ and $A = I$ in Definition 1) verifies (L1). If $A = 0$, any nonnegative function σ verifying the assumptions of Definition 1 will satisfy (L1).

The motivation for condition (L1) rests on part (b) of the proposition below, where we prove that the search direction of the DSG algorithm produces strict improvement of the values of H if and only if the Lagrangian verifies (L1).

Proposition 1. *Consider the notation and definitions of the DSG algorithm:*

(a) $H_{k+1} - H_k \leq s_k(\|A(f_k)\|^2 + \sigma(f_k)^2) + \varepsilon_k \sigma(f_k)^2$.
(b) The following statements hold:

(i) If the augmenting function σ verifies (L1), then the DSG steps produce a sequence $\{H_k\}$ which is strictly increasing. More precisely, if $(u_k, c_k) \notin S(D)$ then $H_{k+1} > H_k$.

(ii) Assume that $\sigma(z)=\sigma(-z)$ for every $z\in\mathbb{R}^m$. If for every problem of the form (P) the DSG steps produce a sequence $\{H_k\}$ which is strictly increasing, then σ must verify (L1).

Proof. Applying Lemma 1(a), we get

$$H_{k+1}-H_k \le \langle u_{k+1}-u_k, -Af_k\rangle + (c_{k+1}-c_k)\sigma(f_k).$$

Using also the definition of the algorithm, the right-hand-side of the expression above can be rewritten as

$$H_{k+1}-H_k \le s_k\|Af_k\|^2 + (s_k+\varepsilon_k)\sigma(f_k)^2,$$

which readily implies the conclusion. This proves item (a). Let us now prove (i) in item (b). Note that the assumption $(u_k,c_k)\notin S(D)$ and Theorem 2 yield $f_k\neq 0$. Therefore $\sigma(f_k)>0$. Using the definition of the algorithm we can write

$$\begin{aligned}
H_{k+1} &= \min_{x\in X} f_0(x) + c_{k+1}\,\sigma(f(x)) - \langle Au_{k+1}, f(x)\rangle \\
&= \min_{x\in X} f_0(x) + c_{k+1}\,\sigma(f(x)) - \langle u_{k+1}, A(f(x))\rangle \\
&= \min_{x\in X} f_0(x) + [c_k + (s_k+\varepsilon_k)\sigma(f_k)]\,\sigma(f(x)) - \langle [u_k - s_k A(f_k)], A(f(x))\rangle \\
&= \min_{x\in X} f_0(x) + c_k\,\sigma(f(x)) - \langle Au_k, f(x)\rangle + (s_k+\varepsilon_k)\sigma(f_k)\,\sigma(f(x)) \\
&\quad + s_k\langle A(f_k), A(f(x))\rangle \\
&\ge \min_{x\in X} f_0(x) + c_k\,\sigma(f(x)) - \langle Au_k, f(x)\rangle + (s_k+\varepsilon_k)\sigma(f_k)\,\sigma(f(x)) \\
&\quad - s_k\sigma(f_k)\,\sigma(f(x)) \\
&= \min_{x\in X} f_0(x) + (c_k+\varepsilon_k\sigma(f_k))\,\sigma(f(x)) - \langle Au_k, f(x)\rangle \\
&= H(u_k, c_k+\varepsilon_k\sigma(f_k)),
\end{aligned}$$

where we used Cauchy–Schwartz and (L1) in the inequality. Let $\hat{x}_k\in X$ be a solution of the minimization problem above. In other words $\hat{x}_k\in X(u_k,c_k+\varepsilon_k\sigma(f_k))$. Assume first that $f(\hat{x}_k)=0$. In this case Theorem 2 yields $(u_k,c_k+\varepsilon_k\sigma(f_k))\in S(D)$. On the other hand, since $(u_k,c_k)\notin S(D)$ we must have $H(u_k,c_k)<H(u_k,c_k+\varepsilon_k\sigma(f_k))\le H_{k+1}$. Therefore the conclusion holds in this case. Assume now that $f(\hat{x}_k)\neq 0$. Then $\sigma(f(\hat{x}_k))>0$ and

$$\begin{aligned}
H_{k+1} &\ge \min_{x\in X} f_0(x) + (c_k+\varepsilon_k\sigma(f_k))\,\sigma(f(x)) - \langle Au_k, f(x)\rangle \\
&= f_0(\hat{x}_k) + (c_k+\varepsilon_k\sigma(f_k))\,\sigma(f(\hat{x}_k)) - \langle Au_k, f(\hat{x}_k)\rangle \\
&= L(\hat{x}_k,(u_k,c_k)) + \varepsilon_k\sigma(f_k)\,\sigma(f(\hat{x}_k)) \\
&\ge \min_{x\in X} L(x,(u_k,c_k)) + \varepsilon_k\sigma(f_k)\,\sigma(f(\hat{x}_k)) = H_k + \varepsilon_k\sigma(f_k)\,\sigma(f(\hat{x}_k)).
\end{aligned}$$

Therefore

$$H_{k+1} \ge H_k + \varepsilon_k\sigma(f_k)\,\sigma(f(\hat{x}_k)) > H_k \tag{4}$$

and the conclusion is proved also for this case. For the converse stated in (b)(ii), we now show that, if (L1) does not hold, then we can find a problem (P) for which the sequence $\{H_k\}$ generated by DSG is not strictly increasing. If (L1) does not hold, then there exists $\bar{z} \in \mathbf{R}^m$ such that $\sigma(\bar{z}) - \|A\bar{z}\| < 0$. Hence $\bar{z} \neq 0$. Let $r := \|\bar{z}\| > 0$. Consider problem (P) with $f_0(x) = \sigma(x) - \|Ax\|$, $X = B(0,r)$ the closed ball of center zero and radius r, and constraint function $f(x) = x$. It is clear that $M_P = 0$. Let $\bar{x} \in \operatorname{argmin}_{x \in B(0,r)} \sigma(x) - \|Ax\|$. By our assumption, we have that $f_0(\bar{x}) \leq f_0(\bar{z}) < 0$. Take now $(u_0, c_0) = (0,0)$, so we can take

$$x_0 := \bar{x} \in \operatorname*{argmin}_{x \in B(0,r)} \sigma(x) - \|Ax\| = X(u_0, c_0),$$

and $H_0 = H(u_0, c_0) = f_0(x_0) < 0 = M_P$. Because $f(x_0) = x_0 \neq 0$ we can perform a DSG step. Let us take s_0, ε_0 such that

$$\frac{s_0}{\varepsilon_0} > \frac{\sigma(x_0)^2}{\|Ax_0\|^2 - \sigma(x_0)^2} > 0.$$

From the DSG step we have $(u_1, c_1) = (-s_0 A x_0, (s_0 + \varepsilon_0)\sigma(x_0))$, so

$$\begin{aligned} H_1 &= \min_{x \in B(0,r)} \sigma(x) - \|Ax\| + (s_0 + \varepsilon_0)\sigma(x_0)\sigma(x) + s_0 \langle Ax_0, Ax \rangle \\ &\leq \sigma(-x_0) - \|A(-x_0)\| + s_0(\sigma(-x_0)^2 - \|Ax_0\|^2) + \varepsilon_0 \sigma(-x_0)\sigma(x_0) \\ &= \sigma(x_0) - \|A(x_0)\| + s_0(\sigma(x_0)^2 - \|Ax_0\|^2) + \varepsilon_0 \sigma(x_0)^2 \\ &< \sigma(x_0) - \|Ax_0\| = H_0, \end{aligned}$$

where we used the fact that $-x_0 \in B(0,r)$ in the first inequality, the assumption $\sigma(z) = \sigma(-z)$ in the second equality, and the condition on s_0, ε_0 in the second inequality. The above expression contradicts the assumption that DSG produces strictly improving dual values, so we must have that (L1) holds.

Remark 4. Note that (4) in the proof above provides a lower bound on the improvement of the value of H. Then the following corollary can be stated. Obviously, when the deflection parameter $\varepsilon_k = 0$, there is no guarantee on the improvement of the value of H, although $\{H_k\}$ would still be nondecreasing.

Corollary 1. *Consider the notation and definitions of the DSG algorithm and assume that (L1) holds. Then for all k we have*

$$\varepsilon_k \sigma(f_k)\, \sigma(f(\hat{x}_k)) \leq H_{k+1} - H_k \leq s_k \|Af_k\|^2 + (s_k + \varepsilon_k)\sigma(f_k)^2,$$

where $\hat{x}_k \in X(u_k, c_k + \varepsilon_k \sigma(f_k))$.

From now on, we always assume that (L1) holds. We establish below a necessary and sufficient condition on the step-sizes s_k and ε_k for guaranteeing boundedness of the dual sequence.

Lemma 2. *Consider the notation and definitions of the DSG algorithm. The following statements are equivalent:*

(a) $\sum_{k=0}^{\infty}(s_k+\varepsilon_k)\sigma(f_k)<\infty$.
(b) The sequence $\{z_k\}$ *is bounded.*

Proof. From the DSG algorithm

$$c_{m+1}-c_0=\sum_{k=0}^{m}(s_k+\varepsilon_k)\sigma(f_k) \quad \text{and} \quad \|u_{m+1}-u_0\|\le\sum_{k=0}^{m}s_k\|Af_k\|. \tag{5}$$

By (L1) we have $\|Af_k\|\le\sigma(f_k)$. If (a) holds then clearly the sequence $\{c_k\}$ is bounded. On the other hand,

$$\|u_{m+1}-u_0\|\le\sum_{k=0}^{m}s_k\|Af_k\|\le\sum_{k=0}^{m}s_k\,\sigma(f_k)<+\infty,$$

which yields boundedness of the sequence $\{u_k\}$. Assume now that (b) holds. By (5) we must have condition (a).

4 Existence and Convergence Results

Conditions that guarantee the existence of dual solutions are often related to properties of the perturbation function associated with the problem under consideration. Unfortunately, the calculation of the perturbation function is very difficult, and hence it makes sense to establish alternative ways of guaranteeing existence of dual solutions. Conditions of this kind can be found in [28–31]. We give below a new existence condition by using the dual sequence generated by the DSG algorithm using a rather general step-size s_k. The proofs of the results in this section and the next one can be deduced (with some suitable modifications) from those in references [8, 11], which were given for $\sigma(\cdot)=\|\cdot\|$ and $A=I$. We include here these proofs, however, for the sake of completeness.

Lemma 3. *Let* $\overline{H}$ *be the optimal value of (P) (i.e.,* $\overline{H}:=M_D=M_P$*). Assume that the sequence* $\{z_k\}$ *generated by the DSG algorithm is bounded and that the step-size* s_k *satisfies*

$$s_k\ge\eta\,\frac{(\overline{H}-H_k)}{\sigma(f_k)^2}, \tag{6}$$

for some fixed $\eta>0$. *Then* $\{H_k\}$ *converges to* $\overline{H}$ *and every accumulation point of* $\{z_k\}$ *is a dual solution. In particular,* $S(D)\neq\emptyset$.

Proof. By (5), boundedness of the sequence implies that

$$\sum_{k=0}^{\infty}s_k\sigma(f_k)<+\infty. \tag{7}$$

Let $(\bar{u},\bar{c})$ be an accumulation point of the sequence $\{(u_k,c_k)\}$, and denote by $\mathcal{K}$ the infinite set of indices such that

$$\lim_{\substack{k\in\mathcal{K},\\ k\to\infty}} (u_k,c_k) = (\bar{u},\bar{c}).$$

We will prove that $(\bar{u},\bar{c}) \in S(D)$. By boundedness of $\{x_k\}$, we can also assume that the whole sequence $\{x_k\}_{k\in\mathcal{K}}$ converges to some $\bar{x}$. If $f(\bar{x}) = 0$, we claim that $\bar{x} \in X(\bar{u},\bar{c})$. In this case, Theorem 2 implies that $(\bar{u},\bar{c}) \in S(D)$. Indeed, by definition of x_k, we have that

$$f_0(x_k) + c_k\sigma(f(x_k)) - \langle Au_k, f(x_k)\rangle \le f_0(x) + c_k\sigma(f(x)) - \langle Au_k, f(x)\rangle,$$

for all $x \in X$ and for all k. Taking limits for $k \in \mathcal{K}$, $k \to \infty$ in the above expression we get

$$f_0(\bar{x}) + \bar{c}\sigma(f(\bar{x})) - \langle A\bar{u}, f(\bar{x})\rangle \le f_0(x) + \bar{c}\sigma(f(x)) - \langle A\bar{u}, f(x)\rangle,$$

for all $x \in X$. Hence $\bar{x} \in X(\bar{u},\bar{c})$ and thus $(\bar{u},\bar{c}) \in S(D)$. Assume now that $f(\bar{x}) \neq 0$. This fact, together with (7), implies that the sequence $\{s_k\}_{k\in\mathcal{K}}$ converges to zero. Using also (6) for $k \in \mathcal{K}$, we conclude that the subsequence of dual values $\{H_k\}_{k\in\mathcal{K}}$ converges to $\overline{H}$. By continuity of H we have that

$$H(\bar{u},\bar{c}) = \limsup_{\substack{k\in\mathcal{K},\\ k\to\infty}} H(u_k,c_k) = \overline{H}.$$

This shows that $H(\bar{u},\bar{c})$ has optimal functional value $\overline{H}$ and hence $(\bar{u},\bar{c}) \in S(D)$. Moreover, from Corollary 1 we know that $\{H_k\}$ is strictly increasing. Since it has a convergent subsequence, the whole sequence must converge to $\overline{H}$. The proof is complete.

The following simple estimate will be useful.

Lemma 4. *Fix $z = (u,c) \in \mathbb{R}^m \times \mathbb{R}_+$. Then*

$$d_{k+1} - d_k \le (s_k^2\|Af_k\|^2 + (s_k+\varepsilon_k)^2\sigma(f_k)^2) + 2[s_k(H_k - H(u,c)) - \varepsilon_k\sigma(f_k)(c-c_k)].$$

Proof. Note that

$$\begin{aligned} d_{k+1} - d_k &= \|z - z_{k+1}\|^2 - \|z - z_k\|^2 \\ &= \|z_k - z_{k+1}\|^2 + 2\langle z - z_k, z_k - z_{k+1}\rangle. \end{aligned}$$

Call $A_k := \|z_k - z_{k+1}\|^2$ and $B_k := \langle z - z_k, z_k - z_{k+1}\rangle$. Using the definition of the DSG algorithm, we can write

$$A_k = \|u_k - u_{k+1}\|^2 + |c_k - c_{k+1}|^2 = s_k^2\,\|Af_k\|^2 + (s_k+\varepsilon_k)^2\sigma(f_k)^2.$$

Combining the two previous expressions, we get

$$d_{k+1} - d_k = s_k^2 \|Af_k\|^2 + (s_k + \varepsilon_k)^2 \sigma(f_k)^2 + 2B_k. \tag{8}$$

The term B_k is written out as follows:

$$\begin{aligned} B_k &= \langle u - u_k, u_k - u_{k+1} \rangle + (c - c_k)(c_k - c_{k+1}) \\ &= \langle u - u_k, s_k A f_k \rangle - (c - c_k)(s_k + \varepsilon_k)\sigma(f_k) \\ &= s_k \left[\langle u - u_k, Af_k \rangle - (c - c_k)\sigma(f_k) \right] - \varepsilon_k (c - c_k)\sigma(f_k). \end{aligned} \tag{9}$$

In order to estimate the expression between brackets, we use the subgradient inequality:

$$H(u,c) \le H(u_k, c_k) + \langle (-Af_k, \sigma(f_k)), (u - u_k, c - c_k) \rangle$$

or

$$[\langle u - u_k, Af_k \rangle - (c - c_k)\sigma(f_k)] \le H_k - H(u,c).$$

Using this in (9), we obtain

$$B_k \le s_k (H_k - H(u,c)) - \varepsilon_k (c - c_k)\sigma(f_k).$$

Equation (8) now yields

$$\begin{aligned} d_{k+1} - d_k \le s_k^2 \|Af_k\|^2 + (s_k + \varepsilon_k)^2 \sigma(f_k)^2 - 2 [s_k (H(u,c) - H_k) \\ + \varepsilon_k (c - c_k)\sigma(f_k)], \end{aligned} \tag{10}$$

which completes the proof.

Lemma 4 allows us to prove that the dual sequence is convergent. For proving that the limit is in fact optimal we will need extra assumptions on the step-sizes (see Theorem 4).

Theorem 3. *If the sequence generated by DSG is bounded, then it is convergent.*

Proof. Assume that the sequence $\{z_k\}$ is bounded and let $\hat{z}$ be an accumulation point of $\{z_k\}$. Call $\{z_{k_j}\}_j$ a subsequence converging to $\hat{z}$. Using Lemma 4 for the choice $z := \hat{z} = (\hat{u}, \hat{c})$, we conclude that the sequence $\{d(\hat{z}, z_k)\}$ verifies

$$\begin{aligned} d(\hat{z}, z_{k+1}) - d(\hat{z}, z_k) \le s_k^2 \|Af_k\|^2 + (s_k + \varepsilon_k)^2 \sigma(f_k)^2 \\ -2s_k(H(\hat{z}) - H_k) - 2\varepsilon_k \sigma(f_k)(\hat{c} - c_k). \end{aligned}$$

By Corollary 1, $\{H_k\}$ is a strictly increasing and therefore $\lim_k H_k = \sup_k H_k = \lim_j H_{k_j} = H(\hat{z})$. Using now the upper-semicontinuity of H we get

$$H(\hat{z}) \ge \limsup_j H_{k_j} = \lim_j H_{k_j} = \sup_j H_j \ge H_k \qquad \text{for every } k \in \mathbb{N}.$$

So $(H(\hat{z}) - H_k) \ge 0$ for all k. Using also that $\{c_k\}$ is a strictly increasing sequence, we have that $(\hat{c} - c_k) \ge 0$ for all k. Hence,

$$d(\hat{z}, z_{k+1}) - d(\hat{z}, z_k) \leq s_k^2 \|Af_k\|^2 + (s_k + \varepsilon_k)^2 \sigma(f_k)^2.$$

Since $\{z_k\}$ is bounded, we can use Lemma 2 to conclude that the series with general term $a_k := s_k^2 \|Af_k\|^2 + (s_k + \varepsilon_k)^2 \sigma(f_k)^2$ is convergent, and this implies (by [23, Lemma 2.2.2]) that the sequence $\{d(\hat{z}, z_k)\}_k$ is convergent. But the subsequence $\{d(\hat{z}, z_{k_j})\}_j$ of this sequence converges to zero, and so the whole sequence converges to zero, yielding the uniqueness of the accumulation point.

It has been established in [8, Example 1] that the sequence x_k generated by DSG may not converge to a primal solution. However, if we consider the slightly perturbed sequence $\{\tilde{x}_k\}$ such that $\tilde{x}_k \in X(u_k, c_k + \beta)$ for a fixed $\beta > 0$, then we can prove that all its accumulation points are primal solutions. We call such a sequence a *primal sequence*.

Theorem 4 (Primal–dual convergence). *Assume that the sequence $\{(u_k, c_k)\}$ generated by the DSG algorithm is bounded. Assume also that for some $\eta > 0$ the step-size s_k satisfies*

$$s_k \geq \eta \frac{(\overline{H} - H_k)}{\sigma(f_k)^2}. \tag{11}$$

Then the limit of the sequence $\{z_k\}$ is a dual solution. Additionally, all accumulation points of $\{\tilde{x}_k\}$ are solutions of (P).

Proof. By Corollary 1 the sequence $\{H_k\}$ is strictly increasing. Moreover, Lemma 3 implies that $H_k \longrightarrow \overline{H}$ and every accumulation point of $\{z_k\}$ is a dual solution. Since $\{z_k\}$ is bounded, Theorem 3 allows us to conclude that $\{z_k\}$ is convergent. Combining these two facts, we get that $\{z_k\}$ converges to a dual solution. Now we will show that all accumulation points of the primal sequence $\{\tilde{x}_k\}$ are solutions of Problem (P). In order to prove this fact, we will show that the numerical sequence $\{\sigma(\tilde{f}_k)\}$ has zero as its unique accumulation point. Fix $\beta > 0$ and take $\tilde{x}_k \in X(u_k, c_k + \beta)$ for all k. Take $a \geq 0$ as an accumulation point of the sequence $\{\sigma(\tilde{f}_k)\}$. So there exists a subsequence $\{\sigma(\tilde{f}_{k_j})\}$ such that $a = \lim_{j\to\infty} \sigma(\tilde{f}_{k_j})$. Then

$$\begin{aligned} H(u_{k_j}, c_{k_j}) = H_{k_j} &\leq H(u_{k_j}, c_{k_j} + \beta) + \langle(-A\tilde{f}_{k_j}, \sigma(\tilde{f}_{k_j})), (\mathbf{0}, -\beta)\rangle \\ &\leq H(u_{k_j}, c_{k_j} + \beta) - \beta\sigma(\tilde{f}_{k_j}). \end{aligned}$$

We can rewrite this as

$$\beta\sigma(\tilde{f}_{k_j}) \leq H(u_{k_j}, c_{k_j} + \beta\sigma(f_{k_j})) - H_{k_j} \leq \overline{H} - H_{k_j}.$$

Using the fact that $\lim_j \overline{H} - H_{k_j} = 0$ we get

$$a = \lim_{j\to\infty} \sigma(\tilde{f}_{k_j}) = 0. \tag{12}$$

Thus the sequence $\{\sigma(\tilde{f}_k)\}$ converges to zero. Take now $\tilde{x}$ as an accumulation point of $\{\tilde{x}_k\}$. Since zero is the limit of $\{\sigma(\tilde{f}_k)\}$, we must have $f(\tilde{x}) = 0$. Without loss of generality, assume the whole sequence $\{\tilde{x}_k\}$ converges to $\tilde{x}$. Then

$$\begin{aligned} M_P \le f_0(\tilde{x}) &= \lim_k f_0(\tilde{x}_k) + (c_k+\beta)\sigma(f(\tilde{x}_k)) - \langle Au_k, f(\tilde{x}_k)\rangle \\ &= \lim_k \min_{x\in X} f_0(x) + (c_k+\beta)\sigma(f(x)) - \langle Au_k, f(x)\rangle \\ &= \lim_k H(u_k, c_k+\beta) \le \bar{H} = M_D, \end{aligned}$$

where we have used the definition of $\tilde{x}_k$ in the second equality. By weak duality, we must have $f_0(\tilde{x}) = M_P$ and since $f(\tilde{x}) = 0$, $\tilde{x}$ is a primal solution.

5 On a Special Choice of s_k

In this section, we study a special choice of the parameter s_k for which nonemptiness of $S(D)$ is equivalent to the boundedness of $\{z_k\}$. The step-size we consider is as follows:

$$\eta\frac{\overline{H}-H_k}{\sigma(f_k)^2} \le s_k \le 2\frac{\overline{H}-H_k}{\sigma(f_k)^2}, \tag{13}$$

with $\eta \in (0,2)$.

For establishing the announced fact, we need an auxiliary result.

Lemma 5. *Assume that $S(D) \neq \emptyset$ and let $\{z_k\}$ be the sequence generated by the DSG algorithm with step-size $\{s_k\}$ satisfying*

$$\liminf_k \left[2\frac{\overline{H}-H_k}{\sigma(f_k)^2} - s_k\right] > -\infty. \tag{14}$$

Then $\{z_k\}$ is bounded.

Proof. Fix a dual solution $(\bar{u},\bar{c}) \in S(D)$. For contradiction purposes, assume that $\{z_k\}$ is unbounded. This means that either $\{u_k\}$ or $\{c_k\}$ are unbounded. If $\{u_k\}$ is unbounded, then

$$\infty = \sum_k s_k\|Af_k\| \le \sum_k s_k\sigma(f_k) \le \sum_k (s_k+\varepsilon_k)\sigma(f_k),$$

where we also used (L1). So $\{c_k\}$ must be unbounded. Hence in either case we must have $\{c_k\}$ unbounded. Since it is a strictly increasing sequence, it tends to infinity. On the other hand, by definition of the DSG algorithm,

$$\begin{aligned} \|\bar{u}-u_{k+1}\|^2 &= \|\bar{u}-u_k+s_kAf_k\|^2 \\ &= \|\bar{u}-u_k\|^2 + 2s_k\langle\bar{u}-u_k, Af_k\rangle + s_k^2\|Af_k\|^2. \end{aligned} \tag{15}$$

In order to estimate the middle term of the expression above we use the subgradient inequality,

$$\overline{H}-H_k \le \langle\bar{u}-u_k, -Af_k\rangle + (\bar{c}-c_k)\sigma(f_k). \tag{16}$$

Multiply both sides by $2s_k$ and rearrange the resulting expression, to get

$$2s_k\langle\bar{u}-u_k, Af_k\rangle \le -2s_k(\overline{H}-H_k) + 2s_k(\bar{c}-c_k)\sigma(f_k).$$

Combine this fact with (L1) and (15) to obtain

$$\begin{aligned}\|\bar{u}-u_{k+1}\|^2 &\leq \|\bar{u}-u_k\|^2 - 2s_k(\overline{H}-H_k) + 2s_k(\bar{c}-c_k)\sigma(f_k) + s_k^2\|Af_k\|^2 \\ &\leq \|\bar{u}-u_k\|^2 + s_k\sigma(f_k)^2\left[s_k - \frac{2(\overline{H}-H_k)}{\sigma(f_k)^2} + \frac{2(\bar{c}-c_k)}{\sigma(f_k)}\right].\end{aligned} \tag{17}$$

Assumption (14) means that there exist a constant $\rho \in \mathrm{R}$ and an index k_0 such that

$$s_k - \frac{2(\overline{H}-H_k)}{\sigma(f_k)^2} \leq \rho,$$

for all $k \geq k_0$. As pointed out above, $\{c_k\}$ tends to infinity and hence there exists an index $k_1 \geq k_0$ such that

$$\rho \leq \frac{2(c_k-\bar{c})}{\sigma(f_k)} \qquad \text{for all } k \geq k_1,$$

where we are also using the fact that the sequence $\{\sigma(f_k)\}$ is bounded. Altogether, we conclude that for all $k \geq k_1$,

$$s_k - \frac{2(\overline{H}-H_k)}{\sigma(f_k)^2} + \frac{2(\bar{c}-c_k)}{\sigma(f_k)} \leq 0.$$

This fact, combined with (17), yields $\|\bar{u}-u_{k+1}\| \leq \|\bar{u}-u_k\|$ for all $k \geq k_1$ and this implies that $\{u_k\}$ is bounded. Using Cauchy–Schartz inequality in (16), we get

$$(c_k-\bar{c})\sigma(f_k) \leq -(\overline{H}-H_k) + \|Af_k\|\,\|\bar{u}-u_k\| \leq \|Af_k\|\|\bar{u}-u_k\| \leq \sigma(f_k)\|\bar{u}-u_k\|,$$

where we used assumption (L1) in the rightmost inequality. Note that, if $f_{k_0} = 0$ for some k_0, then the corresponding $(u_{k_0}, c_{k_0}) \in S(D)$ and the DSG stops at k_0. In this case the sequence is finite and therefore bounded. So the unboundedness assumption implies that $f_k \neq 0$ for all k. Using this fact in the previous expression, we get

$$(c_k-\bar{c}) \leq \|\bar{u}-u_k\|,$$

and hence $\{c_k\}$ must be bounded, a contradiction. This implies that the sequence $\{z_k\}$ must be bounded.

Condition (14) is not practical from an algorithmic point of view, because it cannot be verified during the process. For this reason, and also for simplicity of exposition, we replace it by the right-hand side inequality in (13), which can be effectively checked at each iteration. The latter condition readily implies (14).

Theorem 5. *Assume the step-size in the DSG algorithm is chosen according to* (13), *then the following statements are equivalent:*

(a) The sequence $\{z_k\}$ *is bounded.*
(b) $S(D) \neq \emptyset$.

Proof. The fact that (a) implies (b) is a consequence of Theorem 3 and the left-hand side inequality in (13). Indeed, Theorem 3 implies that every accumulation point of $\{z_k\}$ is a dual solution. In particular, $S(D)$ is nonempty. In order to show that (b) implies (a), observe that this follows from Lemma 5 and the fact that the right-hand side inequality in (13) implies (14).

In the theorem below, we consider again the sequence $\{\tilde{x}_k\}$ such that $\tilde{x}_k \in X(u_k, c_k + \beta)$, where $\beta > 0$. We recover the same convergence results as the ones reported in Theorem 4, but without the assumption of boundedness of $\{z_k\}$.

Theorem 6. *Assume the step-size in the DSG algorithm is chosen according to* (13). *Suppose also that* $S(D) \neq \emptyset$. *Then:*

(i) The dual sequence $\{z_k\}$ converges to a dual solution.
(ii) The sequence of dual values $\{H_k\}$ converges to an optimal dual value.
(iii) All accumulation points of the primal sequence $\{\tilde{x}_k\}$ are solutions of Problem (P).

Proof. By Theorem 5 and the fact that $S(D) \neq \emptyset$, we conclude that $\{z_k\}$ is bounded. Using now the left-hand side of (13) and Theorem 4, we conclude that statements (i)–(iii) hold.

The following simple result is useful for an implementation of the algorithm.

Proposition 2. *Assume that one of the following conditions holds:*

(i) The step-size s_k satisfies (6) *and $\{z_k\}$ is bounded.*
(ii) The step-size s_k satisfies (13) *and $S(D) \neq \emptyset$.*

Then there exists a dual solution $(\bar{u}, \bar{c})$ such that $\bar{c} > c_k$ for all k.

Proof. Under assumption (i), by Theorem 3 it holds that $S(D) \neq \emptyset$. So under either assumption, we must have $S(D) \neq \emptyset$. Now fix a dual solution $(\bar{u}, \bar{c})$. Under assumption (ii), and using Theorem 5, we conclude that the sequence $\{z_k\}$ is bounded. So again under either assumption, we must have $\{z_k\}$ bounded. Thus there exists $\hat{c} \geq c_k$ for all k. Using Lemma 1(b), we have that $(\bar{u}, \hat{c} + \bar{c})$ is also a dual solution, and this dual solution is as in the statement of the proposition.

6 Numerical Implementation

In Sect. 6.1, we select practical step-size parameters for a numerical implementation of the deflected subgradient method. In Sect. 6.2, we list the functional forms we use for the augmenting function σ, and in Sect. 6.3 point to the special case of a penalty function method. In Sect. 6.4, we demonstrate the method with the proposed step-sizes and example choices of σ and A on test problems.

6.1 Step-Size Selection

We assume that the dual sequence $\{z_k\}$ generated by the deflected subgradient method is bounded. Although a wide range of step-sizes can be chosen using (11), we will restrict our attention to the estimate given in Lemma 4 in deriving a step-size, because it reflects the structure of the problem.

For simplicity let s_k and ε_k be related through $s_k = \alpha\,\varepsilon_k$, where $\alpha > 0$. Since the hypotheses of Proposition 2(i) are satisfied, there exists a dual solution $(\bar{u},\bar{c})$ such that $\bar{c} \geq c_k$ for all k. Now, Lemma 4 for $z := (\bar{u},\bar{c})$ and $\varepsilon_k = \alpha s_k$ yields, after trivial manipulations,

$$s_k\left[2\,\frac{(\overline{H}-H_k)+\alpha\,(\bar{c}-c_k)\,\sigma(f_k)}{\|A\,f_k\|+(1+\alpha)^2\sigma(f_k)^2}-s_k\right] \leq \frac{d_k-d_{k+1}}{\|A\,f_k\|+(1+\alpha)^2\sigma(f_k)^2}\,. \tag{18}$$

Taking now s_k such that

$$s_k = \delta\,\frac{(\overline{H}-H_k)+\alpha\,(\bar{c}-c_k)\,\sigma(f_k)}{[1+(1+\alpha)^2]\,\sigma(f_k)^2}\,, \qquad 0<\delta<2\,, \tag{19}$$

and recalling $\sigma(f_k) \geq \|A\,f_k\|$ in Assumption (L1), we see that the left-hand side of the expression in (18) is nonnegative, and therefore the sequence $\{d_k\}$ is nonincreasing, and hence convergent. This choice of s_k forces convergence of the dual values towards the optimal value $\overline{H}$.

The value of the cost function at any point satisfying the constraints constitutes an upper bound $\widehat{H}$ for $\overline{H}$. In the numerical experiments in this chapter, we use $\widehat{H}$ in the place of $\overline{H}$ to illustrate the behaviour of the deflected subgradient method. In some problems, $\overline{H}$ is known exactly, for example, in some formulations of the problem of solving a nonlinear system of equations $\overline{H}=0$. Because in general $\overline{H}$ is not known but is necessary to use for a relatively more efficient implementation of subgradient methods, methods are proposed in the nonlinear programming literature for getting around this difficulty [2, 33].

Recall that any $\widehat{c} \geq \bar{c}$ is also a dual solution. So we can replace $\bar{c}$ in (19) by an upper bound $\widehat{c}$ for $\bar{c}$. So we set the step-size s_k as

$$s_k = \delta\,\frac{(\widehat{H}-H_k)+\alpha\,(\widehat{c}-c_k)\,\sigma(f_k)}{[1+(1+\alpha)^2]\,\sigma(f_k)^2}\,, \qquad 0<\delta<2\,, \quad \alpha>0\,. \tag{20}$$

It must be noted that when $\overline{H}$ is replaced by $\widehat{H} \geq \overline{H}$ the right-hand inequality in (13) does not hold. Therefore, Theorems 5 and 6 cannot be stated. However, the main convergence theorems 3–4 would still hold.

Substitution of the step-size in (20) into the dual variable update formulas in Step k.2 of the DSG algorithm gives

$$c_{k+1} = c_k + \frac{(1+\alpha)\,\delta}{1+(1+\alpha)^2}\left[\frac{\widehat{H}-H_k}{\sigma(f_k)}+\alpha\,(\widehat{c}-c_k)\right]\,, \tag{21}$$

$$u_{k+1} = u_k - \frac{c_{k+1} - c_k}{(1+\alpha)\,\sigma(f_k)}\,A f_k\,, \qquad 0 < \delta < 2\,, \quad \alpha > 0\,. \tag{22}$$

6.2 Augmenting Functional Forms

For testing the proposed method, the following seven functional forms for σ have been chosen:

$$\textit{Norm functional forms}: \ \|\cdot\|_2\,,\ \|\cdot\|_1\,,\ \sqrt{m}\,\|\cdot\|_\infty$$

$$\textit{Exponential functional forms}: \ e^{\|\cdot\|_2} - 1\,,\ e^{\|\cdot\|_1} - 1\,,\ \sqrt{m}\,\left(e^{\|\cdot\|_\infty} - 1\right)$$

$$\textit{Hybrid functional form}: \ \max(\|\cdot\|_2, \|\cdot\|_2^2)$$

In these forms, m is the number of constraints (i.e., m is the dimension of the codomain of the constraint function $f : \mathbf{R}^n \to \mathbf{R}^m$ in problem (P)). All of these forms satisfy the requirement that

$$\sigma(z) \geq \|A\,z\|\,, \quad z \in \mathbf{R}^m\,,$$

where two particular choices of A are considered: $A = I$, the identity matrix, and $A = 0$, the zero matrix. Many other functional forms can be shown to satisfy the above condition. However, we restrict our implementation and comparison to the forms we have chosen above. A much wider choice of functional forms for σ and different choices of matrices for A should be the subject of a comprehensive comparison study.

6.3 A Special Case: Penalty Methods

One should note that the choice of $A = 0$ makes the augmented Lagrangian a penalty function with the penalty term $c\,\sigma(\cdot)$. In this case, any nonnegative function $\sigma(\cdot)$ can be used. The deflected subgradient method then reduces to a penalty method, with the attractive feature that the penalty parameter c is updated automatically.

It is not common in the nonlinear programming literature to find an update rule for the penalty parameter c, which is derived from the structure of the problem. The authors presented an update rule in [10] for c in the penalty function

$$f_c(x) = f_0(x) + c\,\|f(x)\|\,,$$

using the sharp Lagrangian (i.e., when $\sigma(z) = \|z\|$ and $A = I$). The setting we have for the deflected subgradient method prescribes the update rule (21) for the penalty parameter c in the much more general penalty function

$$f_c(x) = f_0(x) + c\,\sigma(f(x))\,.$$

Table 1 Parameters and initial guesses used for the deflected subgradient method

Problem	x_0	c_0	u_0	$\widehat{H}$	$\widehat{c}$	α	δ
PQR-T1-7	$-(5,5,5)$	1.5	$(1,1,1,1,1,1,1)$	0	10	1	0.5
GLR-P1-1	$(0,0,0)$	50	$(0,0,0,0)$	$-20,000$	$10,000$	1	0.5
PPR-P1-3	$(5,5,5,5)$	0.5	$(1,1,1,1,1,1,1,1,1,1)$	20	5	1	0.1
PGR-P1-3	$(0,0,0,0,0)$	0.0001	$(0,0)$	0.5	1	1	0.1
GQR-T1-5	$(1,1,1)$	1.5	$(0,0)$	2	20	1	0.1
QIP [8, 11]	$-(2,2,2,2)$	1	$-(1,1,1,1,1)$	-19	20	3	0.1
Mur-Sau [8, 11]	$0.5(1,1,1,1,1)$	0.5	$(0,1,1)$	0.05	2	1	0.05
Control [8, 11, 17]	$0.5(1,1,1,1,1)$	1.5	$-(1,1,4)$	4	4	2	0.1
SQR-P1-1	$(0,0)$	0.2	$(0,0,0)$	2	5	1	0.1
GQR-P1-1	$(90,10)$	1	$(1,1,1,1,1,1,1)$	0	5	1	0.1
QQR-P1-1	$(0,0,0)$	0.1	$(1,1)$	0	5	1	0.1
QQR-P1-2	$(2,2,2)$	0.1	$(1,1,1,1,1)$	1,000	5	1	0.1
PPR-P1-2	$(1,1,1)$	0.1	$(1,1,1,1)$	10,000	200	1	0.1
LGR-P1-1	$(1,1,1)$	0.1	$(0,0,0,0,0,0,0,0)$	1	2	1	0.1
LPR-P1-1	$(1,1,1,1)$	1	$(0,0,0,0,0,0,0,0,0,0)$	1,000	50,000	1	0.1
GPR-P1-2	$(-2,2,2,-1,-1)$	0.7	$(1,1,1,1,1,1,1,1,1,1,1)$	0.1	3	1	0.1

6.4 Example Applications

We have tested the deflected subgradient method on 16 test problems from the literature [8, 11, 16]. For each problem, a different set of parameters and initial guesses have been used. These are shown in Table 1. All those test problems listed without a citation are from the Hock and Schittkowski collection [16]. We report the results in Tables 2 and 3.

For solving the subproblem we have used MATLAB's `fminsearch`, which implements the Nelder–Mead downhill simplex method. We have replaced the inequality constraints of the form

$$g(x) \leq 0$$

in the test problems by their nonsmooth equivalent

$$\max\{g(x), 0\} = 0,$$

so that all problems were converted into an equality constrained optimization problem.

We ran the numerical experiments under two major cases, namely (1) $A = I$, that is, an augmented Lagrangian case, and (2) $A = 0$, that is, a penalty function case. Under each case, we report the number of iterations and function evaluations it took for the deflected subgradient method to find a solution. The cases when a solution could not be found are indicated by a dash. For each test problem, the least number of function evaluations have been framed by a box.

Table 2 Results of the numerical experiments with the first eight test problems – n_{its} denotes the number of iterations and n_L the number of function evaluations

A	$\sigma(\cdot)$	PQR-T1-7		GLR-P1-1		PPR-P1-3		PGR-P1-3		GQR-T1-5		QIP [8,11]		Mur-Sau [8,11]		Control [8,11,17]	
		n_{its}	n_L	n_{its}	n_L	n_{its}	n_L	n_{its}	n_L	n_{its}	n_L	n_{its}	n_L	n_{its}	n_L	n_{its}	n_L
I	$\\|\cdot\\|_2$	–	–	3	1,332	–	–	3	3,069	–	–	2	1,897	19	9,582	4	1,844
	$\\|\cdot\\|_1$	–	–	3	1,626	–	–	3	2,743	–	–	2	1,954	16	24,014	5	4,772
	$\sqrt{m}\\|\cdot\\|_\infty$	4	4,262	3	1,292	–	–	3	3,535	1	726	2	1,114	17	38,364	6	1,331
	$e^{\\|\cdot\\|_2}-1$	2	1,517	3	1,436	6	3,581	3	7,047	2	527	3	2,718	19	9,122	4	1,941
	$e^{\\|\cdot\\|_1}-1$	2	2,349	3	1,673	5	8,773	3	4,153	2	1,040	3	2,734	14	28,711	3	810
	$\sqrt{m}\left(e^{\\|\cdot\\|_\infty}-1\right)$	1	446	3	1,192	7	14,353	3	3,996	1	939	2	1,562	18	37,359	2	1,762
	$\max(\\|\cdot\\|_2,\\|\cdot\\|_2^2)$	4	1,931	3	1,309	7	4,342	3	4,831	–	–	3	2,151	18	10,315	4	1,330
0	$\\|\cdot\\|_2$	8	2,871	4	1,541	–	–	3	2,745	–	–	–	–	–	–	–	–
	$\\|\cdot\\|_1$	4	3,229	4	1,944	–	–	3	4,750	–	–	–	–	–	–	–	–
	$\sqrt{m}\\|\cdot\\|_\infty$	3	2,814	3	1,272	5	8,664	4	8,220	1	726	2	2,359	–	–	14	10,918
	$e^{\\|\cdot\\|_2}-1$	1	645	4	1,577	3	1,633	3	3,191	2	540	4	3,632	1	1,690	7	1,839
	$e^{\\|\cdot\\|_1}-1$	1	473	4	1,790	2	1,825	4	6,980	2	1,032	3	2,020	2	4,008	5	2,162
	$\sqrt{m}\left(e^{\\|\cdot\\|_\infty}-1\right)$	1	571	3	1,213	4	6,458	4	4,662	1	939	2	1,151	3	11,222	10	9,144
	$\max(\\|\cdot\\|_2,\\|\cdot\\|_2^2)$	1	653	4	1,514	3	3,483	4	3,259	–	–	4	2,474	–	–	–	–

Table 3 Results of the numerical experiments with the remaining test problems – n_{its} denotes the number of iterations and n_L the number of function evaluations

A	$\sigma(\cdot)$	SQR-P1-1		GQR-P1-1		QQR-P1-1		QQR-P1-2		PPR-P1-2		LGR-P1-1		LPR-P1-1		GPR-P1-2	
		n_{its}	n_L	n_{its}	n_L	n_{its}	n_L	n_{its}	n_L	n_{its}	n_L	n_{its}	n_L	n_{its}	n_L	n_{its}	n_L
	$\|\cdot\|_2$	2	653	1	587	2	1,021	–	–	5	1,938	–	–	9	6,109	–	–
	$\|\cdot\|_1$	2	710	1	587	2	1,091	–	–	5	1,938	7	3,709	9	7,204	–	–
	$\sqrt{m}\|\cdot\|_\infty$	2	633	1	499	2	981	–	–	5	1,877	4	2,508	7	6,318	–	–
I	$e^{\|\cdot\|_2}-1$	2	654	1	540	2	1,242	5	1,731	–	–	6	2,416	9	4,764	10	5,550
	$e^{\|\cdot\|_1}-1$	2	680	1	540	2	1,382	4	2,228	–	–	6	3,424	9	6,467	7	17,810
	$\sqrt{m}\left(e^{\|\cdot\|_\infty}-1\right)$	2	639	1	626	2	1,247	4	3,024	–	–	4	2,619	7	5,734	3	10,950
	$\max(\|\cdot\|_2, \|\cdot\|_2^2)$	2	653	1	587	2	993	5	2,030	5	1,911	6	2,690	9	5,372	13	22,239
	$\|\cdot\|_2$	2	671	1	627	5	2,033	–	–	6	2,278	–	–	12	5,973	–	–
	$\|\cdot\|_1$	2	703	1	509	5	2,534	–	–	6	2,278	–	–	10	8,186	–	–
	$\sqrt{m}\|\cdot\|_\infty$	2	631	–	–	4	2,136	–	–	6	2,125	4	2,595	7	5,689	–	–
0	$e^{\|\cdot\|_2}-1$	2	671	1	493	3	1,449	3	1,287	–	–	7	2,649	–	–	1	3,048
	$e^{\|\cdot\|_1}-1$	2	725	1	493	4	2,169	3	1,997	–	–	7	3,867	–	–	1	1,790
	$\sqrt{m}\left(e^{\|\cdot\|_\infty}-1\right)$	2	656	1	503	3	2,538	4	2,431	–	–	4	2,071	7	6,989	–	–
	$\max(\|\cdot\|_2, \|\cdot\|_2^2)$	2	671	1	500	3	1,306	4	1,827	6	2,405	8	3,240	12	6,462	1	3,060

Observations

In order to draw statistical, far reaching, conclusions regarding the augmented Lagrangian and penalty function cases with a sample of augmenting functional forms, one might need a much wider range of problems than the number we have included in this chapter. However, at least for the small-scale problems we have listed, it is worthwhile to make the following observations:

- Overall, 224 ($2 \times 7 \times 16$) experiments have been conducted using the deflected subgradient method with the parameters and initial guesses listed in Table 1. In the augmented Lagrangian case ($A = I$), no solution could be found in 18 experiments (16%). In the penalty function case, 29 of the experiments failed to find a solution (26%). If one does not discriminate amongst the functional forms of σ, then it might be fair to say that overall it is more advantageous to use an augmented Lagrangian function formulation rather than a penalty function one.
- Next we look at the overall failure rates for the three separate groups of augmenting functional forms, in both of the $A = I$ and $A = 0$ cases. The group of norm augmenting functions has a failure rate of 33/96, that is, 34%. The failure rate for the exponential group is only 9/96, that is, 9.4%. The hybrid functional form has failed at the rate 3/16, that is, 19%. One might argue that the number of experiments for the hybrid group may not be large enough to draw conclusions. Nevertheless, the overall percentage failure rates point to the fact that the exponential augmenting functions we have chosen are more successful compared to the norm augmenting functions.
- How efficient are each of the three groups, when they happen to be successful, with respect to one another? In the experiments, the exponential augmenting function group is the winner in 11 of the 16 problems, as opposed to the five times winner norm augmenting function group. The hybrid group has never achieved the least number of function evaluations in any of the problems.
- It is interesting to note that the sharp augmented Lagrangian formulation (where $\sigma(\cdot) = \|\cdot\|_2$ and $A = I$) seems to be amongst the least successful of all of the augmented Lagrangian formulations considered in the experiments. The classical ℓ_1-penalty function (where $\sigma(\cdot) = \|\cdot\|_1$ and $A = 0$) also seems to be the least successful amongst all of the penalty function formulations here.

We should note that a majority of these test problems come originally from differentiable problems and that in their original form they can be solved by simply applying Newton-type methods with far fewer function evaluations. We often reformulated these differentiable problems in a way that the resulting problem was nondifferentiable. In this sense, these problems only serve to test the performance of the different augmented Lagrangian formulations that we have introduced in our chapter.

References

1. d'Antonio, G., Frangioni, A.: Convergence analysis of deflected conditional approximate subgradient methods, SIAM Journal on Optimization **20**, 357–386 (2009)
2. Bazaraa, M.S., Sherali, H.D.: On the choice of step sizes in subgradient optimization, European Journal of Operations Research **7**, 380–388 (1981)
3. Bazaraa, M.S., Sherali, H.D., Shetty, C.M.: Nonlinear Programming: Theory and Algorithms, Wiley, New York (1993)
4. Bertsekas, D.P.: Nonlinear Programming, Athena Scientific, Belmont, MA (1995)
5. Birgin, E.G., Castillo, R.A., Martínez, J.M.: Numerical comparison of augmented Lagrangian algorithms for nonconvex problems. Computational Optimization and Applications **31**, 31–55 (2005)
6. Burachik, R.S.: On primal convergence for augmented Lagrangian duality, Optimization (accepted for publication)
7. Burachik, R.S., Rubinov, A.M.: On the absence of duality gap for Lagrange-type functions, Journal of Industrial Management and Optimization **1**, 33–38 (2005)
8. Burachik, R.S., Gasimov R.N., Ismayilova N.A., Kaya C.Y.: On a modified subgradient algorithm for dual problems via sharp augmented Lagrangian, Journal of Global Optimization **34**, 55–78 (2006)
9. Burachik, R.S., Iusem, A.N., Melo, J.G.: A primal dual modified subgradient algorithm with sharp Lagrangian, Journal of Global Optimization **46**, 347–361 (2010)
10. Burachik, R.S., Kaya, C.Y.: An update rule and a convergence result for a penalty function method, Journal of Industrial and Management Optimization **3**, 381–398 (2007)
11. Burachik, R.S., Kaya, C.Y., Mammadov, M.: An inexact modified subgradient algorithm for nonconvex optimization, Computational Optimization and Applications **45**, 1–24 (2010)
12. Demyanov, V.F.: Algorithms for some minimax problems, Journal of Computer and Systems Sciences **2**, 342–380 (1968)
13. Gasimov, R.N.: Augmented Lagrangian duality and nondifferentiable optimization methods in nonconvex programming, Journal of Global Optimization **24**, 187–203 (2002)
14. Gasimov R.N., Ismayilova, N.A.: The modified subgradient method for equality constrained nonconvex optimization problems. In: Qi, L., Teo, K.L., Yang, X.Q. (eds.) Optimization and Control with Applications, pp. 257–269, Springer, New York (2005)
15. Griva, I., Polyak, R.A.: Primal–dual nonlinear rescaling method with dynamic scaling parameter update, Mathematical Programming A **106**, 237–259 (2006)
16. Hock, W., Schittkowski, K.: Test Examples for Nonlinear Programming Codes, Springer, Berlin (1981)
17. Kaya, C.Y., Noakes, J.L.: Computational method for time-optimal switching control, Journal of Optimization Theory and Applications **117**, 69–92 (2003)
18. Kim, S., Hyunsil, A.: Convergence of a generalized subgradient method for nondifferentiable convex optimization, Mathematical Programming **50**, 75–80 (1991)
19. Konnov, I.V.: On convergence properties of a subgradient method, Optimization Methods and Software **18**, 53–62 (2003)
20. Polyak, B.T.: The conjugate gradient method in extremal problems, Zhurnal Vychislitelnoy Matematiki i Matematicheskoy Fiziki **9**, 94–112 (1969)
21. Polyak, B.T.: Minimization of unsmooth functionals, Zhurnal Vychislitelnoy Matematiki i Matematicheskoy Fiziki **9**, 14–29 (1969)
22. Polyak, B.T.: Iterative methods using Lagrange multipliers for solving extremal problems with constraints of the equation type, Zhurnal Vychislitelnoy Matematiki i Matematicheskoy Fiziki **10**, 1098–1106 (1970)
23. Polyak, B.T.: Introduction to Optimization, Optimization Software Inc., Publications Division, New York (1987)
24. Polyak, R.A.: Nonlinear rescaling vs. smoothing technique in convex optimization, Mathematical Programming A **92**, 197–235 (2002)

25. Polyak, R.A., Teboulle M.: Nonlinear rescaling and proximal-like methods in convex optimization. Mathematical Programming **76**, 265–284 (1997)
26. Ren, Y.-H., Zhang, L.-W., Xiao, X.-T.: A nonlinear Lagrangian based on Fischer–Burmeister NCP function, Applied Mathematics and Computation **188**, 1344–1363 (2007)
27. Rockafellar, R.T., Wets, R.J.-B.: Variational Analysis, Springer, Berlin (1998)
28. Rubinov, A.M.: Abstract Convexity and Global Optimization, Kluwer, Dordrecht (2000)
29. Rubinov, A.M., Gasimov, R.N.: The nonlinear and augmented Lagrangians for nonconvex optimization problems with a single constraint, Applied and Computational Mathematics **1**, 142–157 (2002)
30. Rubinov, A.M., Gasimov, R.N.: Strictly increasing positively homogeneous functions with applications to exact penalization, Optimization **52**, 1–28 (2003)
31. Rubinov, A.M., Glover, B.M., Yang, X.Q.: Decreasing functions with application to penalization, SIAM Journal on Optimization **10**, 289–313 (1999)
32. Rubinov, A.M., Yang, X.Q., Bagirov, A.M., Gasimov, R.N.: Lagrange-type functions in constrained optimization, Journal of Mathematical Sciences **115**, 2437–2505 (2003)
33. Sherali, H.D., Choi, G., Tuncbilek, C.H.: A variable target value method for nondifferentiable optimization, Operations Research Letters **26**, 1–8 (2000)
34. Shor, N.Z.: Minimization Methods for Nondifferentiable Functions, Springer, Berlin (1985)
35. Shor, N.Z.: Dual estimates in multiextremal problems, Journal of Global Optimization **7**, 75–91 (1995)
36. Yang, X.Q., Huang, X.X.: A nonlinear Lagrangian approach to constrained optimization problems, SIAM Journal on Optimization **14**, 1119–1144 (2001)

On Weak Multifunction Equilibrium Problems

Adela Capătă, Gábor Kassay, and Boglárka Mosoni

Abstract This chapter deals with an extended form of the scalar equilibrium problem called the weak multifunction equilibrium problem. New existence results are obtained in the general setting using a well-known separation theorem in infinite dimensional spaces. These results are applied for the particular case of real-valued multifunctions. Furthermore, two gap functions associated with the studied problems are constructed, where, for one of them, the Fenchel duality theory of optimization is used.

1 Introduction

In the last 15 years, the so-called scalar equilibrium problem has been extensively studied within nonlinear analysis especially due to its important particular cases such as scalar and vector optimization problems, saddle point/minimax problems, variational and hemivariational inequalities (see, for instance, [3–5, 7–9, 14, 23–25]). These problems are useful models of many practical situations arising in economics, engineering, physics, chemistry, etc. (see [2]).

In recent years the vector and multifunction form of the equilibrium problem has been studied (see, e.g., [7, 13]). In this chapter, we extend the results from [7], obtained for the vector equilibrium problems, to the so-called weak multifunction equilibrium problems. These problems can be formulated as follows. Recall that a

Adela Capătă
Department of Mathematics, Technical University of Cluj-Napoca, Str. Constantin Daicoviciu 15, 400020 Cluj-Napoca, Romania
e-mail: adela.capata@math.utcluj.ro

Gábor Kassay and Boglárka Mosoni
Faculty of Mathematics and Computer Science, Babeş-Bolyai University, Cluj-Napoca, Romania
e-mail: kassay@math.ubbcluj.ro, mo_bogi@yahoo.com

R.S. Burachik and J.-C. Yao (eds.), *Variational Analysis and Generalized Differentiation in Optimization and Control*, Springer Optimization and Its Applications 47,
DOI 10.1007/978-1-4419-0437-9_6,

subset C of a topological vector space is called *cone* if $\lambda C \subseteq C$ for every $\lambda \geq 0$. The cone C is said to be:

1. *solid*, if $\operatorname{int} C \neq \emptyset$;
2. *pointed*, if $C \cap (-C) = \{0\}$.

Let A be a nonempty subset of a topological vector space X, B a nonempty set, Z a topological vector space, $C \subset Z$ a convex and solid cone, and $\varphi : A \times B \to Z$ be a vector-valued function.

The weak vector equilibrium problem is

$$(WVEP) \qquad \text{find } \bar{a} \in A \text{ such that } \varphi(\bar{a}, b) \notin -\operatorname{int} C \text{ for all } b \in B.$$

Now, if $\varphi : A \times B \to 2^Z$, the weak multifunction equilibrium problem can be defined in two ways:

$$(WWMEP) \qquad \text{find } \bar{a} \in A \text{ such that } \varphi(\bar{a}, b) \nsubseteq -\operatorname{int} C \text{ for all } b \in B,$$

$$(SWMEP) \qquad \text{find } \bar{a} \in A \text{ such that } \varphi(\bar{a}, b) \cap (-\operatorname{int} C) = \emptyset \text{ for all } b \in B.$$

Observe that both problems reduce to $(WVEP)$ when φ is single valued. It is obvious that each solution of $(SWMEP)$ is a solution of $(WWMEP)$, but not vice versa. Indeed, let $A = \mathbb{R}$, $B = [0, \infty)$, $C = \mathbb{R}^2_+$ and $\varphi : A \times B \to 2^{\mathbb{R}^2}$, defined by

$$\varphi(a, b) = \{(a, y) | y \in [-b, b]\}.$$

Take $\bar{a} = -1$. It is easy to see that $\bar{a}$ is a solution of $(WWMEP)$, but not a solution of $(SWMEP)$.

The chapter is organized as follows. In the remaining part of the introduction, we recall some notions and properties considered in the past necessary for our investigations.

In Sect. 2, we state existence theorems for $(WWMEP)$, by applying a separation result of two convex sets in infinite dimensional spaces. The first one (Lemma 4) is very general, but rather technical; we use it to obtain an existence result (Theorem 1) where the technical assumptions of Lemma 4 are replaced by more familiar (generalized) convexity and continuity hypothesis. In the last part of this section, we consider $(WWMEP)$ for a multifunction whose values are subsets of the real line, denoted by (MEP).

Since gap (or merit) functions help us to detect if a point is solution of the problem $(WWMEP)$, in Sect. 3 we introduce a new gap function for multifunction equilibrium problems in infinite dimensional spaces. For a particular form of the multifunction φ, our results recover some of the results of [13]. It turns out that (MEP) is equivalent to a scalar equilibrium problem, therefore using the Fenchel duality theory of optimization we are able to give another gap function for it. Finally, we conclude this section dealing with gap functions associated to (MEP).

Next we recall the following definitions needed in the sequel.

Definition 1 ([20]). Let $T : X \to 2^Z$ be a multifunction. T is said to be upper C-continuous at $x_0 \in X$ if for every open set V in Z, containing $T(x_0)$ there exists an open neighborhood U of x_0 such that $T(x) \subseteq V + C$ for all $x \in U$. If T is upper C-continuous at each point $x_0 \in X$ we say that T is upper C-continuous on X.

The next continuity type definition for multifunctions in metric spaces can be found in [12] and [22].

Definition 2. Let (Z,d) be a metric space, $T : X \to 2^Z$ a multifunction and $x_0 \in X$. T is said to be d-upper semicontinuous at x_0 if, for any $\varepsilon > 0$ there exists a neighborhood U of x_0 such that

$$T(x) \subset B(T(x_0), \varepsilon) \text{ for all } x \in U,$$

where $B(T(x_0), \varepsilon)$ is defined as

$$\left\{ y \in Z | \inf_{z \in T(x_0)} d(y,z) < \varepsilon \right\}.$$

If T is d-upper semicontinuous at each point $x_0 \in X$ we say that T is d-upper semicontinuous on X.

A neighborhood U of zero is said to be *balanced*, if $\lambda U \subseteq U$ for each scalar with $| \lambda | \leq 1$.

Let us recall the following convexity concepts for multifunctions.

Definition 3. Let $T : X \to 2^Z$ be a multifunction, $C \subset Z$ a convex and solid cone. T is said to be:

(i) C-convex if for all $x_1, x_2 \in X$ and $t \in (0,1)$

$$tT(x_1) + (1-t)T(x_2) \subset T(tx_1 + (1-t)x_2) + C;$$

(ii) C-convexlike on X if for all $x_1, x_2 \in X$ and $t \in (0,1)$

$$tT(x_1) + (1-t)T(x_2) \subset T(X) + C;$$

(iii) C-subconvexlike on X if there exists $\theta \in \text{int}C$ such that for all $x_1, x_2 \in X$, $t \in (0,1)$ and $\varepsilon > 0$

$$\varepsilon\theta + tT(x_1) + (1-t)T(x_2) \subset T(X) + C.$$

We say that T is C-subconcavelike on X if $-$T is C-subconvexlike on X.

The first definition has been considered by S.X. Li in [18]. For the vector case (ii) has been introduced by Paeck in [26] and (iii) by Jeyakumar in [16]. These notions have been considered for multifunctions by Z.F. Li in [19].

The next characterizations of C-subconvexlikeness can be found in [19].

Lemma 1. *For the multifunction $T : X \to 2^Z$, the following properties are equivalent:*
(*i*) *T is C-subconvexlike on X.*
(*ii*) *For all $x_1, x_2 \in X$, $t \in (0,1)$ there exists $\theta \in C$ such that for each $\varepsilon > 0$*

$$\varepsilon\theta + tT(x_1) + (1-t)T(x_2) \subset T(X) + C.$$

(*iii*) *For all $\theta' \in \operatorname{int} C$, $x_1, x_2 \in X$ and $t \in (0,1)$*

$$\theta' + tT(x_1) + (1-t)T(x_2) \subset T(X) + \operatorname{int} C.$$

Lemma 2. *A multifunction $T : X \to 2^Z$ is C-subconvexlike (respectively C-convexlike) on X if and only if the set $T(X) + \operatorname{int} C$ (respectively $T(X)+C$) is convex.*

Example 1. For showing that the concept of C-subconvexlike multifunction is more general than that of C-convexlike multifunction we consider $X = \{(x_1,x_2) \mid x_1 \geq 0, x_2 \geq 0, x_1 + x_2 > 1\}$, $Z = \mathbb{R}^2$, $C = \mathbb{R}^2_+$ and $F : X \to 2^Z$, defined by

$$F(x_1,x_2) = \{(x_1,x_2), (0,1), (1,0)\}.$$

It is easy to see that $F(X) + \operatorname{int}\mathbb{R}^2_+$ is a convex set, but the set $F(X) + \mathbb{R}^2_+$ is not convex.

Definition 4. Let A be a nonempty convex subset of the space X, $T : A \to 2^Z$ and $t \in (0,1)$. The multifunction T is said to be (see [11]):
(i) t-Convex if for any $a_1, a_2 \in A$:

$$tT(a_1) + (1-t)T(a_2) \subseteq T(ta_1 + (1-t)a_2);$$

(ii) t-Concave if for any $a_1, a_2 \in A$:

$$T(ta_1 + (1-t)a_2) \subseteq tT(a_1) + (1-t)T(a_2).$$

We denote by X^* the topological dual of X, which is the set of all continuous and linear real functionals, and by $x^*(x)$ the value of $x^* \in X^*$ at $x \in X$. Recall that the *dual cone* of C, denoted by C^* is the set

$$C^* = \{z^* \in Z^* \mid z^*(c) \geq 0 \text{ for all } c \in C\}.$$

In the sequel, we shall need the following simple property.

Lemma 3. *If $z^* \in C^*$ is a nonzero functional, then $z^*(c) > 0$ for all $c \in \operatorname{int} C$.*

We shall conclude this section by recalling some well-known results from the Fenchel–Moreau duality theory.

Definition 5. Let the functions $f_i : X \to \mathbb{R} \cup \{+\infty\}$, $i \in \{1, \ldots, k\}$, be given. The function $f_1 \square \cdots \square f_k : X \to \mathbb{R} \cup \{\pm\infty\}$ defined by

$$f_1\square\cdots\square f_k(x)=\inf\left\{\sum_{i=1}^{k} f_i(x_i) : \sum_{i=1}^{k} x_i = x\right\}$$

is called the infimal convolution function of $f_1,\dots,f_k$. The infimal convolution $f_1\square\cdots\square f_k$ is said to be exact at $x\in X$ if there exist some $x_i\in X$, $i\in\{1,\dots,k\}$, such that $\sum_{i=1}^{k} x_i = x$ and

$$f_1\square\cdots\square f_k(x)=f_1(x_1)+\cdots+f_k(x_k).$$

For a function $f : X\to\mathbb{R}\cup\{+\infty\}$, we denote by $\operatorname{dom} f=\{x\in X : f(x)<\infty\}$ its *effective domain*. The function f is called *proper* if $\operatorname{dom} f\neq\emptyset$.

Definition 6. The Fenchel–Moreau conjugate function of a function $f : X\to\mathbb{R}\cup\{+\infty\}$, where X is a real locally convex space, is $f^* : X^*\to\mathbb{R}\cup\{+\infty\}$ defined by

$$f^*(x^*)=\sup_{x\in X}[x^*(x)-f(x)].$$

We recall, that for a nonempty subset $A\subseteq X$, the *indicator function* is defined by

$$\delta_A(x)=\begin{cases}0, & \text{if } x\in A;\\ +\infty, & \text{otherwise;}\end{cases}$$

while the *support function* is $\sigma_A(x^*)=\sup_{x\in A} x^*(x)$.

Let $f,g : X\to\mathbb{R}\cup\{+\infty\}$ be proper, convex and lower semicontinuous functions such that $\operatorname{dom} f\cap\operatorname{dom} g\neq\emptyset$ and consider the following optimization problem

$$(P)\qquad \inf_{x\in X}\{f(x)+g(x)\}.$$

The Fenchel dual problem for (P) is

$$(D)\qquad \sup_{p\in X^*}\{-f^*(-p)-g^*(p)\}.$$

The existence of strong duality between a convex optimization problem and its Fenchel dual [the optimal value of (P) equals the optimal value of (D) and (D) admits a solution] was established in [6], under the following (weak) regularity condition:

$$(FRC)\ f^*\square g^* \text{ is a lower semicontinuous function and exact at } 0.$$

Theorem 1. *Let (FRC) be fulfilled. Then the value of (P) equals the value of (D) and (D) admits an optimal solution.*

2 Existence Results

By $\mathscr{C}(Z)$ we denote the set of all compact subsets of the space Z.

We need the following technical result whose proof is based on a separation theorem in infinite dimensional spaces.

Lemma 4. *Let $\varphi : A \times B \to \mathscr{C}(Z)$ be a multifunction. For each $b \in B$ and $c \in \operatorname{int} C$ define the set $U_{b,k} = \{a \in A | \varphi(a,b) + k \subseteq -\operatorname{int} C\}$. Suppose that the following assumptions hold:*

(i) If the system $\{U_{b,k} \mid b \in B, k \in \operatorname{int} C\}$ covers A, then it contains a finite subcover.

(ii) For each $a_1, \dots, a_m \in A$, $\lambda_1, \dots, \lambda_m \geq 0$ with $\sum_{i=1}^m \lambda_i = 1$, $b_1, \dots, b_n \in B$, for all $d_j^i \in \varphi(a_i, b_j)$ where $i \in \{1, \dots, m\}$ and $j \in \{1, \dots, n\}$ there exists $u^* \in C^* \setminus \{0\}$ such that

$$\min_{1 \leq j \leq n} \sum_{i=1}^{m} \lambda_i u^*(d_j^i) \leq \sup_{a \in A} \min_{1 \leq j \leq n} \max u^*(\varphi(a, b_j)),$$

where $\max u^*(\varphi(a,b_j))$ is the greatest element of the compact set $u^*(\varphi(a,b_j)) \subseteq \mathbb{R}$.

(iii) For each $b_1, \dots, b_n \in B$ and $z_1^*, \dots, z_n^* \in C^*$ not all zero

$$\sup_{a \in A} \sum_{j=1}^{n} \max z_j^*(\varphi(a, b_j)) \geq 0.$$

Then the equilibrium problem $(WWMEP)$ admits a solution.

Proof. Suppose by contradiction that $(WWMEP)$ has no solution, that is, for each $a \in A$ there exists $b \in B$ with the property $\varphi(a,b) \subseteq -\operatorname{int} C$. Since φ takes compact values, this means that for each $a \in A$ there exist $b \in B$ and $k \in \operatorname{int} C$ such that

$$\varphi(a,b) + k \subseteq -\operatorname{int} C.$$

Hence the family $\{U_{b,k}\}$ covers the set A, and by assumption (i) there exists a finite subcover which covers the same set A, that is, there exist $b_1, \dots, b_n \in B$ and $k_1, \dots, k_n \in \operatorname{int} C$ such that

$$A \subseteq \bigcup_{j=1}^{n} U_{b_j, k_j}. \tag{1}$$

For these $k_1, \dots, k_n \in \operatorname{int} C$, we have that there exist $V_1, \dots, V_n$ balanced neighborhoods of the origin of Z such that $k_j + V_j \subset C$ for all $j \in \{1, \dots, n\}$.

Define $V := V_1 \cap \dots \cap V_n$, thus V is a balanced neighborhood of the origin of the space Z. Let $k_0 \in V \cap \operatorname{int} C$, so we have $-k_0 \in V$. Hence,

$$k_j - k_0 \in k_j + V \subseteq k_j + V_j \subseteq C \text{ for all } j \in \{1, \dots, n\},$$

which gives

$$k_j - k_0 \in C \text{ for all } j \in \{1, \dots, n\}. \tag{2}$$

Now we define the set-valued mapping $F : A \to 2^{Z^n}$ by

$$F(a) := [\varphi(a, b_1) + k_0] \times \cdots \times [\varphi(a, b_n) + k_0].$$

Assert that

$$\operatorname{co} F(A) \cap (\operatorname{int} C)^n = \emptyset, \tag{3}$$

where $\operatorname{co} F(A)$ denotes the convex hull of the set $F(A)$. Supposing the contrary, there exist $z_1, \ldots, z_m \in F(A)$ and $\lambda_1, \ldots, \lambda_m \geq 0$ with $\sum_{i=1}^m \lambda_i = 1$ such that

$$\sum_{i=1}^m \lambda_i z_i \in (\operatorname{int} C)^n. \tag{4}$$

Let $a_1, \ldots, a_m \in A$ such that $z_i \in [\varphi(a_i, b_1) + k_0] \times \cdots \times [\varphi(a_i, b_n) + k_0]$, where $i \in \{1, \ldots, m\}$. This, together with (4) imply the existence of some $d_j^i \in \varphi(a_i, b_j)$ such that

$$\sum_{i=1}^m \lambda_i d_j^i + k_0 \in \operatorname{int} C \text{ for each } j \in \{1, \ldots, n\}. \tag{5}$$

Now, let $u^* \in C^*$ be a nonzero functional for which (ii) holds. Applying u^* to the above relation and taking into account Lemma 3 we obtain that

$$\sum_{i=1}^m \lambda_i u^*(d_j^i) + u^*(k_0) > 0 \qquad \text{for each } j \in \{1, \ldots, n\}.$$

Passing to the minimum over j yields

$$\min_{1 \leq j \leq n} \sum_{i=1}^m \lambda_i u^*(d_j^i) > -u^*(k_0), \tag{6}$$

thus, assumption (ii) and relation (6) imply

$$\sup_{a \in A} \min_{1 \leq j \leq n} \max u^*(\varphi(a, b_j)) > -u^*(k_0). \tag{7}$$

For each $a \in A$, by relation (1) we have that there exists $j_0 \in \{1, \ldots, n\}$ such that $a \in U_{b_{j_0}, k_{j_0}}$, that is, $\varphi(a, b_{j_0}) + k_{j_0} \subseteq -\operatorname{int} C$. This, together with (2) give

$$\varphi(a, b_{j_0}) + k_0 \subseteq -k_{j_0} + k_0 - \operatorname{int} C \subseteq -\operatorname{int} C.$$

By Lemma 3 and the fact that $u^* \in C^* \setminus \{0\}$ we obtain for all $a \in A$ that there exists $j_0 \in \{1, \ldots, n\}$ such that for each $d \in \varphi(a, b_{j_0})$

$$u^*(d) + u^*(k_0) < 0.$$

Using this, we deduce

$$\max u^*(\varphi(a, b_{j_0})) < -u^*(k_0).$$

Passing to minimum over $j \in \{1,\ldots,n\}$ and then to supremum over $a \in A$ the previous relation becomes

$$\sup_{a\in A} \min_{1\le j\le n} \max u^*(\varphi(a,b_j)) \le -u^*(k_0), \tag{8}$$

which is a contradiction to (7). Thus, (3) is true.

By the well-known Hahn–Banach separation theorem (see, e.g., [15, p. 74] or [27, p. 58]), we obtain that there exists a nonzero linear and continuous functional, $z^* \in (Z^n)^*$ such that

$$z^*(u) \le 0, \qquad \text{for all } u \in \operatorname{co} F(A), \tag{9}$$

$$z^*(c) \ge 0, \qquad \text{for all } c \in (\operatorname{int} C)^n. \tag{10}$$

Assert that $z^*(c) \ge 0$, for all $c \in (\operatorname{int} C)^n$, implies $z_j^* \in C^*$ for all $j \in \{1,\ldots,n\}$. Observe that, by the continuity of z^*, (10) holds for every $c \in C^n$. Suppose the contrary, that is, there exist $j_0 \in \{1,\ldots,n\}$ and $w_0 \in C$ such that

$$z_{j_0}^*(w_0) < 0.$$

Fix $c_1,\ldots,c_{j_0-1},c_{j_0+1},\ldots,c_n \in C$. Then $(c_1,\ldots,c_{j_0-1},tw_0,c_{j_0+1},\ldots,c_n) \in C^n$ for any $t > 0$ and by (10) we have that

$$\sum_{j=1, j\ne j_0}^{n} z_j^*(c_j) + t z_{j_0}^*(w_0) \ge 0.$$

Passing with t to infinity within the upper relation we obtain $-\infty \ge 0$, which is false. So, $z_j^* \in C^*$ for all $j \in \{1,\ldots,n\}$.

In particular, by (9), we have $z^*(u) \le 0$ for all $u \in F(A)$. This means that for every $a \in A$ and $h^a := (d_1^a + k_0,\ldots,d_n^a + k_0) \in F(a)$, where $d_j^a \in \varphi(a,b_j)$ for each $j \in \{1,\ldots,n\}$, we have $z^*(h^a) \le 0$, or equivalently,

$$\sum_{j=1}^{n} z_j^*(d_j^a + k_0) \le 0.$$

As not all z_j^* are zero, this implies by Lemma 3 that

$$\sum_{j=1}^{n} z_j^*(d_j^a) \le -\sum_{j=1}^{n} z_j^*(k_0) < 0. \tag{11}$$

In particular, this takes place for those d_j^a for which z_j^* attains its maximal value on the compact set $\varphi(a,b_j)$ for every $j \in \{1,\ldots,n\}$. Passing to supremum over $a \in A$ in (11), we deduce

$$\sup_{a\in A} \sum_{j=1}^{n} \max z_j^*(\varphi(a,b_j)) \le -\sum_{j=1}^{n} z_j^*(k_0) < 0,$$

which is a contradiction to assumption *(iii)*. This completes the proof.

Following the definition of C-subconvexlikeness (Definition 3(iii) and Lemma 1) we introduce a new convexity notion.

Definition 7. Let Y be a topological space, $T : X \times Y \to 2^Z$ be a multifunction and $C \subset Z$ be a convex and solid cone. T is said to be C-subconvexlike in its first variable if for each $\theta \in \mathrm{int}C$, $x_1, x_2 \in X$ and $t \in (0,1)$ there exists an $x_3 \in X$ such that

$$\theta + tT(x_1,y) + (1-t)T(x_2,y) \subset T(x_3,y) + \mathrm{int}C \text{ for all } y \in Y.$$

We say that T is C-subconcavelike in its first variable if $-$T is C-subconvexlike in its first variable.

The next result provides sufficient conditions for the existence of $(WWMEP)$ by means of convexity and continuity assumptions.

Theorem 1. *Let A be a compact set and $\varphi : A \times B \to \mathscr{C}(Z)$ such that:*
(i) $\varphi(\cdot,b)$ is upper $-C$-continuous for all $b \in B$.
(ii) φ is C-subconcavelike in its first variable.
(iii) For each $b_1,\dots,b_n \in B$ and $z_1^*,\dots,z_n^* \in C^*$ not all zero yields

$$\sup_{a\in A} \sum_{j=1}^{n} \max z_j^*(\varphi(a,b_j)) \geq 0.$$

Then the equilibrium problem $(WWMEP)$ admits a solution.

Proof. We shall verify the assumptions of Lemma 4. Let $b \in B$, $k \in \mathrm{int}C$ and $a_0 \in U_{b,k}$. This implies

$$\varphi(a_0,b) + k \subseteq -\mathrm{int}C.$$

Since $V := -\mathrm{int}C - k$ is an open set with the property $\varphi(a_0,b) \subseteq V$ and $\varphi(\cdot,b)$ is upper $-C$-continuous, then there exists an open neighborhood U of a_0 such that

$$\varphi(a,b) \subseteq V - C \quad \text{for all } a \in U.$$

Hence, $U_{b,k}$ is an open set in the space X. This implies that if the family $\{U_{b,k}\}$ where $b \in B$ and $k \in \mathrm{int}C$ covers the compact set A, then it is an open covering of it, thus there exists a finite subcover of the set A. By this, assumption (i) of Lemma 4 is verified.

Now, let $\theta \in \mathrm{int}C$, $a_1,\dots,a_m \in A$, $\lambda_1,\dots,\lambda_m \geq 0$ with $\sum_{i=1}^m \lambda_i = 1$ and $b_1,\dots,b_n \in B$. By the definition of the C-subconcavelikeness in the first variable, we have that there is an element $a_0 \in A$ such that

$$-\theta + \sum_{i=1}^{m} \lambda_i \varphi(a_i,b_j) \subseteq \varphi(a_0,b_j) - \mathrm{int}C \quad \text{for all } j \in \{1,\dots,n\}.$$

Let $d_j^i \in \varphi(a_i,b_j)$, where $i \in \{1,\dots,m\}$ and $j \in \{1,\dots,n\}$. Then there are $d_j^{a_0} \in \varphi(a_0,b_j)$ and $c_j \in \mathrm{int}C$ with

$$-\theta + \sum_{i=1}^{m} \lambda_i d_j^i = d_j^{a_0} - c_j \quad \text{for all } j \in \{1,\dots,n\}. \tag{12}$$

For any $u^* \in C^* \setminus \{0\}$, by relation (12), we obtain

$$-u^*(\theta) + \sum_{i=1}^{m} \lambda_i u^*(d_j^i) < u^*(d_j^{a_0}) \quad \text{for each } j \in \{1,\dots,n\}.$$

This implies

$$-u^*(\theta) + \sum_{i=1}^{m} \lambda_i u^*(d_j^i) < \max u^*(\varphi(a_0,b_j)) \quad \text{for each } j \in \{1,\dots,n\},$$

where passing to minimum over j and then to supremum over a we get

$$\begin{aligned} -u^*(\theta) + \min_{1\le j\le n} \sum_{i=1}^{m} \lambda_i u^*(d_j^i) &< \min_{1\le j\le n} \max u^*(\varphi(a_0,b_j)) \\ &\le \sup_{a\in A} \min_{1\le j\le n} \max u^*(\varphi(a,b_j)). \end{aligned}$$

Since this holds for every $\theta \in \operatorname{int} C$ we deduce

$$\min_{1\le j\le n} \sum_{i=1}^{m} \lambda_i u^*(d_j^i) \le \sup_{a\in A} \min_{1\le j\le n} \max u^*(\varphi(a,b_j)),$$

which is nothing else than assumption (ii) of Lemma 4. Thus the assertion follows from this lemma.

Now, let us consider the particular case $Z = \mathbb{R}$ and $C = \mathbb{R}_+$. Then $\varphi : A \times B \longrightarrow 2^{\mathbb{R}}$ and $(WWMEP)$ becomes

$$(MEP) \text{ find } \bar{a} \in A \text{ such that } \varphi(\bar{a},b) \not\subseteq -\operatorname{int}\mathbb{R}_+ \quad \text{for all } b \in B.$$

For this particular case, using the previous results we obtain the following.

Corollary 1. *Let $\varphi : A \times B \to \mathscr{C}(\mathbb{R})$ be a multifunction. For each $b \in B$ and $k > 0$ define the set $U_{b,k} = \{a \in A | \varphi(a,b) + k \subseteq -\operatorname{int}\mathbb{R}_+\}$. Suppose that the following assumptions hold:*

(i) If the system $\{U_{b,k} \mid b \in B,\, k > 0\}$ covers A, then it contains a finite subcover.

(ii) For each $a_1,\dots,a_m \in A$, $\lambda_1,\dots,\lambda_m \ge 0$ with $\sum_{i=1}^{m} \lambda_i = 1$, $b_1,\dots,b_n \in B$, for all $d_j^i \in \varphi(a_i,b_j)$, where $i \in \{1,\dots,m\}$ and $j \in \{1,\dots,n\}$

$$\min_{1\le j\le n} \sum_{i=1}^{m} \lambda_i d_j^i \le \sup_{a\in A} \min_{1\le j\le n} \max \varphi(a,b_j).$$

(iii) For each $b_1,\dots,b_n \in B$ and $z_1^*,\dots,z_n^* \ge 0$ not all zero

$$\sup_{a\in A} \sum_{j=1}^{n} \max z_j^*(\varphi(a,b_j)) \ge 0.$$

Then the equilibrium problem (MEP) admits a solution.

Corollary 2. *Let A be a compact set and $\varphi : A \times B \to \mathscr{C}(\mathbb{R})$ such that:*
(i) $\varphi(\cdot, b)$ is upper $-\mathbb{R}_+$-continuous for all $b \in B$.
(ii) φ is $\mathbb{R}_+$-subconcavelike in its first variable.
(iii) For each $b_1, \dots, b_n \in B$ and $z_1^*, \dots, z_n^* \geq 0$ not all zero yields

$$\sup_{a \in A} \sum_{j=1}^{n} \max z_j^*(\varphi(a, b_j)) \geq 0.$$

Then the equilibrium problem (MEP) admits a solution.

3 Gap Functions

In connection with the scalar equilibrium problem and its particular cases (e.g., variational inequalities) the so-called *gap functions* play an important role. Namely, they help to analyze if a point is a solution of these problems, by reducing them to an optimization problem (see [1, 10, 21]).

3.1 Gap Functions for (WWMEP)

Definition 8 ([13]). A multifunction $T : A \to 2^Z$ is a gap function for $(WWMEP)$ if:
(i) $T(a) \subseteq -C$ for all $a \in A$.
(ii) $0 \in T(a)$ if and only if a is a solution of $(WWMEP)$.

In what follows, we give an example of gap function for this problem. In this way, we extend a result from [13] to a multifunction that takes values in a topological vector space.

Let $A = B$ and consider the following assumption:

Assumption A.

$$\text{If } a \in A \text{ is a solution of } (WWMEP), \text{ then } \bigcap_{b \in A} \{\varphi(a, b) \cap C\} \neq \emptyset.$$

Theorem 1. *Suppose that C is a pointed cone, $\varphi(a, a) \subseteq -C$ for all $a \in A$ and Assumption A holds. Then the multifunction $T : A \to 2^Z$, defined by $T(a) = \bigcap_{b \in A} \varphi(a, b)$ for each $a \in A$ is a gap function for $(WWMEP)$.*

Proof. Let $a \in A$. Since $T(a) = \bigcap_{b \in A} \varphi(a, b)$, by the assumptions of the theorem we deduce

$$T(a) \subseteq \varphi(a, a) \subseteq -C. \tag{13}$$

Hence, the condition (i) of Definition 8 is verified.

To verify condition (ii) of Definition 8, let $0 \in T(a)$. This is equivalent to $0 \in \bigcap_{b\in A} \varphi(a,b)$, or $0 \in \varphi(a,b)$ for all $b \in A$. By this, we deduce

$$\varphi(a,b) \nsubseteq -\mathrm{int}C \text{ for each } b \in A.$$

Thus, $a \in A$ is a solution of $(WWMEP)$.

For the reverse implication, take $a \in A$ a solution of $(WWMEP)$. Hence, by Assumption A we deduce that

$$T(a) \cap C = \bigcap_{b\in A} \{\varphi(a,b) \cap C\} \neq \emptyset. \tag{14}$$

By relations (13) and (14), we obtain $0 \in T(a)$, and condition (ii) of Definition 8 is satisfied.

Now let us consider a particular case of $(WWMEP)$, which has been studied in [13]. For $n \in \mathbb{N}$, $N = \{1,\dots,n\}$ and $F_l : A \times A \to 2^{\mathbb{R}}$, $l \in N$, consider the problem $(WWMEP)$ for the multifunction defined by $F(a,b) = F_1(a,b) \times \cdots \times F_n(a,b)$, that is, (see [13]):

$$(GFVEP1) \text{ find } \bar{a} \in A \text{ such that } F(\bar{a},b) \nsubseteq -\,\mathrm{int}\mathbb{R}^n_+ \text{ for all } b \in A.$$

Define a multifunction $T_1 : A \to 2^{\mathbb{R}}$ as follows:

$$T_1(a) = \bigcap_{b\in A} \bigcup_{l\in\mathbb{N}} F_l(a,b) \text{ for all } a \in A \tag{15}$$

and consider the following assumption [13]:

Assumption B. Let $a \in A$ be given. If for any $b \in A$ the relation $\bigcup_{l\in N} F_l(a,b) \cap \mathbb{R}_+ \neq \emptyset$ holds, then $\bigcap_{b\in A} \bigcup_{l\in N} \{F_l(a,b) \cap \mathbb{R}_+\} \neq \emptyset$.

Corollary 3 (Theorem 4.4 [13]). *If for any $a \in A$ and each $l \in N, F_l(a,a) \subseteq -\mathbb{R}_+$ and Assumption B holds, then the multifunction T_1 defined by (15) is a gap function for $(GFVEP1)$ in the same sense of Definition 8 where $C = \mathbb{R}_+$.*

Proof. We show that the assumptions of Theorem 1 are satisfied for $\varphi : A \times A \to 2^{\mathbb{R}}$, defined by

$$\varphi(a,b) = \bigcup_{l\in N} F_l(a,b) \qquad \text{for all } a,b \in A.$$

Indeed, by the hypothesis $\varphi(a,a) = \bigcup_{l\in N} F_l(a,a) \subseteq -\mathbb{R}_+$ for each $a \in A$.

Suppose that a is a solution for $(GFVEP1)$, or equivalently

$$\bigcup_{l\in N} F_l(a,b) \cap \mathbb{R}_+ \neq \emptyset \qquad \text{for all } b \in A.$$

Therefore, by Assumption B

$$\bigcap_{b\in A}\varphi(a,b)\cap\mathbb{R}_+=\bigcap_{b\in A}\bigcup_{l\in N}\{F_l(a,b)\cap\mathbb{R}_+\}\neq\emptyset.$$

Hence, by Theorem 1 the multifunction T_1 is a gap function for $(GFVEP1)$. This completes the proof.

3.2 Gap Functions for (MEP)

In what follows, $A=B$ is a nonempty, closed, and convex subset of a real locally convex space and suppose that $\varphi(a,b)$ is a compact subset of $\mathbb{R}$ for each $a,b\in A$. We observe that (MEP) is equivalent to the problem:

$$\text{find } \bar{a}\in A \text{ such that } \max\,\varphi(\bar{a},b)\geq 0 \quad \text{for all } b\in A,$$

or, in order words:

$$(EP)\ \text{find } \bar{a}\in A \text{ such that } \psi(\bar{a},b)\geq 0 \quad \text{for all } b\in A,$$

where $\psi: X\times X\longrightarrow\mathbb{R}\cup\{+\infty\}$, with $A\times A\subseteq \operatorname{dom} f$, defined by $\psi(a,b)=\max\varphi(a,b)$ for all $a,b\in A$. Further, suppose that $\max\varphi(a,a)=0$ for all $a\in A$. Let $a\in X$. According to [1], (EP) can be reduced to the optimization problem

$$P(a) \qquad \inf_{b\in A}\psi(a,b).$$

Is easy to check that $\bar{a}\in A$ is a solution of (EP) if and only if it is a solution of $P(\bar{a})$.

The next definition is a particular case of Definition 8, when $C=-\mathbb{R}_+$. A function $\gamma: X\longrightarrow\mathbb{R}\cup\{\pm\infty\}$ is said to be a gap function for (EP) (see [21]), if it satisfies the properties:

(i) $\gamma(a)\geq 0$ for all $a\in A$.

(ii) $\gamma(a)=0$ and $a\in A$ if and only if a is a solution for (EP).

In what follows, we will give a gap function using the duality theory of optimization. Considering the indicator function δ_A, we can rewrite the optimization problem as

$$P(a) \qquad \inf_{b\in X}\{\psi(a,b)+\delta_A(b)\}.$$

Proposition 1. *If, for each $a\in A$ the multifunction $b\mapsto\varphi(a,b)$ is t-concave for each $t\in(0,1)$ and d-upper semicontinuous, where d is the euclidian metric on $\mathbb{R}$, then the function $b\mapsto\psi(a,b)$ is convex and lower semicontinuous on A.*

Proof. Take $b_1, b_2 \in A$ and $t \in (0,1)$. Since $b \mapsto \varphi(a,b)$ is t-concave in its second variable on A we have

$$\varphi(a, tb_1 + (1-t)b_2) \subseteq t\varphi(a,b_1) + (1-t)\varphi(a,b_2) \quad \text{for all } a \in A.$$

By this, we obtain

$$\begin{aligned}\psi(a, tb_1 + (1-t)b_2) &\leq \max\{t\varphi(a,b_1) + (1-t)\varphi(a,b_2)\} \\ &\leq t\psi(a,b_1) + (1-t)\psi(a,b_2) \quad \text{for any } a \in A,\end{aligned}$$

which is nothing else than the convexity of the function $b \mapsto \psi(a,b)$.

Take $\varepsilon > 0$, $a \in A$ and $b_0 \in A$. By the d-upper semicontinuity of the multifunction $b \mapsto \varphi(a,b)$, there exists a neighborhood U of b_0 such that

$$\varphi(a,b) \subseteq \varphi(a,b_0) + \left(-\frac{\varepsilon}{2}, \frac{\varepsilon}{2}\right),$$

for all $b \in U$. So,

$$\psi(a,b_0) \leq \psi(a,b) + \frac{\varepsilon}{2} \quad \text{for all } b \in U.$$

Hence, $b \mapsto \psi(a,b)$ is convex and lower semicontinuous on A.

Let the assumptions of the previous proposition be satisfied. The Fenchel dual of $P(a)$ is

$$D(a) \qquad \sup_{p \in X^*} \{-\psi_b^*(a,p) - \sigma_A(-p)\},$$

where $\psi_b^*(a,p) = \sup_{b \in X}[\langle p,b\rangle - \psi(a,b)]$. By $v(P(a))$ and $v(D(a))$, we denote the optimal value of $P(a)$ and $D(a)$, respectively.

The regularity condition (FRC) introduced in [6] (see Sect. 2 before), for our problem becomes

$$(FRC;a) \qquad \psi_b^* \Box \sigma_A \text{ is a lower semicontinuous function and exact at } 0,$$

where $(\psi_b^* \Box \sigma_A)(p) = \inf\{\psi_b^*(p_1) + \sigma_A(p_2) \mid p_1 + p_2 = p\}$.

Using Theorem 3.1 from [1], we are able to give gap functions for (MEP) (which is equivalent to (EP)).

Theorem 2 ([6]). *Assume that for all $a \in A$ the regularity condition $(FRC;a)$ be fulfilled. For each $a \in A$, let $b \mapsto \psi(a,b)$ be convex and lower semicontinuous. Then $\gamma_{(EP)}$, defined by*

$$\gamma_{(EP)}(a) = -v(D(a))$$

is a gap function for (EP).

Theorem 3. *Let the regularity condition $(FRC;a)$ be fulfilled for each $a \in A$ and $b \mapsto \varphi(a,b)$ is t-convex for each $t \in (0,1)$ and d-upper semicontinuous on A. Then $\gamma_{(EP)}$ is a gap function for (MEP).*

Proof. By Proposition 1, the function $b \mapsto \psi(a,b)$ is convex and lower semicontinuous. By the weak duality, which always holds we have

$$v(D(a)) \leq v(P(a)) \leq \psi(a,a) = 0.$$

This implies that

$$\gamma_{(EP)}(a) \geq 0 \quad \text{for all } a \in A.$$

Further suppose that $\bar{a}$ is a solution for $P(\bar{a})$. Since the regularity condition $(FRC;a)$ is fulfilled, the strong duality holds. Hence, by Theorem 2 we obtain that $\gamma_{(EP)}$ is a gap function for (EP) and therefore for (MEP).

Acknowledgements The research of Adela Capătă was supported by CNCSIS, Grant PN II, ID 523/2007. Gábor Kassay and Boglárka Mosoni's research was supported by CNCSIS, Grant PN II, ID 543/2008 and co-financed by The Sectoral Operational Programme Human Resources Development, Contract POSDRU 6/1.5/S/3, "Doctoral studies: through science towards society."

The authors are grateful to an anonymous referee for his helpful comments, which greatly improved the presentation of the chapter.

References

1. L. Altangerel, R. I. Boţ, G. Wanka, *On gap functions for equilibrium problems via Fenchel duality*, Pacific Journal of Optimization 2(3) (2006), pp. 667–678.
2. J. P. Aubin, *Optima and Equilibria. An Introduction to Nonlinear Analysis* Series: Graduate Texts in Mathematics 140, Springer, Berlin, 1993.
3. M. Bianchi, R. Pini, *A note on equilibrium problems with properly quasimonotone bifunctions*, Journal of Global Optimization 20(1) (2001), pp. 67–76.
4. M. Bianchi, R. Pini, *Coercivity conditions for equilibrium problems*, Journal of Optimization Theory and Applications 124(1) (2005), pp. 79–92.
5. E. Blum, W. Oettli, *From optimization and variational inequalities to equilibrium problems*, Mathematics Students 63 (1994), pp. 123–145.
6. R. I. Boţ, G. Wanka, *A weaker regularity condition for subdifferential calculus and Fenchel duality in infinite dimensional spaces*, Nonlinear Analysis: Theory Methods and Applications 64(12) (2006), pp. 2787–2804.
7. A. Capătă, G. Kassay, *On vector equilibrium problems and applications*, accepted in Taiwanese Journal of Mathematics, 2009.
8. F. Flores-Bazán, *Existence theorems for generalized noncoercive equilibrium problems: the quasi-convex case*, SIAM Journal of Optimization 11(3) (2000), pp. 675–690.
9. F. Flores-Bazán, *Existence theory for finite dimensional pseudomonotone equilibrium problems*, Acta Applicandae Mathematicae 77(3) (2003), pp. 249–297.
10. F. Giannessi, *On some connections among variational inequalities, combinatorial and continuous optimization*, Annals of Operations Research 58 (1995), pp. 181–200.
11. A. Göpfert, H. Riahi, C. Tammer, C. Zălinescu, *Variational Methods in Partially Ordered Spaces*, Series: CMS Books in Mathematics 17, Springer, New York, 2003.
12. V. G. Gutev, *Weak factorizations of continuous set-valued mappings*, Topology and Applications 102 (2000), pp. 33–51.
13. N. J. Huang, J. Li, S. Y. Wu, *Gap functions for a system of generalized vector quasi-equilibrium problems with set-valued mappings*, Journal of Global Optimization 41(3) (2008), pp. 401–415.

14. A. N. Iusem, G. Kassay, W. Sosa, *On certain conditions for the existence of solutions of equilibrium problems*, Mathematical Programming 116(1,2) (2009), pp. 259–273.
15. J. Jahn, *Vector Optimization: Theory, Applications, and Extensions*, Springer, Berlin, 2004.
16. V. Jeyakumar, *A generalization of a minimax theorem of Fan via a theorem of the alternative*, Journal of Optimization Theory and Applications 48(3) (1986), pp. 525–533.
17. G. Kassay, J. Kolumbán, *On a generalized sup–inf problem*, Journal of Optimization Theory and Applications 91(3) (1996), pp. 651–670.
18. S. X. Li, *Quasiconvexity and nondominated solutions in multiple-objective programming*, Journal of Optimization Theory and Applications 88(1) (1996), pp. 197–208.
19. Z. F. Li, *Benson proper efficiency in the vector optimization of set-valued maps*, Journal of Optimization Theory and Applications 98(3) (1998), pp. 623–649.
20. D. T. Luc, *Theory of Vector Optimization*, Series: Lecture Notes in Economics and Mathematcal Systems, Springer, Berlin, 1989.
21. G. Mastroeni, *Gap functions for equilibrium problems*, Journal of Global Optimization 27(4) (2003), pp. 411–426.
22. E. Michael, *Continuous selections III*, Annals of Mathematics 65(2) (1957), pp. 375–390.
23. B. Mordukhovich, *Optimization and equilibrium problems with equilibrium constraints*, OMEGA 33(5) (2005), pp. 379–384.
24. B. Mordukhovich, *Equilibrium problems with equilibrium constraints via multiobjective optimization*, Optimization Methods and Software 19(5) (2004), pp. 479–492.
25. B. Mordukhovich, *Optimization and equilibrium problems with equilibrium constraints in infinite-dimensional spaces*, Optimization 57(5) (2008), pp. 715–741.
26. S. Paeck, *Convexlike and concavelike conditions in alternative, minimax, and minimization theorems*, Journal of Optimization Theory and Applications 74(2) (1992), pp. 317–332.
27. W. Rudin, *Functional Analysis*, Tata McGraw-Hill, New Delhi, 1973.

Optimality Conditions for a Simple Convex Bilevel Programming Problem

S. Dempe, N. Dinh, and J. Dutta

Abstract The problem to find a best solution within the set of optimal solutions of a convex optimization problem is modeled as a bilevel programming problem. It is shown that regularity conditions like Slater's constraint qualification are never satisfied for this problem. If the lower-level problem is replaced with its (necessary and sufficient) optimality conditions, it is possible to derive a necessary optimality condition for the resulting problem. An example is used to show that this condition in not sufficient even if the initial problem is a convex one. If the lower-level problem is replaced using its optimal value, it is possible to obtain an optimality condition that is both necessary and sufficient in the convex case.

1 Introduction

In this chapter, we are interested in studying optimality conditions for the following bilevel problem (BP)

$$\min f(x) \quad \text{subject to} \quad x \in S,$$

S. Dempe
Department of Mathematics and Computer Science, TU Bergakademie Freiberg, Freiberg, Germany
e-mail: dempe@tu-freiberg.de

N. Dinh
Department of Mathematics, International University, Vietnam National University of Ho Chi Minh city, Ho Chi Minh city, Vietnam
e-mail: ndinh02@yahoo.fr

J. Dutta
Department of Mathematics and Statistics, Indian Institute of Technology, Kanpur, India
e-mail: jdutta@iitk.ac.in

R.S. Burachik and J.-C. Yao (eds.), *Variational Analysis and Generalized Differentiation in Optimization and Control*, Springer Optimization and Its Applications 47, DOI 10.1007/978-1-4419-0437-9_7, where S is given as

$$S = \operatorname{argmin}\{h(x) : x \in \Theta\}.$$

Here f and h are real-valued convex functions on $\mathbb{R}^n$ and Θ is a convex subset of $\mathbb{R}^n$. Thus it is clear that S is a convex set and, hence, problem (BP) is a convex programming problem. We refer to (BP) as a *simple convex bilevel programming problem*. This problem was first studied by Solodov [15], who developed a nice algorithm for the problem and also gave a convergence criterium for the algorithm. In this paper, Solodov also shows that as a special case the problem (BP) contains the standard differentiable convex optimization problem of the form

$$\min f(x) \quad \text{subject to} \quad g(x) \leq 0, Ax = b,$$

where $f : \mathbb{R}^n \longrightarrow \mathbb{R}$ is a differentiable convex function, $g : \mathbb{R}^n \longrightarrow \mathbb{R}^m$ is a differentiable convex function, A is a $l \times n$ matrix and $b \in \mathbb{R}^l$. This problem can be posed as the problem (BP) by simply assuming that the lower-level function h is given as

$$h(x) = ||Ax - b||^2 + ||\max\{0, g(x)\}||^2$$

and the lower-level problem is to minimize the function h over $\mathbb{R}^n$. It is important to note that in the above expression the maximum is taken coordinate-wise.

In this chapter, we want to analyze optimality conditions for the problem (BP). Given the inherent bilevel structure of the problem, it appears that it seems not be straightforward to write down the optimality conditions for (BP). To motivate our study, it might be a good idea to take a brief tour of the usual bilevel programming problem. In general, a bilevel programming problem is given as follows

$$\min_x F(x,y), \quad \text{subject to} \quad x \in X, y \in \Psi(x),$$

where $F : \mathbb{R}^n \times \mathbb{R}^n \longrightarrow \mathbb{R}$, X is a closed subset of $\mathbb{R}^n$ and Ψ is a set-valued map denoting the solution set mapping of the following parametric optimization problem,

$$\min_y f(x,y), \quad \text{subject to} \quad y \in \Theta(x),$$

where $f : \mathbb{R}^n \times \mathbb{R}^m \longrightarrow \mathbb{R}$ and $\Theta(x)$ is a set depending on the parameter x. The bilevel programming problem has been often referred to as an optimization formulation of the Stackelberg game [16]. The leaders problem is the so-called upper-level problem where the minimization is carried out on the variable x while the followers problem is called the lower-level problem where for each of the input x by the leader the follower optimizes his objective function $f(x,y)$ using the variable y depending on some constraints that depend on the input provided. It must be now apparent why we call the problem at hand a *simple bilevel programming problem*. However, an important point is that, unless for each x the lower-level problem or the follower's problem has an unique solution, the upper-level objective is a set-valued map. So in the general scenario a bilevel programming problem can be viewed as a mathematical programming problem with set-valued maps. In order to avoid the set-valued objective, two different approaches, namely the optimistic approach and the pessimistic approach, have been introduced mainly to have the final objective function as a single-valued one. For details on these approaches see, for example, Dempe [2], Dutta and Dempe [10], Dempe et al. [3,4] and the references there in.

Let us note that the overall bilevel programming problem is in general a non-convex problem even if the problem data are convex. Further, a major drawback is that for a bilevel programming problem most standard constraint qualification conditions like the Mangasarian–Fromovitz constraint qualification are never satisfied. Let us note that our problem (BP), which is a convex problem, is also not free from such a drawback. If we assume that the set S is non-empty and put $\alpha = \inf_{x\in\Theta} h$, then the problem (BP) is equivalent to the reformulated problem (RP)

$$\min f(x), \quad \text{subject to} \quad h(x) \le \alpha, x \in \Theta.$$

It is simple to notice that the Slater's constraint qualification does not hold for this problem, and since the Slater's constraint qualification is equivalent to the Mangasarian–Fromovitz constraint qualification for a convex programming problem, thus even for this simple bilevel problem we are faced with the same issues of the usual bilevel programming problem. As we had stated earlier, our main aim is to study optimality conditions for the problem (BP). In fact, our main aim would be to develop a necessary and sufficient optimality condition. In our first approach, we will consider the problem data to be smooth and in fact we will assume that the lower-level function h is twice continuously differentiable while the upper-level objective f is just differentiable and hence continuously differentiable since f is convex. In this setting, we will take the approach as considered for a bilevel programming problem with convex lower-level problems in Dutta and Dempe [10]. However, the optimality condition that we will get is a necessary one and not sufficient even though the problem (BP) is a convex problem. In fact, using this approach we will see that the Lagrange multipliers are related to the coderivative of the normal cone map to the set Θ. This co-derivative appears to be quiet difficult to compute though very recently some advances have been made by Henrion et al. [11]. This approach will be demonstrated in Sect. 2. The natural question is whether it is possible to develop an optimality condition that is both necessary and sufficient for problem (BP). This will be achieved through an alternative approach of reformulating the bilevel program (BP) as the convex programming problem (RP) that never satisfies Slater's constraint qualification. We use very recent results in convex optimization to develop a simple necessary and sufficient optimality condition for the problem (BP) using this reformulation. Our notations are standard. Further, instead of collecting all the preliminary definitions and results in one section, we present the basic tools in the main sections as and when needed.

2 Optimality Conditions: The Bilevel Approach

In this section, we will use techniques from variational analysis to develop a necessary optimality condition for the problem (BP). Assume that the convex function f is differentiable and the convex function h is twice continuously differentiable. We will show through an example that the optimality condition that we develop in this section is necessary but not sufficient. We begin our study by observing that the set S can equivalently be written as

$$S = \{x \in \Theta : 0 \in \nabla h(x) + N_\Theta(x)\}.$$

Let us denote by $F(x) = (x, -\nabla h(x))^{\mathrm{T}}$, where T denotes transpose. Then we can write S as

$$S = \{x \in \Theta : F(x) \in \mathrm{gph} N_\Theta\}, \tag{1}$$

where $\mathrm{gph} N_\Theta$ is the graph of the normal cone map N_Θ, which is given as

$$N_\Theta(x) = \{v \in \mathbb{R}^n : \langle v, y - x\rangle \leq 0, \text{for all } y \in \Theta\}$$

if $x \in \Theta$ and $N_\Theta(x) = \emptyset$ if $x \notin \Theta$. In what follows, we also need to consider the limiting normal cone or the Mordukhovich normal cone to the graph of the normal cone map to the feasible set Θ of the lower-level problem. Hence, we now briefly describe the limiting normal cone. For more details, see Rockafellar and Wets [14].

Given a set $C \subseteq \mathbb{R}^n$ and an element $\bar{x} \in C$ an element $v \in \mathbb{R}^n$ is called a regular normal vector or a Frechét normal vector to C at $\bar{x}$ if

$$v \in \hat{N}_C(\bar{x}) := \{v \in \mathbb{R}^n : \langle v, x - \bar{x}\rangle \leq o(||x - \bar{x}||), \quad \forall x \in C\},$$

where $\frac{o(t)}{t} \to 0$ as $t \to 0$. The set of all regular or Frechét normal vectors to C at $\bar{x}$ is denoted by $\hat{N}_C(\bar{x})$ and forms a cone which is known as the regular of Frechét normal cone to C at $\bar{x}$. This cone is a closed and convex but suffers from the drawback that it can reduce to the trivial cone containing only the zero element at some points of the boundary of C. This problem is eliminated by defining the limiting normal cone or the Mordukhovich normal cone.

Given a set $C \subseteq \mathbb{R}^n$ and $\bar{x} \in C$ a vector $v \in \mathbb{R}^n$ is said to be a limiting normal vector or a Mordukhovich normal vector to C at $\bar{x}$ if there exist sequences $v_k \to v$ and $x_k \to \bar{x}$ with $x_k \in C$ and $v_k \in \hat{N}_C(x_k)$. The set of all limiting normal vectors to the set C at $\bar{x}$ forms a cone and is denoted by $N_C^L(\bar{x})$, which is known as the limiting normal cone or the Mordukhovich normal cone. This cone is not convex in general but it is always closed in our finite dimensional setting. Moreover, the limiting normal cone never reduces to the trivial cone containing only the zero element at the boundary points of C. Of course, if $\bar{x}$ is an interior point of C then $N_C^L(\bar{x}) = \{0\}$. We will now state the main result of this section.

Theorem 1. *Let us consider the bilevel programming problem (BP) where the upper-level objective function f is convex and differentiable, the lower-level objective function is convex and twice continuously differentiable and Θ is a convex* *set. Let $\bar{x}$ be a solution of the problem (BP). Assume that the following qualification condition holds at $\bar{x}$: $(w, v) \in N^L_{gphN_\Theta}(\bar{x}, -\nabla h(\bar{x}))$ satisfying the condition*

$$0 \in w - \nabla^2 h(\bar{x})^T w + N_\Theta(\bar{x}),$$

implies that $w = 0, v = 0$. Then there exists $(\bar{w}, \bar{v}) \in N^L_{gphN_\Theta}(\bar{x}, -\nabla h(\bar{x}))$ such that

$$0 \in \nabla f(\bar{x}) + \bar{w} - \nabla^2 h(\bar{x})^T \bar{w} + N_\Theta(\bar{x}).$$

Proof. Once we have expressed S as in (1) the result follows easily by an application Theorem 6.14 in Rockafellar and Wets [14]. Observe that the qualification condition given in the above theorem is the same as the one given in Theorem 6.14 of Rockafellar and Wets [14].

It is important to note that the above optimality condition is just necessary and not sufficient. We demonstrate this fact through the following example.

Example 1. Consider the problem (BP) with the function $f : \mathbb{R} \to \mathbb{R}$ given by $f(x) = x^2$ and the lower-level objective $h : \mathbb{R} \to \mathbb{R}$ given as follows: $h(x) = x^3$ when $x \geq 0$ and $h(x) = 0$, $x \leq 0$. The lower-level constraint set is $\Theta = [-1, +1]$. Observe that $S = [-1, 0]$. Thus $x = 0$ is the only solution to the problem. However, the optimality condition given in Theorem 1 is satisfied at the point $x = -1$ which we know is not a solution of the problem. This fact can be seen by noting that $(-1, 0)^T \in \text{gph}N_\Theta$ and also observing that $\nabla f(-1) = -2$, $\nabla^2 h(-1) = 0$ and $(4, 0)^T \in N^L_{\text{gph}N_\Theta}(-1, 0)$. Now the optimality condition is satisfied by choosing the element -2 from $N_\Theta(-1) = (-\infty, 0]$.

However, observe that (BP) is overall a convex optimization problem. So naturally one would like to develop necessary and sufficient optimality conditions for the problem (BP). As we have seen in this section that formulation of the problem as done above only produces a necessary optimality condition. In the next section, we demonstrate how we can develop a necessary and sufficient optimality condition.

3 Optimality Conditions: An Alternative Approach

In this section, we proceed to develop necessary and sufficient optimality conditions for the bilevel problem (BP) by reformulating it as the single-level convex optimization problem (RP). We have mentioned that the reformulated problem never satisfies Slater's constraint qualification and, hence, we need modern tools of convex optimization to address this issue. We shall divide this section into three subsections. In the first, one we will describe the tools from convex optimization needed for our study. In the second subsection, we show how this can be used to develop necessary and sufficient optimality conditions for the problem (BP) while in the last one we consider the case where the feasible set Θ of the lower-level problem is described through cone constraints and an abstract constraint.

3.1 Recent Tools from Convex Optimization

We deal with a class of *cone-convex programs* given as

$$\min \vartheta(x) \quad \text{subject to} \quad g(x) \in -D, \text{ and } x \in C, \tag{2}$$

where $\vartheta\colon \mathbb{R}^n \to \overline{\mathbb{R}} := (-\infty, \infty]$ is a *proper, convex, lower semicontinuous* (l.s.c.) function with values in the *extended* real line $\overline{\mathbb{R}}$, $g\colon \mathbb{R}^n \to \mathbb{R}^m$ is a continuous D-convex mapping with D is a closed convex cone in $\mathbb{R}^m$ and $C \subset \mathbb{R}^n$ is a closed and convex subset.

For a set $C \subset \mathbb{R}^n$, the indicator function δ_C is defined as $\delta_C(x) = 0$ if $x \in C$ and $\delta_C(x) = +\infty$ if $x \notin D$. Let us recall that if C is nonempty, closed and convex, then δ_C is a proper l.s.c. convex function. Let $A = \{x \in C \mid g(x) \in -D\}$. Further, let D^+ be the positive dual cone of D, i.e.,

$$D^+ := \{s^* \in \mathbb{R}^m \mid \langle s^*, s\rangle \geq 0, \; \forall s \in D\}.$$

Assume that $\operatorname{dom} f \cap A \neq \emptyset$.

Considering further an extended-real-valued function $\varphi\colon \mathbb{R}^n \to \overline{\mathbb{R}}$ with the *domain* $\operatorname{dom}\varphi := \{x \in \mathbb{R}^n \mid \varphi(x) < \infty\}$, we always assume that it is *proper*, i.e., $\varphi(x) \not\equiv \infty$ on $\mathbb{R}^n$. The *conjugate function* $\varphi^*\colon \mathbb{R}^n \to \overline{\mathbb{R}}$ to φ is defined by

$$\begin{aligned} \varphi^*(x^*) &:= \sup\big\{\langle x^*, x\rangle - \varphi(x) \,\big|\, x \in \mathbb{R}^n\big\} \\ &= \sup\big\{\langle x^*, x\rangle - \varphi(x) \,\big|\, x \in \operatorname{dom}\varphi\big\}. \end{aligned} \tag{3}$$

Definition 1 (Farkas–Minkowski (FM) constraint qualification). We say that problem (2) satisfies the *Farkas–Minkowski constraint qualification*, (FM) in brief, if the cone

$$K := \bigcup_{\lambda \in D^+} \operatorname{epi}(\lambda g)^* + \operatorname{epi}\delta_C^* \tag{4}$$

is closed in the space $\mathbb{R}^n \times \mathbb{R}$.

It is important to note that the set $\bigcup_{\lambda \in D^+} \operatorname{epi}(\lambda g)^*$ is a closed convex cone. This was shown in [13].

Definition 2 ((CC) constraint qualification). We say that problem (2) satisfies the *(CC) constraint qualification*, if the

$$\operatorname{epi}(\vartheta)^* + K \tag{5}$$

is closed in the space $\mathbb{R}^n \times \mathbb{R}$, where K is given in (4).

Remark 1. The constraint qualification conditions (CC) and (FM) are often known as *closedness conditions* and *closed cone constraint qualification*. The first one was introduced in [1] for the first time and the second one was proposed in the unpublished manuscript [12] (it is called (CCCQ) condition), all for dealing with the problem of model (2) in Banach space setting. They have been used extensively in [6, 8, 9], and others (see the references in the papers listed above) to establish optimality conditions, duality results for DC programs subject to cone constraints, and extended to infinite convex and DC problems (infinitely many convex constraints, in infinite dimensional spaces) in [5, 7]. These are also used to establish necessary conditions for bilevel programs in [6]. It was shown in the papers mentioned above that (FM) condition [(CC) condition] is much weaker than the classical constraint qualification of Slater-type (weaker than many combinations of Slater-type conditions and qualification conditions such as the requirement that the cost functional f must be continuous at some point in the feasible set of the problem in consideration, respectively). As we will see below, when reformulated the simple bilevel program as an optimization problem, the new problems never satisfies the Slater's condition while the (CC) and (FM) may hold.

Remark 2. It is worth noting that if f is continuous at one point in A then [5]

$$\operatorname{epi}(f+\delta_A)^* = \operatorname{cl}\{\operatorname{epi} f^* + \operatorname{epi}\delta_A^*\} = \operatorname{epi} f^* + \operatorname{epi}\delta_A^* = \operatorname{epi} f^* + \operatorname{cl} K,$$

where cl$A$ denotes the closure of the set $A \subset \mathbb{R}^n$. So, if (FM) holds (i.e., K is closed), then (CC) holds.

The following optimality condition for (2) was established in [8].

Theorem 2 ([8] Necessary and sufficient optimality conditions for cone-convex programs). *Let the qualification condition (CC) hold for the convex program* (2). *Then* $\bar{x} \in A \cap \operatorname{dom}\vartheta$ *is a (global) solution to* (2) *if and only if there is* $\lambda \in D^+$ *such that*

$$0 \in \partial\vartheta(\bar{x}) + \partial(\lambda g)(\bar{x}) + N_C(\bar{x}), \tag{6}$$

$$\lambda g(\bar{x}) = 0. \tag{7}$$

Remark 3. It is worth noting that the above necessary and sufficient optimality was also proved in the infinite dimensional setting by Burachik and Jeyakumar [1] provided that both the conditions (FM) and (CC) hold. It can be shown [9] that (CC) is not the same as both (CC) and (FM). Note also that in the case where $(\vartheta)^*$ is continuous at one point in A then (CC) is equivalent to both (CC) and (FM).

3.2 Applications to the Simple Bilevel Problem

Let us consider the simple bilevel programming problem (BP) given in Sect. 1. We have already stated in Sect. 1 that the problem (BP) is equivalent to the following convex optimization problem (RP)

$$\min f(x) \quad \text{subject to} \quad h(x) - \alpha \le 0,\ x \in \Theta.$$

We would just like to recall that $\alpha = \inf_{x\in\Theta} h$ and further we also assume that α is a finite real number.

It is worth noting that Slater's constraint qualification (and other interior-types of constraint qualification conditions) never hold for the problem (RP).

We now give an optimality condition for (RP) [which is an optimality condition for problem (BP) as well] that is a consequence of Theorem 2.

Theorem 3. *For the problem (RP), assume that*

$$\{\operatorname{cone}\{(0,1)\} \cup \operatorname{cone}[(0,\alpha)+\operatorname{epi} h^*]\} + \operatorname{epi}\delta_\Theta^*$$

is closed. Then $\bar{x} \in \Theta$ is a (global) solution to (RP) if and only if there is $\lambda \in \mathbb{R}_+$ such that

$$0 \in \partial f(\bar{x}) + \lambda\partial h(\bar{x}) + N_\Theta(\bar{x}), \tag{8}$$

$$\lambda[h(\bar{x}) - \alpha] = 0. \tag{9}$$

Proof. We observe firstly that problem (RP) is of the type (2) with $D = D^+ = \mathbb{R}_+$ and $C = \Theta$. Secondly, for each $u^* \in \mathbb{R}^n$, and $\mu \in \mathbb{R}_+$

$$(\mu(h(.)-\alpha))^*(u^*) = \mu\alpha + (\mu h)^*(u^*).$$

It then follows that

$$\operatorname{epi}(\mu(h(.)-\alpha))^* = (0,\mu\alpha) + \operatorname{epi}(\mu h)^*.$$

Now observe that when $\mu = 0$ we have

$$\operatorname{epi}(\mu h)^* = \operatorname{cone}\{(0,1)\},$$

and when $\mu > 0$ we have

$$\operatorname{epi}(\mu h)^* = \mu\operatorname{epi} h^*.$$

Thus we have

$$\operatorname{epi}(\mu(h(.)-\alpha))^* = \operatorname{cone}\{(0,1)\} \cup \left\{\bigcup_{\mu>0} \mu[(0,\alpha)+\operatorname{epi} h^*]\right\}.$$

Noting that $\operatorname{cone}\{(0,1)\} \cup \{(0,0)\} = \operatorname{cone}\{(0,1)\}$ we have

$$\operatorname{epi}(\mu(h(.)-\alpha))^* = \operatorname{cone}\{(0,1)\} \cup \operatorname{cone}[(0,\alpha)+\operatorname{epi} h^*].$$

Now from the hypothesis of the theorem it is clear that the problem (RP) satisfies (FM) and hence, it satisfies (CC) since f is continuous (see Remark 2). It now follows from Theorem 2 that there is $\lambda \in \mathbb{R}_+$ such that

$$\begin{aligned}&0 \in \partial f(\bar{x}) + \lambda \partial[h(.) - \alpha](\bar{x}) + N_\Theta(\bar{x}),\\&\lambda[h(\bar{x}) - \alpha] = 0.\end{aligned} \tag{10}$$

Since $\partial[h(.) - \alpha](\bar{x}) = \partial h(\bar{x})$, the conclusion follows.

Example 2. Let us consider the bilevel problem of the model (2) where $f(x) = x^2 + 1$, $\Theta = [-1, 1]$, and $h(x) = \max\{0, x\}$.

It is easy to see that $\operatorname{epi} \delta^*_\Theta = \operatorname{epi}|.|$, $S = [-1, 0]$, and $\alpha = 0$. The optimization problem reformulated from this bilevel problem is

$$\min f(x) := x^2 + 1 \;\text{ subject to } h(x) = \max\{0, x\} \leq 0,\; x \in [-1, 1]. \tag{11}$$

Note that for each $u \in \mathbb{R}$,

$$h^*(u) = \begin{cases} +\infty & \text{if } \; u < 0 \; \text{ or } \; u > 1, \\ 0 & \text{if } \; u \in [0, 1]. \end{cases}$$

We have

$$\operatorname{epi} h^* = \{(u, r) \mid u \in [0, 1], r \geq 0\} = [0, 1] \times \mathbb{R}_+,$$

and

$$\operatorname{cone}\{\operatorname{epi} h^*\} + \operatorname{epi} \delta^*_\Theta = \mathbb{R}^2_+ \cup \{(u, r) \mid u \leq 0,\; r \geq -u\}$$

is a closed subset of $\mathbb{R}^2$. This shows that for the problem (11), (FM) holds since $\operatorname{cone}\{(0, 1)\} \subset \operatorname{epi} h^*$. Since f is continuous, (CC) holds as well [note that the Slater's condition fails to hold for (11)]. It is easy to see that $\bar{x} = 0$ is a solution of the bilevel problem. Since $N_\Theta(0) = \{0\}$, $\partial f(0) = \{0\}$, and $\partial h(0) = [0, 1]$, (8)–(9) are satisfied with $\lambda = 0$.

3.3 Lower-Level Problem with Explicit Constraints

We now consider the bilevel problem of the type (2),

$$\inf_{x \in S} f(x), \tag{12}$$

where S is the solution set of the lower-level problem:

$$\min h(x) \quad \text{subject to} \quad g_1(x) \in -D_1,\; x \in C. \tag{13}$$

Here the data are as in Sect. 3.1. Concretely, $h \colon \mathbb{R}^n \to \overline{\mathbb{R}} := (-\infty, \infty]$ is a *proper, convex, lower semicontinuous* (l.s.c.) function, and $g_1 : \mathbb{R}^n \to \mathbb{R}^m$ is a D_1-convex mapping with D_1 is a closed convex cone in $\mathbb{R}^m$ and $C \subset \mathbb{R}^n$ is a closed and convex set. Further, assume that the mapping g_1 is continuous on $\mathbb{R}^n$.

Suppose that

$$\alpha = \inf_{g_1(x) \in -D_1, x \in C} h(x) < +\infty.$$

For the sake of simplicity, assume that $\alpha = 0$. This can be achieved by setting $h(x) := h(x) - \alpha$. Then the bilevel program (12) is equivalent to the following optimization problem:

$$\min f(x) \quad \text{subject to} \quad h(x) \leq 0,\ g_1(x) \in -D_1,\ x \in C. \tag{14}$$

Now let

$$D := \mathbb{R}_+ \times D_1, \quad g : \mathbb{R}^n \to \mathbb{R}^{m+1}, \quad g(x) = (h(x), g_1(x)).$$

Then, problem (14) can be reformulated as

$$\min f(x) \quad \text{subject to} \quad g(x) \in -D,\ x \in C, \tag{15}$$

which is of the type (2).

The next theorem gives necessary and sufficient conditions for optimality of the bilevel programming problem (12).

Theorem 4. *For the problem* (12), *assume that*

$$\operatorname{cone}\operatorname{epi} h^* + \bigcup_{\lambda \in D_1^+} \operatorname{epi}(\lambda g_1)^* + \operatorname{epi} \delta_C^*$$

is closed. Then $\bar{x} \in g^{-1}(-D) \cap C$ is a (global) solution to (12) *if and only if there is $r \in \mathbb{R}_+$ and $\lambda \in D_1^+$ such that*

$$0 \in \partial f(\bar{x}) + r\partial h(\bar{x}) + \partial(\lambda g_1)(\bar{x}) + N_C(\bar{x}), \tag{16}$$

$$rh(\bar{x}) = 0 \text{ and } \lambda g_1(\bar{x}) = 0. \tag{17}$$

Proof. Observe that $D^+ = \mathbb{R}_+ \times D_1^+$ and for any $\widetilde{\lambda} = (r, \lambda) \in D^+$,

$$(\widetilde{\lambda} g)(x) = rh(x) + (\lambda g_1)(x).$$

Moreover,

$$\begin{aligned} \operatorname{epi}(\widetilde{\lambda} g)^* &= \operatorname{cl}\{\operatorname{epi}(rh)^* + \operatorname{epi}(\lambda g_1)^*\} \\ &= \operatorname{epi}(rh)^* + \operatorname{epi}(\lambda g_1)^* \\ &= r \cdot \operatorname{epi} h^* + \operatorname{epi}(\lambda g_1)^*. \end{aligned}$$

Therefore,

$$\bigcup_{\widetilde{\lambda} \in D^+} \operatorname{epi}(\widetilde{\lambda} g)^* + \operatorname{epi} \delta_C^* = \operatorname{cone}\operatorname{epi} h^* + \bigcup_{\lambda \in D_1^+} \operatorname{epi}(\lambda g_1)^* + \operatorname{epi} \delta_C^*.$$

By assumption, this set is closed. Hence, (FM) holds for the problem (15). Since f is continuous, (CC) holds for (15) as well (see Remark 2). Since the problem (12) is equivalent to (15) [also (14)], by Theorem 2, $\bar{x} \in A$ is an optimal solution of (12) if and only if there exists $\widetilde{\lambda} = (r, \lambda) \in D^+$ such that

$$0 \in \partial f(\bar{x}) + \partial(\widetilde{\lambda g})(\bar{x}) + N_C(\bar{x}), \tag{18}$$

$$(\widetilde{\lambda g})(\bar{x}) = 0. \tag{19}$$

It is obvious that

$$\partial(\widetilde{\lambda g})(\bar{x}) = r\partial h(\bar{x}) + \partial(\lambda g_1)(\bar{x}).$$

On the other hand,

$$(\widetilde{\lambda g})(\bar{x}) = rh(\bar{x}) + \lambda g_1(\bar{x}) = 0.$$

Since $r \geq 0$, $h(\bar{x}) \leq 0$, $\lambda g_1(\bar{x}) \leq 0$, we get $rh(\bar{x}) = 0$ and $\lambda g_1(\bar{x}) = 0$. It then follows from (18)–(19) that

$$0 \in \partial f(\bar{x}) + r\partial h(\bar{x}) + \partial(\lambda g_1)(\bar{x}) + N_C(\bar{x}),$$
$$rh(\bar{x}) = 0 \quad \text{and} \quad \lambda g_1(\bar{x}) = 0,$$

which is desired.

It will be interesting to revisit Example 1 in the light of the above theorem. This is what we present below.

Example 3 (Example 1 revisited). Recall that in Example 1 we consider the problem (BP) with the function $f : \mathbb{R} \to \mathbb{R}$ given by $f(x) = x^2$ and the lower-level objective $h : \mathbb{R} \to \mathbb{R}$ given as follows: $h(x) = x^3$ when $x \geq 0$ and $h(x) = 0$, $x \leq 0$. The lower-level constraint set is $C = [-1, +1]$. This problem is of the model (12)–(13), where $g_1 \equiv 0$.

A direct calculation gives $\operatorname{epi} \delta_C^* = \operatorname{epi} |\cdot|$ while for $u \in \mathbb{R}$,

$$h^*(u) = \begin{cases} +\infty & \text{if } \ u < 0, \\ 0 & \text{if } \ u \geq 0. \end{cases}$$

Therefore, $\operatorname{epi} h^* = \operatorname{cone} \operatorname{epi} h^* = \mathbb{R}^2_+$ and hence,

$$\operatorname{cone} \operatorname{epi} h^* + \operatorname{epi} \delta_C^* = \mathbb{R}^2_+ + \operatorname{epi} |\cdot|$$

is closed in $\mathbb{R}^2$ which shows that the closedness condition in Theorem 4 holds for (BP). It is easy the see that the system (16) [applies to (BP)] leads to the unique solution $x = 0$. So, by Theorem 4, $x = 0$ is the solution of (BP).

4 Conclusion

The problem (BP) of finding a "best" optimal solution of a convex optimization problem

$$\min\{h(x) : x \in \Theta\}, \tag{20}$$

where Θ is a convex set and h a convex function defined on Θ with respect to a convex function f is modeled as a bilevel programming problem. Using the necessary and sufficient optimality conditions

$$0 \in \nabla h(x) + N_{\Theta}(x)$$

for the lower-level problem, optimality conditions for problem (BP) can be derived. An example shows that these optimality conditions are necessary but not sufficient in general, even if problem (BP) is a convex optimization problem.

To formulate necessary and sufficient optimality conditions for problem (BP), this needs to be transformed using the optimal function value of problem (20). Then, using tools from cone-convex optimization optimality conditions of Karush–Kuhn–Tucker type can be developed provided some weak constraint qualification is satisfied.

The results presented again show that the reformulation of the bilevel programming problem using the optimal value function for the lower-level problem is more promising than using the (necessary and sufficient) optimality conditions of the lower-level problem itself, see also [4, 10].

Acknowledgements We thank the two anonymous referees whose constructive suggestions have improved the presentation of the chapter. The second author, N. Dinh, expresses his sincere thanks to Vietnam National University, HCM city, and to NAFOSTED, Vietnam, for their support.

References

1. R. S. Burachik and V. Jeyakumar, A dual condition for the convex subdifferential sum formula with applications. *Journal of Convex Analysis* **12** (2005), 279–290.
2. S. Dempe, *Foundations of Bilevel Programming*, Kluwer, Dordrecht, 2000.
3. S. Dempe, J. Dutta, and S. Lohse, Optimality conditions for bilevel programming problems. *Optimization* **55** (2006), 505–524.
4. S. Dempe, J. Dutta, and B. Mordukhovich, New necessary optimality conditions in optimistic bilevel programming. *Optimization* **56** (2007), 577–604.
5. N. Dinh, M.A. Goberna, M.A. López, and T.Q. Son, New Farkas-type results with applications to convex infinite programming. *ESAIM: Control, Optimisation and Calculus of Variations* **13** (2007), 580–597.
6. N. Dinh, B.S. Mordukhovich, and T.T.A. Nghia, Subdifferentials of value functions and optimality conditions for some classes of DC and bilevel infinite and semi-infinite programs. *Mathematical Programming* **123** (2010), 101–138.
7. N. Dinh, B.S. Mordukhovich, and T.T.A. Nghia, Qualification and optimality conditions for DC programs with infinite constraints. *Acta Mathematica Vietnamica* **34** (2009), 123–153.
8. N. Dinh, T.T.A. Nghia, and G. Vallet, A closedness condition and its applications to DC programs with convex constraints. *Optimization* **59**(4) (2010), 541–560.
9. N. Dinh, G. Vallet, and T.T.A. Nghia, Farkas-type results and duality for DC programs with convex constraints. *Journal of Convex Analysis* **15**(2) (2008), 235–262.
10. J. Dutta and S. Dempe, Bilevel programming with convex lower level problems. In: S. Dempe and V. Kalashnikov (eds.) Optimization with Multivalued Mappings, 51–71, *Springer Optim. Appl.*, 2, Springer, New York, 2006.

11. R. Henrion, B. Mordukhovich, and N.M. Nam, Second-order analysis of polyhedral systems in finite and infinite dimensions with application to robust stability of variational inequalities. SIAM Journal on Optimization **20**(5) (2010), 2199–2227.
12. V. Jeyakumar, N. Dinh, and G.M. Lee, A new closed cone constraint qualification for convex optimization, Applied Mathematics Report, UNSW, AMR04/8, 2004.
13. V. Jeyakumar, A.M. Rubinov, B.M. Glover, and Y. Ishizuka, Inequality systems and global optimization. *Journal of Mathematical Analysis and Applications* **202**(3) (1996), 900–919.
14. R.T. Rockafellar and R.J.-B. Wets, Variational Analysis, Springer, Berlin, 1998.
15. M.V. Solodov, An explicit descent method for bilevel convex optimization. *Journal of Convex Analysis* **14**(2) (2007), 227–237.
16. H.v. Stackelberg, Marktform und Gleichgewicht, Springer, Berlin, 1934 (Engl. transl.: The Theory of the Market Economy, Oxford University Press, Oxford, 1952).

Smooth Representations of Optimal Solution Sets of Piecewise Linear Parametric Multiobjective Programs

Ya Ping Fang and Xiao Qi Yang

Abstract In this chapter, we investigate the smooth representation of the (weakly) efficient solution set of a piecewise linear parametric multiobjective program. We show that if the data of a piecewise linear multiobjective program are smooth functions of a parameter then the (weakly) efficient solution set of the problem can be locally represented as a union of finitely many polyhedra whose vertices and directions are smooth functions of the parameter.

1 Introduction

Multiobjective programs have been extensively studied and applied to various decision-making problems in economics, management science, and engineering (see, e.g., [1, 2, 5–7, 10, 13, 14, 24, 25, 27–29]). One of the most important topics in multiobjective programs is the study of the structure of optimal solution sets (see, e.g., [1, 5, 10, 14, 25–27, 29]). Arrow, Barankin, and Blackwell, in their pioneering paper [1], proved that if the feasible set and the ordering cone are polyhedral, respectively, then the (weakly) efficient solution set of a linear multiobjective program in the setting of finite-dimensional spaces is connected and the union of finitely many polyhedra. This result has been known as the ABB Theorem. Recently, the ABB Theorem has been generalized to the piecewise linear case. Zheng and Yang [27] proved that if the objective function is a piecewise linear function between two normed spaces and if the ordering cone and feasible set are polyhedral then the set

Ya Ping Fang
Department of Mathematics, Sichuan University, Chengdu, Sichuan, China
e-mail: fabhcn@yahoo.com.cn

Xiao Qi Yang
Department of Applied Mathematics, The Hong Kong Polytechnic University, Hung Hum, Kowloon, Hong Kong
e-mail: mayangxq@polyu.edu.hk

R.S. Burachik and J.-C. Yao (eds.), *Variational Analysis and Generalized Differentiation in Optimization and Control*, Springer Optimization and Its Applications 47,
DOI 10.1007/978-1-4419-0437-9_8,

of all weak Pareto solutions is the union of finitely many polyhedra and that if the objective function is a convex piecewise linear function, the feasible set is polyhedral and the ordering cone is a convex and closed cone, not necessarily a polyhedra one, then the set of all weak Pareto solutions is connected and the union of finitely many polyhedra. Yang and Yen [25] further generalized Zheng and Yang's result to the efficient solution set case. Yang and Yen also [25] showed that the set of all efficient solutions of a nonconvex piecewise linear multiobjective program is not necessarily the union of finitely many polyhedra, but the union of finitely many semiclosed polyhedra. Very recently, Fang and Yang [8] showed that the (weakly) efficient solution set of a discontinuously piecewise linear multiobjective program is the union of finitely many semiclosed polyhedra. For more works on the extensions on the ABB Theorem, we refer the reader to [29] and the references therein.

Another important topic in multiobjective programs is the study of stability. When the data of a multiobjective program depends on a parameter, we obtain a parametric multiobjective program. The stability problem of multiobjective programs deals with the continuity properties of their solution maps. There exists an extensive literature devoted to the study of stability in multiobjective programs. For details, we refer the reader to [2,13,17,18,20,23] and the references therein. Sensitivity analysis arising in single-objective programs deals with the differentiability of the solution map and has been extensively studied (see, e.g., [4,9,11,12]). Different from the single-objective case, the study of sensitivity in multiobjective programs is very limited because its solution map is set-valued in general. The techniques developed for single-objective programs cannot be applied to the study of the sensitivity of multiobjective programs. Up to now, there exist two approaches to study the sensitivity of multiobjective programs. The first approach is based on the use of generalized derivatives for set-valued maps (see, e.g., [3, 12, 19, 21]). The second approach, developed for the linear case in the works [15,16], is based on the smooth representation of a parametric polyhedron by vertices and recession directions. Using the second approach, Thuan and Luc [24] proved that if the data of a linear multiobjective program are smooth functions of a parameter then the (weakly) efficient solution set of the problem can be locally represented as a union of finitely many polyhedra whose vertices and directions are smooth functions of the parameter. Thuan and Luc's result is an interesting extension of the ABB Theorem.

The purpose of this chapter is to extend Thuan and Luc's result to the piecewise linear case. We shall show that if the data of a convex piecewise linear multiobjective program are smooth functions of a parameter then the (weakly) efficient solution set of the problem can be locally represented as a union of finitely many polyhedra whose vertices and directions are smooth functions of the parameter. The convexity requirement plays an important role here as the establishment of the main results will use a linear scalarization approach. Our results can be regarded as extensions of the corresponding results in [24,25,27].

2 Preliminaries

Let C be a closed, convex and pointed cone of $\mathbb{R}^m$ with $int\,C \neq \emptyset$, where $int\,C$ denotes the topological interior of C. Denote by C^* the dual cone of C, i.e.,

$$C^* = \{\xi \in \mathbb{R}^m : \langle \xi, x \rangle \geq 0, \quad \forall x \in C\}.$$

Given a nonempty set $A \subset \mathbb{R}^m$, a point $a \in A$ is called an *efficient point* of A (with respect to C) if there is no $a' \in A$ such that $a - a' \in C \backslash \{0\}$. Similarly, a point $a \in A$ is called a *weakly efficient point* of A (with respect to C) if there is no $a' \in A$ such that $a - a' \in int\,C$. The sets of all efficient points and all weakly efficient points of A are denoted by $MinA$ and $WMinA$, respectively.

Definition 1. A subset P of $\mathbb{R}^n$ is called a *polyhedron* if it is the intersection of finitely many closed half-spaces, i.e., $\exists \{x_1^*, x_2^*, \ldots, x_l^*\} \subset \mathbb{R}^n$ and $\{c_1, c_2, \ldots, c_l\} \subset \mathbb{R}$ such that

$$P = \{x \in \mathbb{R}^n : \langle x_i^*, x \rangle \leq c_i, i = 1, 2, \ldots, l\}.$$

For the definitions of face, extreme point and extreme direction of a polyhedron, please refer to [22].

The following lemma is important to establish our main results, whose proof can be found in [13, 29].

Lemma 1. *Let $A = \cup_{i=1}^l A_i$ with A_i being a polyhedron in $\mathbb{R}^m$ and C be a pointed, closed, and convex cone of $\mathbb{R}^m$ with $int\,C \neq \emptyset$. If $A + C$ is convex, then the following conclusions hold:*

(i) $x \in MinA$ (resp. $WMinA$) if and only if there exists $\xi \in int\,C^*$ (resp. $C^* \backslash \{0\}$) such that

$$x \in L_\xi(A) := \left\{ y \in A : \xi(y) = \min_{y' \in A} \xi(y') \right\}.$$

(ii) There exist $\Lambda_1, \Lambda_2, \ldots, \Lambda_l$ with $\Lambda_i \subset int\,C^*$ (resp. $C^* \backslash \{0\}$) being a finite set such that

$$MinA \ (resp.\ WMinA) = \cup_{i=1}^l \left[\cup_{\xi \in \Lambda_i} A_i \cap L_\xi(A) \right].$$

(iii) $MinA$ (resp. $WMinA$) is a union of a finite family consisting of some faces of $A_i, i = 1, 2, \ldots, l$.

Lemma 2. *Let $A = \cup_{i=1}^l A_i$ with A_i being a polyhedron in $\mathbb{R}^m$, C be a pointed, closed, and convex cone of $\mathbb{R}^m$ with $int\,C \neq \emptyset$ and $A + C$ be convex. Suppose that F_i is a face of A_i. Then, $F_i \subset MinA$ if and only if there exists $\hat{x} \in ri\,F_i$ such that $\hat{x} \in MinA$, where ri denotes the relative interior of a set.*

Proof. If $F_i \subset MinA$, then every point of its relative interior is an efficient one of A by definition.

Conversely, suppose that there exists $\hat{x} \in riF_i$ such that $\hat{x} \in MinA$. By (i) of Lemma 1, there exists $\xi \in int\,C^*$ such that $\hat{x} \in L_\xi(A)$. Then,

$$\xi(x) \geq \xi(\hat{x}), \quad \forall x \in F_i.$$

It is clear that $\hat{x}$ belongs to the face of the convex set $A+C$, which is determined by the hyperplane $\xi(x) = \xi(\hat{x})$. As $\hat{x}$ is a relative interior point of $F_i \subseteq A+C$, we deduce that F_i lies in that hyperplane too. In other words, every point of F_i minimizes $\xi(\cdot)$ on $A+C$, hence on A as well. By Lemma 1, it is efficient. □

Given a nonempty set $X \subset \mathbb{R}^n$ and a vector-valued function $f : \mathbb{R}^n \to \mathbb{R}^m$, the multiobjective program,

$$\text{(MP)} \qquad C - \min_{x \in X} f(x)$$

consists of finding a point $x_0 \in X$ such that $f(x_0)$ is an efficient point or a weakly efficient point of $f(X)$ (with respect to C). Such x_0 is called an *efficient solution* or a *weakly efficient solution* of (MP). The sets of all efficient solutions and all weakly efficient solutions of (MP) are denoted by $S(f,X)$ and $WS(f,X)$, respectively.

Definition 2. See [25, 27, 28]. A function $f : \mathbb{R}^n \to \mathbb{R}^m$ is said to be *piecewise linear* if there exist polyhedra $P_1, P_2, \ldots, P_l$ in $\mathbb{R}^n$, matrices $T_1, T_2, \ldots, T_l$ in $\mathbb{R}^{m\times n}$ and vectors $b_1, b_2, \ldots, b_l$ in $\mathbb{R}^m$ such that

$$\mathbb{R}^n = \cup_{i=1}^l P_i \quad \text{and} \quad f(x) = T_i x + b_i, \quad \forall x \in P_i \text{ and } 1 \leq i \leq l. \tag{1}$$

Definition 3. Let C be a pointed, closed, and convex cone of $\mathbb{R}^m$ and $f : \mathbb{R}^n \to \mathbb{R}^m$. f is said to be *C-convex* if

$$f(tx_1 + (1-t)x_2) - tf(x_1) - (1-t)f(x_2) \in -C, \quad \forall t \in [0,1] \text{ and } x_1, x_2 \in \mathbb{R}^n.$$

3 Sensitivity Analysis

In this section, we study the sensitivity of piecewise linear multiobjective programs by using the approach developed in [15, 16, 24]. First we recall some results on parametric polyhedra.

Set the parametric polyhedron as follows:

$$N(\omega) = \{x \in \mathbb{R}^m : \langle a_i(\omega), x\rangle \leq c_i(\omega), i = 1, 2, \ldots, \tau\}, \tag{2}$$

where a_i and c_i are functions of the parameter ω on $\mathbb{R}^r$. For convenience, we always say that $N(\omega)$ defined by (2) is of *class* C^d if a_i and c_i are of class C^d for all i.

Lemma 3 ([15]). *Let $N(\omega)$ be a parametric polyhedron of class C^d. Then, for any open set $W \subset \mathbb{R}^r$, there exists an open nonempty set $W_0 \subset W$ such that either*

$$N(\omega) = \emptyset, \quad \forall \omega \in W_0$$

or there exist functions $v_1, v_2, \ldots, v_s$ *of class* C^d *and integer* $0 < k \leq s$ *such that, for any* $w \in W_0$,

$$N(\omega) = \left\{ x \in \mathbb{R}^m : x = \sum_{i=1}^{s} \lambda_i v_i(\omega), \sum_{i=1}^{k} \lambda_i = 1, \lambda_i \geq 0, i = 1, 2, \ldots, s \right\}.$$

Lemma 4 ([24]). *Let* $N(\omega)$ *be of class* C^d. *Then, for any open set* $W \subset \mathbb{R}^r$, *there exist an open nonempty set* $W_0 \subset W$, *an integer* $\kappa \geq 1$ *and* 2κ *finite index sets* $I_1, \ldots, I_\kappa$ *and* $J_1, \ldots, J_\kappa$, 2κ *families* $\mathscr{U}_1, \ldots, \mathscr{U}_\kappa$ *and* $\mathscr{D}_1, \ldots, \mathscr{D}_\kappa$ *with* $\mathscr{U}_i = \{u_k \in C^d : k \in I_i\}$ *and* $\mathscr{D}_i = \{d_{k'} \in C^d : k' \in J_i\}$ *such that for any* $\omega \in W_0$,

$$N(\omega) = \cup_{i=1}^{\kappa} N_i(\omega),$$

where

$$N_i(\omega) = co\{u_k(\omega) : k \in I_i\} + cone\{d_{k'}(\omega) : k' \in J_i\}$$

is the face of $N(\omega)$, $u_k(\omega)$ *is the vertex of* $N(\omega)$ *and* $d_{k'}(\omega)$ *is the extreme direction of* $N(\omega)$.

As a direct consequence of Lemma 4, we have

Lemma 5. *Let* $A(\omega) = \cup_{i=1}^{l} A_i(\omega)$ *with* $A_i(\omega)$ *being a parametric polyhedron of class* C^d. *Then, for any set* $W \subset \mathbb{R}^r$, *there exist an open nonempty subset* $W_0 \subset W$, *index sets* I_j^i *and* J_j^i, $j = 1, \ldots, \kappa_i$ *(each* κ_i *is a natural number);* $i = 1, \ldots, l$ *and families* $\mathscr{U}_j^i$ *and* $\mathscr{D}_j^i$ *with* $\mathscr{U}_j^i = \{u_k(\omega) \in C^d : k \in I_j^i\}$ *and* $\mathscr{D}_j^i = \{d_{k'}(\omega) \in C^d : k' \in J_j^i\}$ *such that for any* $\omega \in W_0$,

$$A(\omega) = \cup_{i=1}^{l} \cup_{j=1}^{\kappa_i} F_j^i(\omega),$$

where

$$F_j^i(\omega) = co\{u_k(\omega) : k \in I_j^i\} + cone\{d_{k'}(\omega) : k' \in J_j^i\}$$

is the face of $A_i(\omega)$, $u_k(\omega)$ *with* $k \in I_j^i$ *is the vertex of* $A_i(\omega)$, *and* $d_{k'}(\omega)$ *with* $k' \in J_j^i$ *is the extreme direction of* $A_i(\omega)$.

Lemma 6. *Let* $A(\omega), W_0$ *and* $F_j^i(\omega)$ *be as in Lemma 5. Suppose that* $A(\omega) + C$ *is convex for all* $\omega \in \mathbb{R}^r$ *and* $F_j^i(\omega_0) \subset MinA(\omega_0)$ *for some* $\omega_0 \in W_0$. *Then there exists an open neighborhood* $W_1 \subset W_0$ *of* ω_0 *such that*

$$F_j^i(\omega) \subset MinA(\omega), \quad \forall \omega \in W_1.$$

Proof. Let $\hat{x}_0 \in ri F_j^i(\omega_0)$. Then $\hat{x}_0 \in MinA(\omega_0)$ since $F_j^i(\omega_0) \subset MinA(\omega_0)$. By the Minkowski's theorem [7], there exist $\{\lambda_k > 0 : k \in I_j^i\}$ with $\sum_{k \in I_j^i} \lambda_k = 1$ and $\{t_{k'} > 0 : k' \in J_j^i\}$ such that

$$\hat{x}_0 = \sum_{k \in I_j^i} \lambda_k u_k(\omega_0) + \sum_{k' \in J_j^i} t_{k'} d_{k'}(\omega_0).$$

Define

$$\hat{x}(\omega) = \sum_{k \in I_j^i} \lambda_k u_k(\omega) + \sum_{k' \in J_j^i} t_{k'} d_{k'}(\omega).$$

Clearly, $\hat{x}(\omega)$ is of class C^d and $\hat{x}(\omega_0) = \hat{x}_0$.

Set

$$I = \cup_{i'=1}^{l} \cup_{j'=1}^{\kappa_{i'}} I_{j'}^{i'} \quad \text{and} \quad J = \cup_{i'=1}^{l} \cup_{j'=1}^{\kappa_{i'}} J_{j'}^{i'}.$$

By Lemma 1, there exists $\xi_0 \in int\, C^*$ such that

$$\hat{x}_0 \in A_i(\omega_0) \cap L_{\xi_0}(A(\omega_0)).$$

Take

$$\hat{I} = \{k \in I : \langle \xi_0, u_k(\omega_0)\rangle = \langle \xi_0, \hat{x}_0\rangle\} \quad \text{and} \quad \hat{J} = \left\{k' \in J : \langle \xi_0, d_{k'}(\omega_0)\rangle = 0\right\}.$$

Clearly, $I_j^i \subset \hat{I}$ and $J_j^i \subset \hat{J}$. Consider the following two systems for $w \in \mathbb{R}^r$:

$$\begin{cases} \langle \xi, u_k(\omega) - \hat{x}(\omega)\rangle = 0, & k \in \hat{I}, \\ \langle \xi, d_{k'}(\omega)\rangle = 0, & k' \in \hat{J}, \end{cases} \tag{3}$$

and

$$\begin{cases} \langle \xi, u_k(\omega) - \hat{x}(\omega)\rangle > 0, & k \in I \backslash \hat{I}, \\ \langle \xi, d_{k'}(\omega)\rangle > 0, & k' \in J \backslash \hat{J}. \end{cases} \tag{4}$$

Then systems (3) and (4) have a common solution ξ_0 when $\omega = \omega_0$. By Lemma 3, there exist an open neighborhood $W_1 \subset W_0$ of ω_0, and functions $\xi_k(\omega)$, $k = 1, \ldots, p$, of C^d such that every solution of system (3) can be written in the form

$$\xi(\omega) = \sum_{k=1}^{p} \beta_k \xi_k(\omega), \quad \beta_k \geq 0,$$

for all $\omega \in W_1$. Since $\xi_0 = \xi(\omega_0) \in int\, C^*$, $\xi(\omega) \in int\, C^*$ when ω is sufficiently near to ω_0.

Since

$$\begin{cases} \langle \xi(\omega_0), u_k(\omega_0) - \hat{x}(\omega_0)\rangle > 0, & k \in I \backslash \hat{I}, \\ \langle \xi(\omega_0), d_{k'}(\omega_0)\rangle > 0, & k' \in J \backslash \hat{J}, \end{cases}$$

and $\xi, u_k, \hat{x}, d_{k'}$ are continuous, $\xi(\omega)$ is a solution of system (4) when ω is sufficiently near to ω_0.

So, without loss of generality, we can suppose that $\xi(\omega)$ is a common solution of systems (3) and (4), and $\xi(\omega) \in int\, C^*$ for all $\omega \in W_1$.

Next we prove that for any $\omega \in W_1$,

$$\hat{x}(\omega) \in L_{\xi(\omega)}(A(\omega)).$$

Assume to the contrary that there exists $x \in A(\omega)$ such that

$$\langle \xi(\omega), x\rangle < \langle \xi(\omega), \hat{x}(\omega)\rangle.$$

By the Minkowski's theorem [7], there exist index sets $I^{i_0}_{j_0}$ and $J^{i_0}_{j_0}$, $\{\lambda_k \geq 0 : k \in I^{i_0}_{j_0}\}$ with $\sum_{k\in I^{i_0}_{j_0}} \lambda_k = 1$ and $\{t_{k'} \geq 0 : k' \in J^{i_0}_{j_0}\}$ such that

$$x = \sum_{k\in I^{i_0}_{j_0}} \lambda_k u_k(\omega) + \sum_{k'\in J^{i_0}_{j_0}} t_{k'} d_{k'}(\omega).$$

It follows that

$$\begin{aligned}\langle \xi(\omega), x\rangle &= \sum_{k\in I^{i_0}_{j_0}} \lambda_k \langle \xi(\omega), u_k(\omega)\rangle + \sum_{k'\in J^{i_0}_{j_0}} t_{k'} \langle \xi(\omega), d_{k'}(\omega)\rangle \\ &\geq \sum_{k\in I^{i_0}_{j_0}} \lambda_k \langle \xi(\omega), \hat{x}(\omega)\rangle = \langle \xi(\omega), \hat{x}(\omega)\rangle,\end{aligned}$$

a contradiction. So $\hat{x}(\omega) \in L_{\xi(\omega)}(A(\omega))$ for all $\omega \in W_1$. By Lemma 1, $\hat{x}(\omega) \in MinA(\omega)$. Since $\hat{x}(\omega) \in ri F^i_j(\omega)$, it follows from Lemma 2 that $F^i_j(\omega) \subset MinA(\omega)$ for all $\omega \in W_1$. □

Lemma 7. *Let $A(\omega), W_0 \subset W$ and $F^i_j(\omega)$ be as in Lemma 5. Suppose that $A(\omega)+C$ is convex for all $\omega \in \mathbb{R}^r$ and $F^i_j(\omega_0) \not\subset WMinA(\omega_0)$ for some $\omega_0 \in W_0$. Then there exists an open neighborhood $W_1 \subset W_0$ of ω_0 such that*

$$F^i_j(\omega) \not\subset WMinA(\omega), \quad \forall \omega \in W_1.$$

Proof. Since $F^i_j(\omega_0) \not\subset WMinA(\omega_0)$, there exists $x_0 \in F^i_j(\omega_0)$ and $x \in A(\omega_0)$ such that $x_0 - x \in int C$. By Lemma 5, there exists an open neighborhood $W_1' \subset W_0$ of ω_0 such that for any $\omega \in W_1'$,

$$F^i_j(\omega) = co\{u_k(\omega) : k \in I^i_j\} + cone\{d_{k'}(\omega) : k' \in J^i_j\}$$

and

$$A(\omega) = \cup^{l}_{i'=1} \cup^{K_{i'}}_{j'=1} F^{i'}_{j'}(\omega),$$

where $u_k(\omega)$ and $d_{k'}(\omega)$ are functions defined as in Lemma 5. Since $x_0 \in F^i_j(\omega_0)$, there exist $\{\lambda_k \geq 0 : k \in I^i_j\}$ with $\sum_{k\in I^i_j} \lambda_k = 1$ and $\{t_{k'} \geq 0 : k' \in J^i_j\}$ such that

$$x_0 = \sum_{k\in I^i_j} \lambda_k u_k(\omega_0) + \sum_{k'\in J^i_j} t_{k'} d_{k'}(\omega_0).$$

Define

$$x_0(\omega) = \sum_{k\in I^i_j} \lambda_k u_k(\omega) + \sum_{k'\in J^i_j} t_{k'} d_{k'}(\omega), \quad \forall \omega \in \mathbb{R}^r.$$

Then $x_0(\omega)$ is of class C^d such that for any $\omega \in W_1'$,

$$x_0(\omega) \in F_j^i(\omega) \quad \text{and} \quad x_0(\omega_0) = x_0.$$

Since $x \in A(\omega_0)$ and

$$A(\omega) = \cup_{i'=1}^{l} \cup_{j'=1}^{\kappa_{i'}} F_{j'}^{i'}(\omega),$$

by same arguments we can find a function $x(\omega)$ of class C^d such that for any $\omega \in W_1'$,

$$x(\omega) \in A(\omega) \quad \text{and} \quad x(\omega_0) = x.$$

It follows that $\lim_{\omega \to \omega_0} x_0(\omega) = x_0$ and $\lim_{\omega \to \omega_0} x(\omega) = x$. Since $x_0 - x \in int C$, $x_0(\omega) - x(\omega) \in int C$ when ω is sufficiently near to ω_0. Therefore, there exists an open neighborhood $W_1 \subset W_0$ of ω_0 such that $F_j^i(\omega) \not\subset WMinA(\omega)$ for all $\omega \in W_1$. □

Remark 1. The conclusion of Lemma 6 is not true for the weakly efficient point set, while the conclusion of Lemma 7 is not true for the efficient point set. This has been pointed out by Thuan and Luc [24] in the linear case.

With the above lemmas in hand, we are ready to establish the main results of this chapter. In the case where the feasible set X and the objective function f of (MP) depend on a parameter ω in $\mathbb{R}^r$, we obtain the following parametric multiobjective program:

$$(\text{MP})_\omega \qquad C\text{-}\min_{x \in X(\omega)} f(\omega, x).$$

In the sequel we need the following assumption.

Hypothesis (H): Let $f : \mathbb{R}^r \times \mathbb{R}^n \to \mathbb{R}^m$ be such that:

(i) There exist parametric polyhedra $P_i(\omega), i = 1, \ldots, l$, of class C^d such that

$$\cup_{i=1}^{l} P_i(\omega) = \mathbb{R}^n, \quad \forall \omega \in \mathbb{R}^r$$

and

$$f(\omega, x) = T_i(\omega)x + b_i(\omega), \quad \forall x \in P_i(\omega) \text{ and } i \in \{1, 2, \ldots, l\},$$

where $T_i(\omega) \in \mathbb{R}^{m \times n}$ is a matrix with its entries being of class C^d, and $b_i(\omega) \in \mathbb{R}^m$ is of class C^d.

(ii) $f(\omega, \cdot) : \mathbb{R}^n \to \mathbb{R}^m$ is C-convex for all $\omega \in \mathbb{R}^r$.

(iii) $X(\omega) \subset \mathbb{R}^n$ is a parametric polyhedron of class C^d on $\mathbb{R}^r$.

Remark 2. Assume that Hypothesis (H) holds. It follows from Lemma 2.2 of [27] that $(\text{MP})_\omega$ is equivalent to the following program:

$$\begin{aligned} &C\text{-}\min\ y \\ &\quad s.t.\ x \in X(\omega), \\ &\qquad T_i(\omega)x + b_i(\omega) - y \in -C, \quad 1 \le i \le l. \end{aligned}$$

In addition, if C is a polyhedral cone, i.e., $C = \{y \in \mathbb{R}^m : Dy \in -\mathbb{R}^k\}$ for some $k \times m$-matrix D, then $(\mathrm{MP})_\omega$ collapses to the following linear parametric multiobjective program:

$$\begin{aligned} &C\text{-}\min\ y \\ &\quad s.t.\ \ x \in X(\omega), \\ &\qquad\quad D(T_i(\omega)x + b_i(\omega)) - Dy \in -\mathbb{R}^k, \quad 1 \le i \le l. \end{aligned}$$

The sensitivity analysis of the above linear parametric multiobjective program has been investigated in [24].

Theorem 1. *Assume that Hypothesis (H) holds and* $V(\omega) = Min\, f(\omega, X(\omega))$. *Then, for any open nonempty set* $W \subset \mathbb{R}^r$, *there exist an open set* $W_0 \subset W$, *two finite index sets* I *and* J, *and two families* $\{u_k(\omega) \in C^d : k \in I\}$ *and* $\{d_{k'}(\omega) \in C^d : k' \in J\}$ *such that:*

(i) Either $V(\omega) = \emptyset$ *for all* $\omega \in W_0$;
(ii) Or there exist an integer $l_* \ge 1$ *and* $2l_*$ *index sets,*

$$I_1, \ldots, I_{l_*} \subset I, \quad J_1, \ldots, J_{l_*} \subset J$$

such that for any $\omega \in W_0$,

$$V(\omega) = \cup_{j=1}^{l_*} N_j(\omega),$$

where

$$N_j(\omega) = co\{u_k(\omega) : k \in I_j\} + cone\{d_{k'}(\omega) : k' \in J_j\}$$

is a face of the set $T_i(\omega)(X(\omega) \cap P_i(\omega)) + b_i(\omega)$ *for some* $i \in \{1, \ldots, l\}$.

Proof. Set

$$A(\omega) = f(\omega, X(\omega)), \quad A_i(\omega) = T_i(w)(X(\omega) \cap P_i(\omega)) + b_i(\omega), i = 1, \ldots, l.$$

Then each $A_i(\omega)$ is a parametric polyhedron of class C^d and

$$A(\omega) = \cup_{i=1}^{l} A_i(\omega).$$

By Lemma 5, there exist an open set $W_1 \subset W$, index sets I_j^i and J_j^i, $j = 1, \ldots, \kappa_i$; $i = 1, \ldots, l$ and families $\mathscr{U}_j^i$ and $\mathscr{D}_j^i$ with $\mathscr{U}_j^i = \{u_k(\omega) \in C^d : k \in I_j^i\}$ and $\mathscr{D}_j^i = \{d_{k'}(\omega) \in C^d : k' \in J_j^i\}$ such that for any $\omega \in W_1$,

$$A(\omega) = \cup_{i=1}^{l} \cup_{j=1}^{\kappa_i} F_j^i(\omega),$$

where

$$F_j^i(\omega) = co\{u_k(\omega) : k \in I_j^i\} + cone\{d_{k'}(\omega) : k' \in J_j^i\}$$

is the face of $A_i(\omega)$, $u_k(\omega)$ with $k \in I_j^i$ is the vertex of $A_i(\omega)$, and $d_{k'}(\omega)$ with $k' \in J_j^i$ is the extreme direction of $A_i(\omega)$.

If $V(\omega) = \emptyset$ for all $\omega \in W_1$, the proof is completed. Now we suppose that

$$V(\omega_0) \neq \emptyset, \qquad \text{for some } \omega_0 \in W_1.$$

Since $f(\omega_0, \cdot)$ is C-convex, $A(\omega_0) + C$ is convex. By (iii) of Lemma 1, $V(\omega_0)$ is a union of a finite family consisting of some faces of $A_i(\omega_0), i = 1, \dots, l$. By Lemma 6, there exists an open neighborhood $W_2 \subset W_1$ of ω_0 such that

$$V(\omega) \neq \emptyset, \quad \forall \omega \in W_2.$$

For any $\omega \in W_2$, let $l(\omega)$ be the number of the sets $F_j^i(\omega)$ with $F_j^i(\omega) \subset V(\omega)$. It is easy to see that there exists $\omega_* \in W_2$ such that

$$l_* := l(\omega_*) = \max_{\omega \in W_2} l(\omega).$$

Let $N_1(\omega_*), \dots, N_{l_*}(\omega_*)$ be the sets $F_j^i(\omega_*)$ which are contained in $V(\omega_*)$, i.e.,

$$V(\omega_*) = \cup_{j=1}^{l_*} N_j(\omega_*).$$

By Lemma 6, there exists an open neighborhood $W_0 \subset W_2$ of ω_* such that

$$\cup_{j=1}^{l_*} N_j(\omega) \subset V(\omega), \quad \forall \omega \in W_0.$$

By the definition of l_*,

$$\cup_{j=1}^{l_*} N_j(\omega) = V(\omega), \quad \forall \omega \in W_0.$$

□

Corollary 1. *Assume that Hypothesis (H) holds and $S(\omega) = \{x \in X(\omega) : f(\omega, x) \in V(\omega)\}$. Then, for any open set $W \subset \mathbb{R}^r$, there exist an open set $W_0 \subset W$, two index sets $\bar{I}$ and $\bar{J}$, and two families $\{v_k(\omega) \in C^d : k \in \bar{I}\}$ and $\{h_{k'}(\omega) \in C^d : k' \in \bar{J}\}$ such that:*

(i) Either $S(\omega) = \emptyset$ for all $\omega \in W_0$;
(ii) Or there exist an integer $l^* \geq 1$ and $2l^*$ index sets,

$$I_1, \dots, I_{l^*} \subset \bar{I}, \quad J_1, \dots, J_{l^*} \subset \bar{J}$$

such that for any $\omega \in W_0$,

$$S(\omega) = \cup_{j=1}^{l^*} F_j(\omega),$$

where

$$F_j(\omega) = co\{v_k(\omega) : k \in I_j\} + cone\{h_{k'}(\omega) : k' \in J_j\}.$$

Proof. By Theorem 1, there exist an open set $W_0 \subset W$, two index sets I and J, and two families $\{u_k(\omega) \in C^d : k \in I\}$ and $\{d_{k'}(\omega) \in C^d : k' \in J\}$ such that:

(a) Either $V(\omega) = \emptyset$ for all $\omega \in W_0$;
(b) Or there exist a number $l_* \geq 1$ and $2l_*$ index sets,

$$I_1, \dots, I_{l_*} \subset I, \quad J_1, \dots, J_{l_*} \subset J$$

such that for any $\omega \in W_0$,

$$V(\omega) = \cup_{j=1}^{l_*} N_j(\omega),$$

where

$$N_j(\omega) = co\{u_k(\omega) : k \in I_j\} + cone\{u_{k'}(\omega) : k' \in J_j\}.$$

Conclusion (i) holds trivially if (a) holds. So we suppose that (b) holds. By Hypothesis (H),

$$\begin{aligned} S(\omega) &= X(\omega) \cap f^{-1}(\omega, \cdot)(V(\omega)) \\ &= X(\omega) \cap f^{-1}(\omega, \cdot)(\cup_{j=1}^{l_*} N_j(\omega)) \\ &= X(\omega) \cap [\cup_{i=1}^{l} P_i(\omega) \cap f^{-1}(\omega, \cdot)(\cup_{j=1}^{l_*} N_j(\omega))] \\ &= \cup_{i=1}^{l} \cup_{j=1}^{l_*} [X(\omega) \cap P_i(\omega) \cap T_i^{-1}(\omega, \cdot)(N_j(\omega) - b_i(\omega))], \quad \forall \omega \in W_0. \end{aligned} \tag{5}$$

It is easy to verify that each $X(\omega) \cap P_i(\omega) \cap T_i^{-1}(\omega, \cdot)(N_j(\omega) - b_i(\omega))$ is a parametric polyhedron of class C^d. Therefore, the conclusion follows directly from (5) and Lemma 5. □

Theorem 2. *Assume that Hypothesis (H) holds and $V^w(\omega) = WMin\, f(\omega, X(\omega))$. Then, for any open set $W \subset \mathbb{R}^r$, there exist an open set $W_0 \subset W$, two index sets I and J, and two families $\{u_k(\omega) \in C^d : k \in I\}$ and $\{d_{k'}(\omega) \in C^d : k' \in J\}$ such that:*

(i) Either $V^w(\omega) = \emptyset$ for all $\omega \in W_0$;
(ii) Or there exist a number $l_* \geq 1$ and $2l_*$ index sets,

$$I_1, \dots, I_{l_*} \subset I, \quad J_1, \dots, J_{l_*} \subset J$$

such that for any $\omega \in W_0$,

$$V^w(\omega) = \cup_{j=1}^{l_*} N_j(\omega),$$

where

$$N_j(\omega) = co\{u_k(\omega) : k \in I_j\} + cone\{u_{k'}(\omega) : k' \in J_j\}$$

is a face of the set $T_i(\omega)(X(\omega) \cap P_i(\omega)) + b_i(\omega)$ for some $i \in \{1, \dots, l\}$.

Proof. Let $A(\omega)$ and $A_i(\omega)$ be the same as in the proof of Theorem 1. By the same arguments, we can find an open set $W_1 \subset W$, index sets I_j^i and J_j^i, $j = 1, \dots, \kappa_i$; $i = 1, \dots, l$ and families $\mathscr{U}_j^i$ and $\mathscr{D}_j^i$ with $\mathscr{U}_j^i = \{u_k(\omega) \in C^d : k \in I_j^i\}$ and $\mathscr{D}_j^i = \{d_k(\omega) \in C^d : k \in J_j^i\}$ such that for any $\omega \in W_1$,

$$A(\omega) = \cup_{i=1}^{l} \cup_{j=1}^{\kappa_i} F_j^i(\omega), \tag{6}$$

where

$$F_j^i(\omega) = co\{u_k(\omega) : k \in I_j^i\} + cone\{d_{k'}(\omega) : k' \in J_j^i\}$$

is the face of $A_i(\omega)$, $u_k(\omega)$ with $k \in I_j^i$ is the vertex of $A_i(\omega)$, and $d_{k'}(\omega)$ with $k' \in J_j^i$ is the extreme direction of $A_i(\omega)$.

If $V^w(\omega_0) = \emptyset$ for some $\omega_0 \in W_1$, then $V^w(\omega) = \emptyset$ for all ω sufficiently near to ω_0. So we can suppose that

$$V^w(\omega) \neq \emptyset, \quad \forall \omega \in W_1.$$

Since $f(\omega, \cdot)$ is C-convex, $A(\omega) + C$ is convex. By (iii) of Lemma 1, $V^w(\omega_0)$ is a union of a finite family consisting of some faces $F_j^i(\omega_0)$ of $A_i(\omega_0), i = 1, \ldots, l$. Let s_* denote the number of the sets $F_j^i(\omega)$ presented in the right set of (6). For any $\omega \in W_1$, let $l(\omega)$ be the number of all the sets $F_j^i(\omega)$ contained in $V^w(\omega)$. Then there exists $\omega_* \in W_1$ such that

$$s_* - l_* := s_* - l(\omega_*) = \max\{s_* - l(\omega) : \omega \in W_1\}.$$

Let $\hat{N}_1(\omega_*), \ldots, \hat{N}_{s_*-l_*}(\omega_*)$ be the sets $F_j^i(\omega_*)$ which are not contained in $V^w(\omega_*)$, i.e.,

$$\hat{N}_j(\omega_*) \not\subset V^w(\omega_*), \quad j = 1, 2, \ldots, s_* - l_*.$$

By Lemma 7, there exists an open neighborhood $W_0 \subset W_1$ of ω_* such that for any $\omega \in W_0$,

$$\hat{N}_j(\omega) \not\subset V^w(\omega), \quad j = 1, 2, \ldots, s_* - l_*.$$

By the definition of $s_* - l_*$, $V^w(\omega)$ is a union of a family consisting of l_* faces of $A_i(\omega), i = 1, 2, \ldots, l$. □

Corollary 2. *Assume that Hypothesis (H) holds and $S^w(\omega) = \{x \in X(\omega) : f(\omega, x) \in V^w(\omega)\}$. Then, for any open set $W \subset \mathbb{R}^r$, there exist an open set $W_0 \subset W$, two index sets $\bar{I}$ and $\bar{J}$, and two families $\{v_k(\omega) \in C^d : k \in \bar{I}\}$ and $\{h_{k'}(\omega) \in C^d : k' \in \bar{J}\}$ such that:*

(i) Either $S^w(\omega) = \emptyset$ for all $\omega \in W_0$;
(ii) Or there exist an integer $l^* \geq 1$ and $2l^*$ index set,

$$I_1, \ldots, I_{l^*} \subset \bar{I}, \quad J_1, \ldots, J_{l^*} \subset \bar{J}$$

such that for any $\omega \in W_0$,

$$S^w(\omega) = \cup_{j=1}^{l^*} F_j(\omega),$$

where

$$F_j(\omega) = co\{v_k(\omega) : k \in I_j\} + cone\{h_{k'}(\omega) : k' \in J_j\}.$$

Proof. The conclusion follows from Theorem 2 and the same argument as in the proof of Corollary 1. □

Acknowledgements This work was partially supported by the Research Grants Council of Hong Kong (PolyU 5317/07E) and the National Natural Science Foundation of China (10831009 and 11001187).

References

1. K.J. Arrow, E.W. Barankin and D. Blackwell, Admissible points of convex sets. In: *Contribution to the Theory of Games*, Edited by H.W. Kuhn and A.W. Tucker, Princeton University Press, Princeton, NJ (1953), pp. 87–92.
2. H. Attouch and H. Riahi, Stability results for Ekeland's ε-variational principle and cone extremal solutions. Math. Oper. Res. **18**(1), 173–201 (1993).
3. J.P. Aubin and H. Frankowska, *Set-Valued Analysis*, Birkhäuser, Boston, MA (1990).
4. B. Bank, J. Guddat, D. Klatte, B. Kummer, and K. Tammer, *Nonlinear Parametric Optimization*, Academic, Berlin (1982).
5. G.R. Bitran and T.L. Magnanti, The structure of admissible points with respect to cone dominance. J. Optim. Theory Appl. **29**(4), 573–614 (1979).
6. S. Deng and X.Q. Yang, Weak sharp minima in multicriteria programming. SIAM J. Optim. **15**, 456–460 (2004).
7. M. Ehrgott, *Multicriteria Optimization*, second edition, Springer, Berlin (2005).
8. Y.P. Fang and X.Q. Yang, Multiple objective and discontinuously piecewise linear programs (submitted).
9. A.V. Fiacco, *Introduction to Sensitivity and Stability Analysis in Nonlinear Programming*, Academic, New York (1983).
10. X.H. Gong, Connectedness of the efficient solution set of a convex vector optimization in normed spaces. Nonlinear Anal. **23**(9), 1105–1114 (1994).
11. E.S. Levitin, *Perturbation Theory in Mathematical Programming and Its Applications*, Wiley, Chichester (1994).
12. A.B. Levy and R.T. Rockafellar, Sensitivity analysis of solutions to generalized equations. Trans. Am. Math. Soc. **345**(2), 661–671 (1994).
13. D.T. Luc, *Theory of Vector Optimization*, Springer, Berlin (1989).
14. D.T. Luc, Contractibility of efficient point sets in normed spaces. Nonlinear Anal. **15**(6), 527–535 (1990).
15. D.T. Luc, Smooth representation of a parametric polyhedral convex set with application to sensitivity in optimization. Proc. Am. Math. Soc. **125**(2), 555–567 (1997).
16. D.T. Luc and P.H. Dien, Differentiable selection of optimal solutions in parametric linear programming. Proc. Am. Math. Soc. **125**(3), 883–892 (1997).
17. D.T. Luc, R. Lucchetti, and C. Maliverti, Convergence of the efficient sets. Set-Valued Anal. **2**(1–2), 207–218 (1994).
18. R.E. Lucchetti and E. Miglierina, Stability for convex vector optimization problems. Optimization **53**(5–6), 517–528 (2004).
19. B.S. Mordukhovich, Generalized differential calculus for nonsmooth and set-valued mappings. J. Math. Anal. Appl. **183**(1), 250–288 (1994).
20. P.H. Naccache, Stability in multicriteria optimization. J. Math. Anal. Appl. **68**(2), 441–453 (1979).
21. S.M. Robinson, An implicit-function theorem for a class of nonsmooth functions. Math. Oper. Res. **16**(2), 292–309 (1991).
22. R.T. Rockafellar, *Convex Analysis*. Princeton University Press, Princeton, NJ (1970).
23. T. Tanino and Y. Sawaragi, Stability of nondominated solutions in multicriteria decision-making. J. Optim. Theory Appl. **30**(2), 229–253 (1980).
24. L.V. Thuan and D.T. Luc, On sensitivity in linear multiobjective programming. J. Optim. Theory Appl. **107**(3), 615–626 (2000).

25. X.Q. Yang and N.D. Yen, Structure and weak sharp minimum of the pareto solution set for piecewise linear multiobjective optimization. J. Optim. Theory Appl. **147**(1), 113–124.
26. P.L. Yu and M. Zeleny, The set of all nondominated solutions in linear cases and a multicriteria simplex method. J. Math. Anal. Appl. **49**, 430–468 (1975).
27. X.Y. Zheng and X.Q. Yang, The structure of weak Pareto solution sets in piecewise linear multiobjective optimization in normed spaces. Sci. China Ser. A **51**(7), 1243–1256 (2008).
28. X.Y. Zheng and X.Q. Yang, Weak sharp minima for piecewise linear multiobjective optimization in normed spaces. Nonlinear Anal. **68**(12), 3771–3779 (2008).
29. X.Y. Zheng, Pareto solutions of polyhedral-valued vector optimization problems. Set-valued and Variational Analysis **17**, 389–408 (2009).

Global Optimality Conditions for Classes of Non-convex Multi-objective Quadratic Optimization Problems

V. Jeyakumar, G.M. Lee, and G. Li

Abstract We present necessary and sufficient conditions for identifying *global weak minimizers* of non-convex multi-objective quadratic optimization problems. We derive these results by exploiting the hidden convexity of the joint range of (non-convex) quadratic functions. We also present numerical examples to illustrate our results.

1 Introduction

Consider the following multi-objective non-convex quadratic optimization problem

$$\begin{aligned} \text{(MP)} \ \min \ & (f_1(x), \ldots, f_p(x)) \\ \text{s.t.} \ & g_j(x) \leq 0, \, j = 1, \ldots, m, \end{aligned}$$

where $f_i(x) = \frac{1}{2}x^T A_i x + a_i^T x$, $g_j(x) = \frac{1}{2}x^T B_j x + b_j^T x + c_j$, $A_i, B_j \in S^n$, the space of all $(n \times n)$ symmetric matrices, $a_i, b_j \in R^n$ and $c_j \in R$.

One of the fundamental studies in the area of multi-objective optimization is to develop dual conditions for identifying global (weak) minimizers of (MP) in terms of Lagrange multipliers. Such results in multi-objective optimization have useful economic interpretations (e.g., see Arrow [1]) and hence have attracted a great deal of researchers (see [4] for a comprehensive and excellent survey).

Over the years, significant advances have been made in identifying solutions of multi-objective *convex* optimization problems. However, the development of global

V. Jeyakumar, G. Li
Department of Applied Mathematics, University of New South Wales, Sydney 2052, Australia
e-mail: v.jeyakumar@unsw.edu.au, g.li@unsw.edu.au

G.M. Lee
Department of Applied Mathematics, Pukyong National University, Busan 608-737, Korea
e-mail: gmlee@pknu.ac.kr

R.S. Burachik and J.-C. Yao (eds.), *Variational Analysis and Generalized Differentiation in Optimization and Control*, Springer Optimization and Its Applications 47,
DOI 10.1007/978-1-4419-0437-9_9, © Springer Science+Business Media, LLC 2010

optimality conditions for non-convex quadratic optimization problems has so far been limited to some classes of *single-objective* quadratic optimization problems (see, e.g., [3, 7–9, 11] and the references therein). Notably, one of the main ingredient in developing such global optimality conditions for *single-objective* non-convex quadratic optimization problems has been the hidden convexity of the joint range of quadratic functions.

The purpose of this chapter is to derive Lagrange multiplier conditions for global weak minimizers of multi-objective optimization problems (MP). We derive necessary as well as sufficient Lagrange multiplier conditions for global weak minimizers of (MP) by first establishing joint range convexity results for systems of quadratic functions.

2 Joint-Range Convexity Conditions

We begin this section by fixing the notation and definitions that will be used throughout this chapter. The real line is denoted by R and the n-dimensional Euclidean space is denoted by R^n. The set of all non-negative vectors of R^n is denoted by R^n_+, and the interior of R^n_+ is denoted by $\text{int}R^n_+$. The space of all $(n \times n)$ symmetric matrices is denoted by S^n. The notation $A \succeq B$ means that the matrix $A - B$ is positive semidefinite. Moreover, the notation $A \succ B$ means the matrix $A - B$ is positive definite. The positive semidefinite cone is defined by $S^n_+ := \{M \in S^n : M \succeq 0\}$. The n-simplex $\{(x_1, \dots, x_n) \in R^n : x_i \geq 0, \sum_{i=1}^n x_i = 1\}$ is denoted by Δ_n.

Let us give the following generalized Gordan alternative theorem, which can be established by following a similar argument as in [6, Theorem 2.1]. For the reader's convenience, we also provide the proof here. This alternative theorem plays a key role in deriving dual global optimality conditions.

Theorem 1. *Let h_i, $i = 1, \dots, p$, be real-valued functions on R^n such that $(h_1, \dots, h_p)(R^n) + \text{int}R^p_+$ is convex where $(h_1, \dots, h_p)(R^n) = \{(h_1(x), \dots, h_p(x)) : x \in R^n\}$. Then, exactly one of the following two statements holds:*

(i) $\exists x \in R^n$, $h_i(x) < 0$, $i = 1, \dots, p$.
(ii) $(\exists \lambda \in R^p_+ \backslash \{0\})$ $(\forall x \in R^n)$ $\sum_{i=1}^p \lambda_i h_i(x) \geq 0$.

Proof. It suffices to show Not(i) $\Rightarrow$ (ii) as the converse implication holds always. Suppose that (i) does not hold. Then we have

$$0 \notin (h_1, \dots, h_p)(R^n) + \text{int}R^p_+.$$

As $(h_1, \dots, h_p)(R^n) + \text{int}R^p_+$ is a convex set, by the convex separation theorem, there exists $\lambda = (\lambda_1, \dots, \lambda_p) \in R^p \backslash \{0\}$ such that

$$\sum_{i=1}^{p} \lambda_i a_i \geq 0, \qquad \text{for all } a = (a_1, \dots, a_n) \in (h_1, \dots, h_p)(R^n) + \text{int}R^p_+. \qquad (1)$$

In particular, we have each $\lambda_i \geq 0$, $i = 1,\dots,p$, and so, $\lambda = (\lambda_1,\dots,\lambda_p) \in R^p_+ \setminus \{0\}$. Moreover, fix an arbitrary $x \in R^n$. Then, for any $\varepsilon > 0$, we have

$$(h_1(x)+\varepsilon,\dots,h_p(x)+\varepsilon) \in (h_1,\dots,h_p)(R^n) + \text{int}R^p_+,$$

and (1) implies that

$$\sum_{i=1}^{p} \lambda_i (h_i(x)+\varepsilon) \geq 0.$$

Letting $\varepsilon \to 0$, we have $\sum_{i=1}^{p} \lambda_i h_i(x) \geq 0$. Hence, (ii) holds. □

It is clear that if each h_i, $i = 1,\dots,p$ is convex, then $(h_1,\dots,h_p)(R^n) + \text{int}R^p_+$ is a convex set.

Now, we give some sufficient conditions ensuring the convexity of $(h_1,\dots,h_p)(R^n) + \text{int}R^p_+$ when h_i, $i = 1,\dots,p$ are non-convex quadratic functions.

Proposition 1. *The set $(h_1,\dots,h_p)(R^n) + \text{int}R^p_+$ is convex if any one of the following conditions holds:*

(1) *$p = 2$, $h_i(x) = \frac{1}{2}x^T A_i x$, $i = 1,2$, where $A_i \in S^n$.*
(2) *$n \geq 2$, $p = 2$, $h_i(x) = \frac{1}{2}x^T A_i x + b_i^T x + c_i$, $i = 1,2$, where $A_i \in S^n$ and there exists $(\mu_1,\mu_2) \in R^2$ such that*

$$\mu_1 A_1 + \mu_2 A_2 \succ 0.$$

(3) *$n \geq 3$, $p = 3$, $h_i(x) = \frac{1}{2}x^T A_i x$, $i = 1,2,3$, where $A_i \in S^n$ and there exists $(\mu_1,\mu_2,\mu_3) \in R^3$ such that*

$$\mu_1 A_1 + \mu_2 A_2 + \mu_3 A_3 \succ 0.$$

(4) *$p \in N$, $h_i(x) = \frac{1}{2}x^T A_i x$, $i = 1,\dots,p$, where $A_i \in S^n$ and the matrices $A_1,\dots,A_p$ commute.*
(5) *$p \in N$, $h_i(x) = \frac{1}{2}x^T A_i x + b_i^T x + c_i$, $i = 1,\dots,p$, where $A_i \in S^n_+$, $i = 1,\dots,p-1$, $A_p \in S^n$, $b_i \in R^n$, $c_i \in R$ and there exists $v \in \bigcap_{i=1}^{p-1}\{d \in R^n : A_i d = 0, b_i^T d \leq 0\}$ such that $v^T A_p v < 0$.*

Proof.
Proof of (1): By Dine's Theorem (cf. [5]), $(h_1,h_2)(R^n)$ is convex in R^2 and hence $(h_1,h_2)(R^n) + \text{int}R^2_+$ is also convex.

Proof of (2): By [12, Theorem 2.2], $(h_1,h_2)(R^n)$ is convex in R^2 and hence $(h_1,h_2)(R^n) + \text{int}R^2_+$ is also convex.

Proof of (3): By [12, Theorem 2.1], $(h_1,h_2,h_3)(R^n)$ is convex in R^3 and hence $(h_1,h_2,h_3)(R^n) + \text{int}R^3_+$ is also convex.

Proof of (4): By [12, Proposition 3.7], $(h_1,\dots,h_p)(R^n)$ is convex in R^p and hence $(h_1,\dots,h_p)(R^n) + \text{int}R^p_+$ is also convex.

Proof of (5): Let $M := (h_1,\dots,h_{p-1})(R^n) + \text{int}R^{p-1}_+$. Since h_i, $i = 1,\dots,p-1$, are convex (by the fact that $A_i \succeq 0$, $i = 1,\dots,p-1$), M is a convex set. Indeed, let

$(x_1,\dots,x_{p-1}) \in M$, $(y_1,\dots,y_{p-1}) \in M$ and $\lambda \in (0,1)$. Then, there exist $u,v \in R^n$ such that

$$h_i(u) < x_i \quad \text{and} \quad h_i(v) < y_i, \quad i = 1,\dots,p-1.$$

Thus, by the convexity of h_i $i = 1,\dots,p-1$, for all $i = 1,\dots,p-1$

$$h_i(\lambda u + (1-\lambda)v) \leq \lambda h_i(u) + (1-\lambda)h_i(v) < \lambda x_i + (1-\lambda)y_i.$$

Thus $\lambda(x_1,\dots,x_{p-1}) + (1-\lambda)(y_1,\dots,y_{p-1}) \in M$ and hence M is convex in R^{p-1}. Now we verify that $N := (h_1,\dots,h_p)(R^n) + \mathrm{int}R^p_+$ is convex in R^p. Let $(x_1,\dots,x_p) \in N$, $(y_1,\dots,y_p) \in N$ and $\lambda \in (0,1)$. From the above argument,

$$(\lambda x_1 + (1-\lambda)y_1,\dots,\lambda x_{p-1} + (1-\lambda)y_{p-1}) \in M := (h_1,\dots,h_{p-1})(R^n) + \mathrm{int}R^{p-1}_+.$$

Thus, there exists $u_0 \in R^n$ such that for all $i = 1,\dots,p-1$

$$h_i(u_0) < \lambda x_i + (1-\lambda)y_i.$$

Consider $u_t := u_0 + tv$, where v is defined as in the assumption in (5) and $t \geq 0$. Then, for $i = 1,\dots,p-1$ and $t \geq 0$,

$$\begin{aligned} h_i(u_t) &= \frac{1}{2}u_t^T A_i u_t + b_i^T u_t + c_i \\ &= \frac{1}{2}(u_0 + tv)^T A_i (u_0 + tv) + b_i^T (u_0 + tv) + c_i \\ &\leq \frac{1}{2}u_0^T A_i u_0 + b_i^T u_0 + c_i \\ &= h_i(u_0) < \lambda x_i + (1-\lambda)y_i. \end{aligned}$$

Moreover, note that

$$\begin{aligned} h_p(u_t) &= \frac{1}{2}u_t^T A_p u_t + b_p^T u_t + c_p \\ &= \frac{1}{2}(u_0 + tv)^T A_p (u_0 + tv) + b_p^T (u_0 + tv) + c_p \\ &= \left(\frac{1}{2}v^T A_p v\right) t^2 + \left(v^T A_p u_0 + b_p^T v\right) t + \left(\frac{1}{2}u_0^T A_p u_0 + b_p^T u_0 + c_p\right). \end{aligned}$$

Thus since $v^T A_p v < 0$, $\lim_{t\to+\infty} h_p(u_t) = -\infty$, and hence there exists $t_0 > 0$ such that

$$h_p(u_{t_0}) < \lambda x_p + (1-\lambda)y_p.$$

Therefore, $h_i(u_{t_0}) < \lambda x_i + (1-\lambda)y_i$, $i = 1,\dots,p$. It follows that $\lambda(x_1,\dots,x_p) + (1-\lambda)(y_1,\dots,y_p) \in N$. Hence N is convex in R^p. □

It is worth noting that, in our condition (5), we require the existence of a vector v such that h_i, $i = 1,\dots,p-1$, is non-increasing along this direction and h_p approaches $-\infty$ along this direction. As we will see later (e.g., Example 4), the condition (5) can be used as a sufficient condition to verify the convexity of $(h_1,\dots,h_p)(R^n)+\mathrm{int}R^p_+$ when $h_1,\dots,h_p$ are not all convex, and so, plays an important role in verifying the global optimality condition for the corresponding non-convex quadratic optimization problem.

To end this section, we give an example showing that the set $(h_1,\dots,h_p)(R^n)+\mathrm{int}R^p_+$ may not be convex when h_i, $i=1,\dots,p$ are general quadratic functions.

Example 1. Let $h_1(x_1,x_2)=x_1$, $h_2(x_1,x_2)=-x_1^2+x_2^2$ and $h_3(x_1,x_2)=-x_1$. Let $M=(h_1,h_2,h_3)(R^2)+\mathrm{int}R^3_+$. Since $h_1(0,0)=h_2(0,0)=h_3(0,0)=0$, $h_1(1,0)=1$ and $h_2(1,0)=h_3(1,0)=-1$, $a:=(\frac{1}{10},\frac{1}{10},\frac{1}{10})\in M$ and $b:=(\frac{11}{10},\frac{-9}{10},\frac{-9}{10})\in M$. Consider the point $\frac{a+b}{2}=(\frac{12}{20},\frac{-8}{20},\frac{-8}{20})$. Note that the following system has no solution:

$$x_1<\frac{12}{20},\quad -x_1^2+x_2^2<-\frac{8}{20},\quad -x_1<-\frac{8}{20}.$$

Thus $\frac{a+b}{2}\notin M$. So, M is not convex in this example.

3 Necessary and Sufficient Optimality Conditions

Recall the multi-objective non-convex quadratic optimization problem

$$\begin{aligned}\text{(MP)}\ \min\ & (f_1(x),\dots,f_p(x))\\ \text{s.t.}\ & g_j(x)\le 0,\ j=1,\dots,m,\end{aligned}$$

where $f_i(x)=\frac{1}{2}x^TA_ix+a_i^Tx$, $g_j(x)=\frac{1}{2}x^TB_jx+b_j^Tx+c_j$, $A_i,B_j\in S^n$, the space of all $(n\times n)$ symmetric matrices, $a_i,b_j\in R^n$ and $c_j\in R$. Denote $F_P=\{x\in R^n: g_j(x)\le 0,\ j=1,\dots,m\}$. A point $\bar{x}\in F_P$ is called a *weak minimizer* of (MP) if there do not exist $x\in F_P$ such that $f_i(x)<f_i(\bar{x})$, $i=1,\dots,p$.

In this section, we derive Lagrange multiplier conditions for a weak minimizer of (MP).

Theorem 2 (Necessary optimality theorem). *Suppose that $(f_1,\dots,f_p,g_1,\dots,g_m)(R^n)+\mathrm{int}R^{p+m}_+$ is convex in R^{p+m} and there exists $x_0\in R^m$ such that $g_j(x_0)<0$ $j=1,\dots,m$. Assume that $\bar{x}\in F_P$ is a weak minimizer of (MP). Then there exists $(\bar{\lambda},\bar{\mu})\in\Delta_p\times R^m_+$ such that for all $x\in R^n$*

$$\sum_{i=1}^p\bar{\lambda}_i(f_i(x)-f_i(\bar{x}))+\sum_{j=1}^m\bar{\mu}_jg_j(x)\ge 0. \tag{2}$$

In particular, we have

$$\sum_{i=1}^p\bar{\lambda}_i\nabla f_i(\bar{x})+\sum_{j=1}^m\bar{\mu}_j\nabla g_j(\bar{x})=0, \tag{3}$$

$$\bar{\mu}_jg_j(\bar{x})=0,\ j=1,\dots,m, \tag{4}$$

$$\sum_{i=1}^{p} \overline{\lambda}_i A_i + \sum_{j=1}^{m} \overline{\mu}_j B_j \succeq 0. \tag{5}$$

Proof. Let $\overline{x}$ be a weak minimizer of (MP). Then the following system has no solution:

$$f_i(x) - f_i(\overline{x}) < 0, i = 1, \ldots, p, \quad g_j(x) \leq 0, j = 1, \ldots, m.$$

Hence the following system also has no solution:

$$f_i(x) - f_i(\overline{x}) < 0, i = 1, \ldots, p, \quad g_j(x) < 0, j = 1, \ldots, m.$$

So, by Theorem 1, there exists $(\lambda_1, \ldots, \lambda_p, \mu_1, \ldots, \mu_m) \in R_+^{p+m} \backslash \{0\}$ such that for all $x \in R^n$

$$\sum_{i=1}^{p} \lambda_i (f_i(x) - f_i(\overline{x})) + \sum_{j=1}^{m} \mu_j g_j(x) \geq 0.$$

If $(\lambda_1, \ldots, \lambda_p) = (0, \ldots, 0)$, then $\sum_{j=1}^{m} \mu_j g_j(x) \geq 0$ for all $x \in R^n$. Since there exists $x_0 \in R^n$ such that $g_j(x_0) < 0$, $j = 1, \ldots, m$ this is impossible. Thus $(\lambda_1, \ldots, \lambda_p) \neq (0, \ldots, 0)$ and hence $\sum_{i=1}^{p} \lambda_i > 0$. Therefore, define $\overline{\lambda}_i = \frac{\lambda_i}{\sum_{i=1}^{p} \lambda_i}$ and $\overline{\mu}_j = \frac{\mu_j}{\sum_{i=1}^{p} \lambda_i}$. Then $\overline{\lambda} = (\overline{\lambda}_1, \ldots, \overline{\lambda}_p) \in \Delta_p$, $\overline{\mu} = (\overline{\mu}_1, \ldots, \overline{\mu}_m) \in R_+^m$ and (2) holds. From (2), one has $\sum_{j=1}^{m} \overline{\mu}_j g_j(\overline{x}) \geq 0$. Since $\overline{x} \in F_P$, $\overline{\mu}_j g_j(\overline{x}) \leq 0$. Thus $\overline{\mu}_j g_j(\overline{x}) = 0$, $j = 1, \ldots, m$. Define $L(x, \overline{\lambda}, \overline{\mu}) = \sum_{i=1}^{p} \overline{\lambda}_i f_i(x) + \sum_{j=1}^{m} \overline{\mu}_j g_j(x)$. It follows from (2) that $\nabla_x L(x, \overline{\lambda}, \overline{\mu}) = 0$ and $\nabla_{xx}^2 L(x, \overline{\lambda}, \overline{\mu}) \succeq 0$. Thus (3)–(5) hold. □

Theorem 3 (Sufficient optimality theorem). *Let $\overline{x} \in F_P$.*

(1) *If there exists $(\overline{\lambda}, \overline{\mu}) \in \Delta_p \times R_+^m$ such that for all $x \in R^n$ (2) holds, then $\overline{x}$ is a weak minimizer of (MP).*
(2) *If there exists $(\overline{\lambda}, \overline{\mu}) \in \Delta_p \times R_+^m$ such that (3)–(5) hold, then $\overline{x}$ is a weak minimizer of (MP).*

Proof.

Proof of (1): Suppose that there exist $\overline{\lambda} \in \Delta_p$ and $\overline{\mu} \in R_+^m$ such that for all $x \in R^n$ (2) holds. Then for any $x \in F_P$,

$$\sum_{i=1}^{p} \overline{\lambda}_i f_i(x) \geq \sum_{i=1}^{p} \overline{\lambda}_i f_i(\overline{x}). \tag{6}$$

Now, suppose on the contrary that $\overline{x}$ is not a weak minimizer of (MP). Then, there exists $\hat{x} \in F_P$ such that

$$f_i(\hat{x}) < f_i(\overline{x}), i = 1, \ldots, p.$$

Thus, from $\overline{\lambda} \in \Delta_p$ one has

$$\sum_{i=1}^{p} \overline{\lambda}_i f_i(\hat{x}) < \sum_{i=1}^{p} \overline{\lambda}_i f_i(\overline{x}).$$

This contradicts the inequality (6).

Proof of (2): Note that (3)–(5) hold if and only if (2) holds. Thus (2) follows. □

The following example shows that our sufficient optimality condition in the preceding theorem can be used for finding weak minimizer even when the range condition "$(h_1,\ldots,h_p)(R^n)+\text{int}R^p_+$ is convex" does not hold.

Example 2. Consider the following multi-objective non-convex quadratic optimization problem

$$\begin{array}{ll}(\text{MP}) & \min\ (x_1,-x_1^2+x_2^2)\\ & \text{s.t.}\quad -x_1\le 0.\end{array}$$

Let $f_1(x_1,x_2)=x_1$, $f_2(x_1,x_2)=-x_1^2+x_2^2$ and $g_1(x_1,x_2)=-x_1$. Let $F_P=\{(x_1,x_2)\in R^2 : g_1(x_1,x_2)\le 0\}$. Then $F_P=\{(x_1,x_2) : x_1\ge 0\}$ and $M:=(f_1,f_2,g_1)(R^2)+\text{int}R^3_+$ is not convex (see Example 1). Thus our necessary optimality theorem cannot be applied to this example. However, our sufficient optimality theorem still can be applied. Let $(\bar{x}_1,\bar{x}_2)\in\{(0,x_2) : x_2\in R\}$. Then $\nabla f_1(\bar{x}_1,\bar{x}_2)=(1,0)$, $\nabla f_2(\bar{x}_1,\bar{x}_2)=(0,2\bar{x}_2)$, $\nabla g_1(\bar{x}_1,\bar{x}_2)=(-1,0)$ and hence the following holds with $(\overline{\lambda}_1,\overline{\lambda}_2,\overline{\mu}_1)=(1,0,1)$:

$$\overline{\lambda}_1\nabla f_1(\bar{x}_1,\bar{x}_2)+\overline{\lambda}_2\nabla f_2(\bar{x}_1,\bar{x}_2)+\overline{\mu}_1\nabla g_1(\bar{x}_1,\bar{x}_2)=0 \quad\text{and}\quad \overline{\mu}_1 g_1(\bar{x}_1,\bar{x}_2)=0.$$

Let $A_1=\nabla^2 f_1(\bar{x}_1,\bar{x}_2)$, $A_2=\nabla^2 f_2(\bar{x}_1,\bar{x}_2)$ and $B_1=\nabla^2 g_1(\bar{x}_1,\bar{x}_2)$. Then, one has

$$A_1=\begin{pmatrix}0&0\\0&0\end{pmatrix},\quad A_2=\begin{pmatrix}-2&0\\0&2\end{pmatrix}\quad\text{and}\quad B_1=\begin{pmatrix}0&0\\0&0\end{pmatrix},$$

and hence $\overline{\lambda}_1A_1+\overline{\lambda}_2A_2+\overline{\mu}_1B_1=A_1+B_1\succeq 0$. Hence, it follows from our sufficient optimality theorem that $(\bar{x}_1,\bar{x}_2)$ is a weak minimizer of (MP). In fact, since $(f_1,f_2)(F_p)=\{(x_1,x_2)\in R^2 : x_1\ge 0, x_2\ge -x_1^2\}$, the set of all weak minimizers is

$$\{(0,x_2) : x_2\in R\}\cup\{(x_1,0) : x_1>0\}.$$

Now, we define a Karush–Kuhn–Tucker (KKT) point of (MP) as follows.

Definition 1. A point $\bar{x}\in F_P$ is called a KKT point of (MP) if there exists $(\overline{\lambda},\overline{\mu})\in\Delta_p\times R^m_+$ such that (3)-(4) hold.

Next, we present an example showing that a KKT point of (MP) need not be a weak minimizer of (MP).

Example 3. Consider the following quadratic multi-objective optimization problem

$$\begin{array}{ll}(\text{MP}) & \min\ (x_1,-x_1^2-x_2^2)\\ & \text{s.t.}\quad x_1^2+x_2^2\le 1.\end{array}$$

Let $f_1(x_1,x_2)=x_1$, $f_2(x_1,x_2)=-x_1^2-x_2^2$ and $g_1(x_1,x_2)=x_1^2+x_2^2-1$. Then the feasible set is $F_P=\{(x_1,x_2)\in R^2 : x_1^2+x_2^2\le 1\}$. Since $(f_1,f_2)(F_P)=\{(x_1,x_2) : -1\le$

$x_2 \le -x_1^2\}$, the set consisting of all the weak minimizer is $\{(x_1,x_2) : x_1^2 + x_2^2 = 1\}$. Let $(\bar{x}_1,\bar{x}_2) = (0,0)$ and $(\bar{\lambda}_1,\bar{\lambda}_2,\bar{\mu}_1) = (0,1,0)$. Then, one has $\bar{\mu}_1 g(\bar{x}_1,\bar{x}_2) = 0$ and

$$\bar{\lambda}_1 \nabla f_1(\bar{x}_1,\bar{x}_2) + \bar{\lambda}_2 \nabla f_2(\bar{x}_1,\bar{x}_2) + \bar{\mu}_1 \nabla g_1(\bar{x}_1,\bar{x}_2) = 0.$$

Thus $(\bar{x}_1,\bar{x}_2)$ is a KKT point but not a weak minimizer of (MP).

Now, we consider a special case of multi-objective non-convex quadratic optimization problem in which any non-zero KKT point is a weak minimizer.

Consider the following multi-objective non-convex quadratic problem with homogeneous quadratic objective functions and convex quadratic constraint functions:

$$\begin{aligned} \text{(HMP)}\ \min\ & \left(\frac{1}{2}x^T A_1 x, \ldots, \frac{1}{2}x^T A_p x\right) \\ \text{s.t.}\ & \frac{1}{2}x^T B_j x \le 1,\ j = 1,\ldots,m, \end{aligned}$$

where $A_i \in S^n$, $i = 1,\ldots,p$, $B_j \in S^n_+$, $j = 1,\ldots,m$. We denote the feasible set of (HMP) by F_{HP}. Let $A \in S^n$. We denote the ith eigenvalue of A by $\sigma_i(A)$. The eigenvalues are ordered as follows $\sigma_1(A) \le \sigma_2(A) \cdots \le \sigma_n(A)$.

Theorem 4. *Assume that $\bar{x} \in F_{HP}$ is a non-zero KKT point of (HMP) with $(\bar{\lambda},\bar{\mu}) \in \Delta_p \times R^m_+$. If $\sigma_2(\sum_{i=1}^p \bar{\lambda}_i A_i) > 0$, then $\sum_{i=1}^p \bar{\lambda}_i A_i + \sum_{j=1}^m \bar{\mu}_j B_j \succeq 0$ and so, $\bar{x}$ is a weak minimizer of (HMP).*

Proof. Since $\bar{x} \in F_{HP}$ is a KKT point of (HMP), one has

$$\left(\sum_{i=1}^p \bar{\lambda}_i A_i + \sum_{j=1}^m \bar{\mu}_j B_j\right)\bar{x} = 0.$$

Note that $\bar{x} \ne 0$. It follows that 0 is an eigenvalue of $\sum_{i=1}^p \bar{\lambda}_i A_i + \sum_{j=1}^m \bar{\mu}_j B_j$. On the other hand, note that $\sum_{j=1}^m \bar{\mu}_j B_j \succeq 0$ (since $B_j \in S^n_+$ and $\mu_j \ge 0$ $j = 1,\ldots,m$) and $\sigma_k(\sum_{i=1}^p \bar{\lambda}_i A_i) \ge \sigma_2(\sum_{i=1}^p \bar{\lambda}_i A_i)$ $k = 2,\ldots,n$. Thus, for all $k = 2,\ldots,n$ we have

$$\begin{aligned} \sigma_k\left(\sum_{i=1}^p \bar{\lambda}_i A_i + \sum_{j=1}^m \bar{\mu}_j B_j\right) &\ge \sigma_k\left(\sum_{i=1}^p \bar{\lambda}_i A_i\right) + \sigma_1\left(\sum_{j=1}^m \bar{\mu}_j B_j\right) \\ &\ge \sigma_2\left(\sum_{i=1}^p \bar{\lambda}_i A_i\right) > 0, \end{aligned}$$

where the first inequality follows from [10, Lemma 2] and [2, III.2.2]. Therefore, we have for all $k = 1,\ldots,n$

$$\sigma_k\left(\sum_{i=1}^p \bar{\lambda}_i A_i + \sum_{j=1}^m \bar{\mu}_j B_j\right) \ge 0,$$

and so, $\sum_{i=1}^{p} \overline{\lambda}_i A_i + \sum_{j=1}^{m} \overline{\mu}_j B_j \succeq 0$. Thus, it follows from Theorem 3(2) that $\overline{x}$ is a weak minimizer of (HMP). □

Finally, we give an example illustrating Proposition 1(5) and Theorems 2 and 3.

Example 4. Consider the following multi-objective indefinite quadratic optimization problem:

$$\begin{aligned} \text{(MP)} \ \min \ & (x_1, x_1^2 - x_2^2) \\ \text{s.t.} \ & x_1^2 - x_1 \leq 0. \end{aligned}$$

Let $f_1(x_1,x_2) = x_1$, $f_2(x_1,x_2) = x_1^2 - x_2^2$ and $g_1(x_1,x_2) = x_1^2 - x_1$. The feasible set is $F_P = \{(x_1,x_2) : x_1^2 - x_1 \leq 0\}$. Then $f_i(x) = \frac{1}{2}x^T A_i x + a_i^T x$, $i = 1,2$ and $g_1(x) = \frac{1}{2}x^T B_1 x + b_1^T x$, where $x = (x_1,x_2) \in R^2$, $a_1 = (1,0)$, $a_2 = (0,0)$, $b_1 = (-1,0)$ and

$$A_1 = \begin{pmatrix} 0 & 0 \\ 0 & 0 \end{pmatrix}, \quad A_2 = \begin{pmatrix} 2 & 0 \\ 0 & -2 \end{pmatrix}, \quad \text{and} \quad B_1 = \begin{pmatrix} 2 & 0 \\ 0 & 0 \end{pmatrix}.$$

(1) Note that $A_1 \succeq 0$ and $B_1 \succeq 0$. Since

$$\begin{aligned} \Omega &:= \{d = (d_1,d_2) \in R^2 : A_1 d = 0, a_1^T d \leq 0, B_1 d = 0, b_1^T d \leq 0\} \\ &= \{(0,d_2) : d_2 \in R\}, \end{aligned}$$

$v := (0,1) \in \Omega$. It can be verified that $v^T A_2 v = -2 < 0$. Thus, all the assumptions in Proposition 1(5) are satisfied. Thus $(f_1, f_2, g_1)(R^2) + \text{int}R^3_+$ is convex.

(2) It can be verified that $\{(x_1,x_2) \in F_P : \exists(\overline{\lambda},\overline{\mu}) \in \Delta_2 \times R_+$ such that (x_1,x_2) is a KKT point of (MP) with $\sum_{i=1}^{2} \overline{\lambda}_i A_i + \overline{\mu}_1 B_1 \succeq 0\} = \{(0,x_2) : x_2 \in R\}$. It follows from Theorems 2 and 3 that the set consisting of all weak minimizers is $\{(0,x_2) : x_2 \in R\}$.

Acknowledgements The authors are grateful to the referees and the guest editor for their helpful comments and valuable suggestions contributed to the final preparation of the chapter. The second author was supported by the Korea Science and Engineering Foundation (KOSEF) NRL Program grant funded by the Korean government (MEST) (No. ROA-2008-000-20010-0).

References

1. K. J. Arrow, An extension of the basic theorems of classical welfare economics, in J. Neyman (ed), Proceedings of the Second Berkley Symposium on Mathematical Statistics and Probability, University of California Press, Berkeley, CA, 1951, pp. 507–532.
2. R. Bhatia, *Matrix Analysis*, Graduate Texts in Mathematics, **169**, Springer, Berlin, 1996.
3. X. Chen and Y. X. Yuan, On local solutions of the Celis–Dennis–Tapia subproblem, *SIAM J. Optim.* **10** (1999), 359–383.
4. A. Chinchuluun and P. M. Pardalos, A survey of recent developments in multiobjective optimization, *Ann. Oper. Res.* **154** (2007), 29–50.
5. L. L. Dines, On the mapping of quadratic forms, *Bull. Am. Math. Soc.* **47** (1941), 494–498.

6. V. Jeyakumar, A generalization of a minimax theorem of Fan via a theorem of the alternative. *J. Optim. Theory Appl.* **48** (1986), 525–533.
7. V. Jeyakumar, G. M. Lee, and G. Li, Alternative theorems for quadratic inequality systems and global quadratic optimization, *SIAM J. Optim.* **20** (2009), 983–1001.
8. V. Jeyakumar, A. M. Rubinov and Z. Y. Wu, Non-convex quadratic minimization problems with quadratic constraints: global optimality conditions, *Math. Program. A* **110** (2007), 521–541.
9. V. Jeyakumar and S. Srisatkunarajah, Lagrange multiplier necessary conditions for global optimality for non-convex minimizations over a quadratic constraint via S-lemma, *Optim. Lett.* **3**(1) (2009), 23–33.
10. A. Jourani and J. J. Ye, Error bounds for eigenvalue and semidefinite matrix inequality, *Math. Program. B* **104** (2005), 525–540.
11. J. M. Peng and Y. X. Yuan, Optimality conditions for the minimization of a quadratic with two quadratic constraints, *SIAM J. Optim.* **7** (1997), 579–594.
12. B. T. Polyak, Convexity of quadratic transformation and its use in control and optimization, *J. Optim. Theory Appl.* **99** (1998) 563–583.

Application of Variational Analysis and Control Theory to Nonparametric Maximum Likelihood Estimation of a Density Function

Ilya Shvartsman

Abstract In this chapter, we propose a new approach to the estimation of the probability density function based on the maximum likelihood method if it is known that the underlying density function is Lipschitz. We treat this problem as an optimal control problem and prove convergence results using techniques of variational analysis.

1 Introduction and Preliminaries

The probability density function is a fundamental concept in probability and statistics. Consider a random variable X that has a probability density function f. Then probabilities associated with X can be found from the relation

$$P(\alpha < x < \beta) = \int_{\alpha}^{\beta} f(x)\,\mathrm{d}x \quad \text{for } \alpha < \beta.$$

Let $x_1, x_2, \ldots, x_n$ be independent realizations of the random variable X with unknown density function f. *Density estimation* is the construction of an estimate of the density function from the observed data.

An overview of methods of density estimation can be found, for example, in [16]. The simplest method of density estimation is by means of a histogram, which is widely used in applied statistical problems. The price paid for simplicity of this method is discontinuity of the histogram and its dependence on the choice of the bin width. In 1957, the so-called *kernel method* was introduced in [12] and has been thoroughly investigated since then (see, e.g., [13, 14, 20] and references therein).

Ilya Shvartsman
Department of Mathematics and Computer Science, Penn State Harrisburg, Middletown, PA 17057, USA
e-mail: ius13@psu.edu

R.S. Burachik and J.-C. Yao (eds.), *Variational Analysis and Generalized Differentiation in Optimization and Control*, Springer Optimization and Its Applications 47,
DOI 10.1007/978-1-4419-0437-9_10,

A kernel density estimator for f at an arbitrary fixed x is

$$\hat{f}(x) = \frac{1}{nh}\sum_{i=1}^{n}\varphi\left(\frac{x-x_i}{h}\right), \tag{1}$$

where the kernel φ is such that $\int_{-\infty}^{\infty}\varphi(x)\,\mathrm{d}x = 1$ determines the shape of the "bumps" centered at each data point, and h is the smoothing parameter: as h tends to zero, $\hat{f}$ tends to a sum of Dirac delta functions with spikes at the observations, while as h becomes large, the features of the distribution become obscured. A histogram is a special case of a kernel estimator with h equal to the bin width and

$$\varphi(x) = \begin{cases} 1, & |x| \le 1/2, \\ 0, & |x| > 1/2. \end{cases}$$

If φ is smooth, a kernel estimator for f can be viewed as a smoothed version of a histogram.

Both the histogram and the kernel methods are derived in an ad hoc way from the definition of density. The *maximum likelihood method* is a commonly used statistical technique, which we describe below. Assume for illustration purposes that f belongs to a parametric family $f = f(x,\theta)$, where θ is a finite-dimensional parameter to be estimated. The likelihood of θ as a function of $x_1, x_2, \ldots, x_n$ is defined as

$$\mathrm{lik}\,(\theta) = \prod_{i=1}^{n} f(x_i,\,\theta).$$

The *maximum likelihood estimator (mle)* of θ is a value of θ that maximizes the likelihood – that is, makes the observed data "most probable." Rather than maximizing the likelihood itself, it is usually easier to maximize its natural logarithm, (the so-called *log-likelihood*)

$$l(\theta) = \sum_{i=1}^{n}\ln[f(x_i,\,\theta)].$$

Assume, for example, that the sample is taken from the normal distribution with mean μ and variance σ:

$$f(x,\mu,\sigma) = \frac{1}{\sigma\sqrt{2\pi}}\mathrm{e}^{-\frac{(x-\mu)^2}{2\sigma^2}}, \qquad -\infty < x < \infty,$$

where the parameters μ and σ are to be estimated. In this case, the log-likelihood is

$$l(\mu,\sigma) = -n\ln\sigma - \frac{n}{2}\ln 2\pi - \frac{1}{2\sigma^2}\sum_{i=1}^{n}(x_i-\mu)^2.$$

Setting the partial derivatives with respect to μ and σ equal to zero and solving the resulting equations, we obtain

$$\hat{\mu} = \frac{1}{n}\sum_{i=1}^{n}x_i =: \bar{x}, \quad \hat{\sigma}^2 = \frac{1}{n}\sum_{i=1}^{n}(x_i-\bar{x})^2.$$

Finding parameters of interest is not always possible in a closed form (e.g., for gamma distribution), but there are numerical algorithms for finding the maximum likelihood estimator built in a lot of statistical software.

What makes the maximum likelihood method very common and useful in parametric density estimation is its asymptotic property: Under certain smoothness assumptions on f, maximum likelihood estimator $\hat{\theta}$ converges to the true parameter θ in probability. Moreover, the probability distribution of $\sqrt{nI(\theta)}(\hat{\theta}-\theta)$ tends to standard normal as $n \to \infty$, where $I(\theta) = E[\frac{\partial}{\partial\theta} \ln f(X|\theta)]^2$ is the so-called *Fisher information* (e.g., [2]), that is, asymptotically,

$$\hat{\theta} - \theta \sim N\left(0, \frac{1}{nI(\theta)}\right). \tag{2}$$

Can the maximum likelihood method be similarly applied to problems of *nonparametric* density estimation? It is easy to see that the likelihood

$$\text{lik}\,(x_1, \ldots, x_n; f) = \prod_{i=1}^{n} f(x_i)$$

has no finite maximum over the class of all (even continuous) densities because by taking f to be the sum of Dirac-type spikes at the observations, the likelihood can be made arbitrarily large, while still satisfying the constraint $\int f(x)\,\mathrm{d}x = 1$. To get around this difficulty, a method of *penalized likelihood* was originally introduced in [5] and developed in a series of works (see, e.g., [1, 3, 9]). The essence of this method is that a penalized likelihood function

$$l_\alpha(f) := \sum_{i=1}^{n} \ln f(x_i) - \alpha R(f) \tag{3}$$

is maximized. Here α is the *smoothing parameter* and $R(f)$ is the *roughness penalty*. The penalized likelihood is a way of quantifying the trade-off between smoothness and goodness-of-fit to the data. Possible choices of $R(f)$ suggested in [5] are, for $\gamma = \sqrt{f}$,

$$R(f) = \int \gamma'^2(x)\,\mathrm{d}x \quad \text{and} \quad R(f) = \int \gamma''^2(x)\,\mathrm{d}x$$

or a linear combination of those. The first choice penalizes for the slope, while the second choice penalizes for the curvature.

To apply the maximum likelihood method directly (without penalization), the class of potential estimators must be large enough to be of interest, and, at the same time, small enough so that the maximum likelihood estimator exists. To the best of the author's knowledge, direct maximum likelihood method has been applied in the context of monotone densities (e.g., [7]), densities with monotone hazard rate $r(x) := f(x)/(1 - \int_{-\infty}^{x} f(t)\,\mathrm{d}t)$ [8], unimodal densities (e.g., [4, 18]) (a density function is said to be unimodal if it has exactly one local maximum), and to densities which are finite linear combinations of indicator functions of disjoint intervals [19]. In this chapter, we suggest application of this method to *Lipschitz functions* with a known bound on the Lipschitz constant. This is a natural assumption because we often do not expect the underlying density function to change much on a small interval, and it seems to be less restrictive than, say, unimodality.

In what follows, we will denote the true unknown density by $f_0 : [a,b] \to \boldsymbol{R}$, the maximum likelihood estimator by $\hat{f}$ or $\hat{f}_n$ to emphasize its dependence on n where relevant. We assume that f_0 and $\hat{f}_n$ belong to the class $\mathscr{F}$ of density functions with Lipschitz constant l:

$$\mathscr{F} := \left\{ f : [a,b] \to \boldsymbol{R} \,|\, f \geq 0, \int_a^b f(x)\,\mathrm{d}x = 1, |f(\gamma_1) - f(\gamma_2)| \right.$$
$$\left. \leq l|\gamma_1 - \gamma_2| \;\forall \gamma_1, \gamma_2 \in [a,b] \right\}. \tag{4}$$

We consider the following two problems:

(P_1) Find a maximum likelihood estimator $\hat{f}_n$ of f_0
(P_2) Investigate asymptotic properties of $\hat{f}_n$ as $n \longrightarrow \infty$

Although these problems are statistical in nature, problem (P_1), as we will see in Sect. 2, can be reduced to and analysed as an optimal control problem. Problem (P_2) will be dealt with in Sect. 3 applying methodology typical to variational analysis – by introducing approximations and passing to the limit.

The main result of this chapter is the following theorem.

Theorem 1.

(a) *The sequence of maximum likelihood estimators $\hat{f}_n$ converges to f_0 uniformly in probability on $[a,b]$. This means that for any positive ε and δ there exists a number N such that for all $n > N$*

$$P\left(\max_{x \in [a,b]} \left| \hat{f}_n(x) - f_0(x) \right| > \varepsilon \right) < \delta. \tag{5}$$

(Here, and below, $P(A)$ stands for the probability of event A.)

(b) *There exists a constant $K > 0$ such that for any $x \in [a,b]$ where $f_0(x) > 0$, for any $z > 0$, the following estimate holds asymptotically:*

$$P\left(\left| \hat{f}_n(x) - f_0(x) \right| > z \right) \leq 2 \left[1 - \Phi\left(\frac{K\sqrt{f_0(x)}}{l(l+1)^3} \frac{\sqrt{n}}{\ln n} z^4 \right) \right], \tag{6}$$

where

$$\Phi(y) = \frac{1}{\sqrt{2\pi}} \int_{-\infty}^{y} \mathrm{e}^{-\frac{x^2}{2}}\,\mathrm{d}x$$

is the cumulative standard normal distribution.

(c) *For any $x \in [a,b]$ such that f_0 is equal to zero on some open interval containing x, the confidence interval is narrower for $0 < z < 1$:*

$$P\left(\left| \hat{f}_n(x) - f_0(x) \right| > z \right) \leq 2 \left[1 - \Phi\left(\frac{K}{l} \frac{\sqrt{n}}{\ln n} z^2 \right) \right]. \tag{7}$$

Note that $\Phi(y) \approx 1$ for $y \geq 3$ and $\Phi(0) = 1/2$. It is natural to observe that the convergence slows down as l increases.

Convergence of the estimator to the true density (*consistency*) is essential for any useful estimator. It is shown in [15] that the kernel estimator (1) is consistent in the sense that

$$\sup_x \left|\hat{f}_n(x) - f_0(x)\right| \to 0 \text{ in probability as } n \to \infty$$

under very mild assumptions of uniform continuity of f_0 and the following relationship between h and n:

$$h \to 0 \quad \text{and} \quad \frac{\ln n}{nh} \to 0 \text{ as } n \to \infty.$$

The rate of this convergence was shown in [6] to be of the order $O((\ln n/nh)^{1/2})$. This estimate was obtained under the assumption of integrability of f_0'', which is a standard assumption for kernel estimators, although it does not hold for simplest nonsmooth functions. For the penalized maximum likelihood estimator, under a rather restrictive assumption on integrability of f_0'' and $f_0'^2/f_0$ the convergence rate was shown to be of the order $O(n^{-2/5})$ if the smoothing parameter $\alpha(n)$ in (3) is of the order $O(n^{-1/5})$. Theorem 1 establishes convergence at the much quicker *normal* convergence rate, generalizing, to some extend, the finite-dimensional convergence result (2).

The structure of the chapter is as follows. In Sect. 2, we show that the estimator can be found as a solution of a finite-dimensional mathematical program, and we consider a few simple examples. Section 3 is devoted to the proof of Theorem 1. In this manuscript, we are not suggesting a numerical method for finding the estimator – this may be addressed in future work. At the same time, it is common in variational analysis (in mathematical programming and optimal control in particular) to establish certain desirable properties of optimizers (like Lipschitz continuity or differentiability) only from the conditions of optimality, without finding optimizers explicitly.

2 Problem (P_1) as an Optimal Control Problem

The problem (P_1) can be stated as

$$\begin{cases} \text{Maximize } \varphi(f) := \sum_{i=1}^{n} \ln f(x_i) \\ \text{over } f \in W^{1,1}[a,b] \text{ satisfying} \\ f(x) \geq 0, \\ \int_a^b f(x)\,\mathrm{d}x = 1, \\ |f'(x)| \leq l \ \text{ for a.a. } x \in [a,b]. \end{cases} \tag{8}$$

Denote

$$y_1(x) := \int_a^x f(s)\,\mathrm{d}s,$$
$$y_2(x) := f(x), \quad x \in [a,b],$$

and introduce the control function u as $u(x) := f'(x) = y_2'(x)$. Then system (8) can be represented as

$$\begin{cases} \text{Minimize } \tilde{\varphi}(u) := -\sum_{i=1}^{n} \ln y_2(x_i) \\ \text{over measurable } u : [a,b] \to \boldsymbol{R} \text{ subject to contraints} \\ y_1' = y_2, \\ y_2' = u, \\ y_2(x) \geq 0, \\ y_1(a) = 0, \\ y_1(b) = 1, \\ |u(x)| \leq l \ \text{ for a.a. } x \in [a,b]. \end{cases} \tag{9}$$

This is an optimal control problem where the objective function includes interior points and this problem also contains a state constraint. Such problems are studied, in a more general nonsmooth framework, in [10], [11], and [17].

Set $x_0 := a$ and $x_{n+1} := b$. The necessary optimality conditions for u imply the following on each of the intervals (x_i, x_{i+1}), $i = 0, \ldots, n$ where $y_2(x) > 0$.

There exist adjoint functions $p_1, p_2 : [a,b] \to \boldsymbol{R}$ satisfying the equations

$$\begin{aligned} p_1' &= 0, \\ p_2' &= -p_1. \end{aligned} \tag{10}$$

Furthermore, the function p_2 may have jumps at the points x_i, $i = 1, \ldots, n$ satisfying $p_2(x_i + 0) - p_2(x_i - 0) \leq 0$, while the function p_1 is continuous on $[a,b]$. It follows from (9) that the function p_2 is linear on each of the intervals (x_i, x_{i+1}) (if the state constraint is inactive), hence it changes its sign at most once on these intervals. The optimal control u must maximize the function $H(p_2, u) = p_2 u$ for a.a. $x \in [a,b]$, i.e., is given by

$$u(x) = l \operatorname{sgn} p_2(x).$$

Therefore, the optimal control u is bang-bang with at most one point of switch on (x_i, x_{i+1}). A typical graph of $y_2(x) = \hat{f}(x)$ corresponding to an optimal control u is shown below.

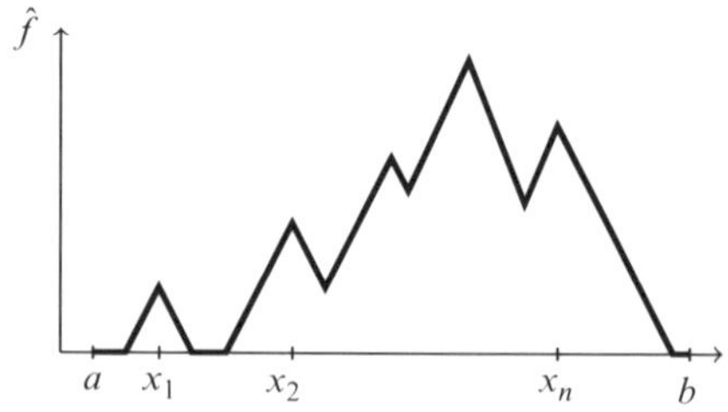

Thus the infinite-dimensional problem (P_1) is reduced to a finite-dimensional mathematical program of maximizing $\sum \ln f(x_i)$ (with $f(x_i)$, $i = 1,\dots,n$ being the parameters) with a constraint that the area below the curve (broken line) is a one. Below we consider simple examples with just two and three sample points and show that the solution is not intuitively obvious.

Examples.

Two-point problem. We have the data x_1, x_2. Denote

$$\tau = |x_2 - x_1|.$$

If x_1 and x_2 are sufficiently far from each other and the interval endpoints (precisely, if $\tau \geq \sqrt{2/l}$), then the graph of the maximum likelihood estimator $\hat{f}$ contains two "triangles," each with area 1/2 and height $\sqrt{l/2}$.

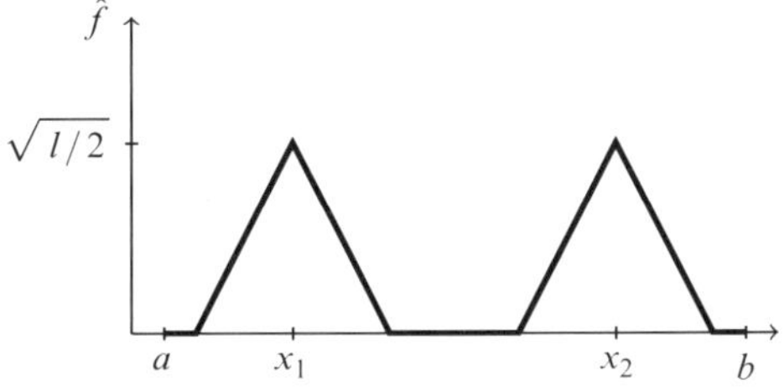

If these points are close to each other ($\tau < \sqrt{2/l}$), then the graph of $\hat{f}$ contains "clustered triangles" with common height $h := \hat{f}(x_1) = \hat{f}(x_2)$, whose total area can be shown to be equal to $h^2/l + \tau h - l\tau^2/4$.

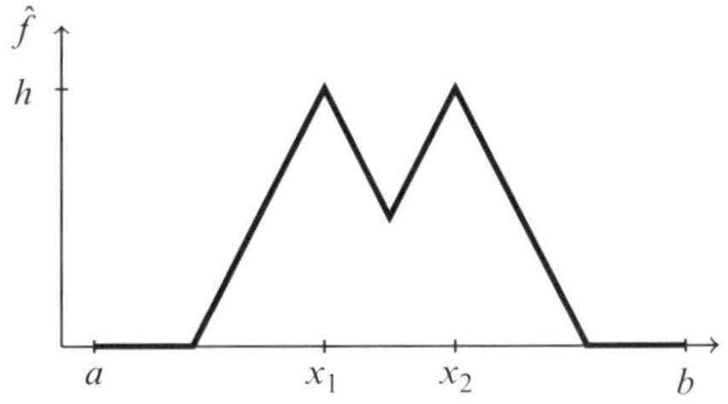

Therefore, h is the solution of the quadratic equation

$$\frac{h^2}{l} + \tau h - \frac{l\tau^2}{4} = 1.$$

Three-point problem. It may seem natural to conjecture that the maximum likelihood estimator $\hat{f}_n$ is such that each data point x_i contributes equally $1/n$ to the cumulative distribution (that is, $\hat{F}_n(x_{i+1}) - \hat{F}_n(x_i) = 1/n$), which is the case for the empirical cumulative distribution function

$$\hat{F}_n = \frac{1}{n}(\#x_i \leq x).$$

However, it turns out not to be the case as we will see.

Consider the situation when two of the three points are close to each other (so that the corresponding triangles overlap) and the third one is far from them, as shown in Fig. 1.

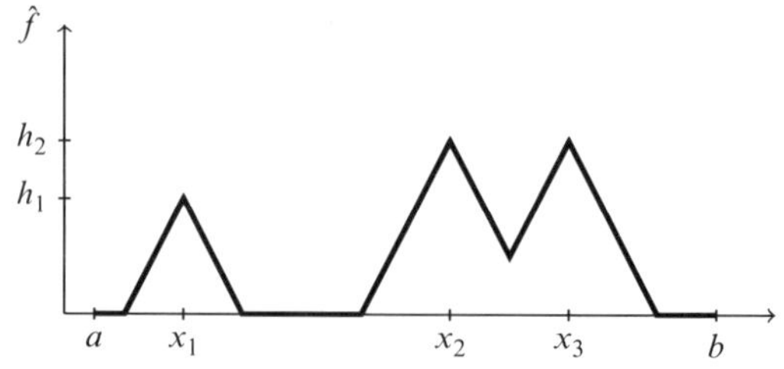

Fig. 1 Graph of $\hat{f}$ for three observation points

Set $h_1 := \hat{f}(x_1)$, $h_2 := \hat{f}(x_2) = \hat{f}(x_3)$ and $\tau = |x_3 - x_2|$. The total area under the curve of $\hat{f}$ is $h_1^2/l + h_2^2/l + \tau h_2 - l\tau^2/4$, which implies the constraint on h_1 and h_2

$$\frac{h_1^2}{l} + \frac{h_2^2}{l} + \tau h_2 - \frac{l\tau^2}{4} = 1. \tag{11}$$

The log-likelihood to be maximized is $\mathscr{L}(h_1, h_2) = \ln h_1 + 2\ln h_2$, therefore the Lagrangian is

$$L = \ln h_1 + 2\ln h_2 - \lambda\left(\frac{h_1^2}{l} + \frac{h_2^2}{l} + \tau h_2 - \frac{l\tau^2}{4} - 1\right),$$

where λ is the Lagrange multiplier.

Setting the partial derivatives L_{h_1}, L_{h_2} equal to zero we get a system

$$\frac{1}{h_1} - \frac{2\lambda h_1}{l} = 0,$$
$$\frac{2}{h_2} - \frac{2\lambda h_2}{l} - \lambda\tau = 0.$$

From the first equation, we obtain $\lambda = l/(2h_1^2)$. Substituting it into the second equation, we get

$$\frac{2}{h_2} - \frac{h_2}{h_1^2} - \frac{\tau l}{2h_1^2} = 0,$$

or, equivalently,

$$4h_1^2 - 2h_2^2 - h_2\tau l = 0,$$

which, together with constraint (11) gives a system with respect to h_1 and h_2 if τ is sufficiently small (precise range can be shown to be $\tau \le 2/\sqrt{3l}$).

The solution of this system is

$$h_2 = \frac{1}{12}\left(\sqrt{49l^2\tau^2 + 96l} - 5l\tau\right),$$

$$h_1^2 = \frac{11}{72}l^2\tau^2 + \frac{l}{3} - \frac{l\tau}{72}\sqrt{49l^2\tau^2 + 96l}.$$

The graph of h_1^2/l (the area of the single triangle in Fig. 1) as a function of τ is shown in the case $l = 1$.

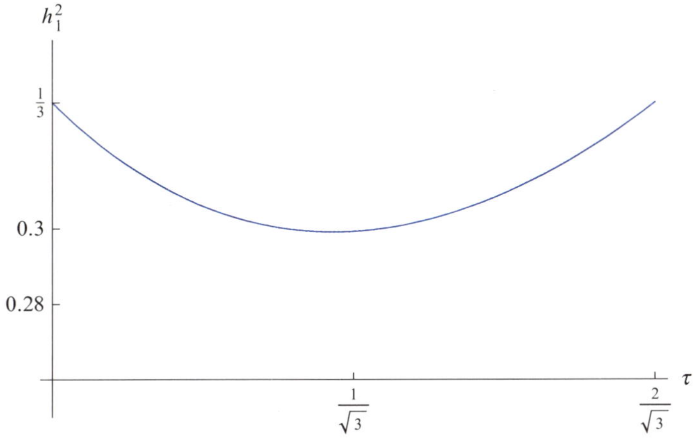

We can see that the "single" point x_1 contributes less than a third of what the "clustered" points x_2 and x_3 do to the cumulative distribution, and the minimum of this contribution occurs close to the mid-interval. This is a very nonintuitive result.

If we have more than three data points, the problem of finding the maximum likelihood estimator, in general, can no longer be solved analytically; numerical algorithms have to be developed.

3 Asymptotic Properties of the Maximum Likelihood Estimators

In this section, we will prove Theorem 1. Below, unless stated otherwise, by convergence $a_n \to a$ we mean convergence in probability, that is, if $P(|a_n - a| > \varepsilon) \to 0$ for any $\varepsilon > 0$ as $n \to \infty$.

Proof of Theorem 1. Recall that $\hat{f}_n$ is a maximizer of the log-likelihood functional $\sum_{i=1}^n \ln f(x_i)$ or, equivalently, of the averaged log-likelihood functional

$$\mathscr{L}_n(f) := \frac{1}{n}\sum_{i=1}^n \ln f(x_i)$$

over the class $\mathscr{F}$ given by (4). For fixed $f \in \mathscr{F}$, the expected value of $\mathscr{L}_n(f)$ with respect to the realization of $x_1, \ldots, x_n$ is given by the functional

$$J(f) := E[\ln f(X)] = \int_a^b \ln f(x) f_0(x)\,\mathrm{d}x, \tag{12}$$

where $f_0 \in \mathscr{F}$ is the true unknown density function.

Set

$$I := \{x \in [a,b] \,|\, f_0(x) > 0\} \quad \text{and} \quad \int_{[a,b]\setminus I} \ln f(x) f_0(x)\,\mathrm{d}x := 0$$

even if f is zero at some points in $[a,b] \setminus I$. Then

$$J(f) = \int_I \ln f(x) f_0(x)\,\mathrm{d}x$$

and the domain of J consists of all f for which this integral is finite (in particular, for positive f bounded above and separated below from zero).

Our idea is to carry out the proof along the following lines. First, we would want to show that

$$\mathscr{L}_n(f) \to J(f) \qquad \text{as } n \to \infty \tag{13}$$

uniformly over a class of functions that contains all possible estimators $\hat{f}_n$. Then we would consider the identity

$$J(f_0) - J(\hat{f}_n) = [J(f_0) - \mathscr{L}_n(f_0)] + \left[\mathscr{L}_n(f_0) - \mathscr{L}_n(\hat{f}_n)\right] + \left[\mathscr{L}_n(\hat{f}_n) - J(\hat{f}_n)\right]. \tag{14}$$

The second difference on the right side of (14) is nonpositive, and if (13) is true, then the first and the third differences on the right side converge to zero in probability. On the other hand, it is possible to show that f_0 is the unique maximizer of the functional J over $\mathscr{F}$, therefore the left side of (14) is nonnegative. This would imply that $J(\hat{f}_n) \to J(f_0)$ and we would be able to deduce uniform convergence $\hat{f}_n \to f_0$ and other assertions of Theorem 1.

There is, however, an obstacle on this way: if it happens that $\hat{f}_n$ is equal to zero on a set where $f_0 \neq 0$ (which is possible), then $J(\hat{f}_n) = \int_I \ln \hat{f}_n(x) f_0(x)\,\mathrm{d}x$ is undefined, so we cannot talk about the convergence (13). To get around this difficulty, for $c > 0$ (to be specified later) we define

$$\hat{f}_n^c := \max\{\hat{f}_n, c\}.$$

Since $\hat{f}_n^c$ is separated from zero and bounded, the value of $J(\hat{f}_n^c)$ is finite.

Let $\mathscr{F}_c$ be the class of nonnegative functions with Lipschitz constant $l(1 + c(b-a))$ and whose integral over $[a,b]$ does not exceed $1 + c(b-a)$, that is,

$$\mathscr{F}_c := \Big\{ f : [a,b] \to R \,|\, f \geq 0, |f(\gamma_1) - f(\gamma_2)| \leq l(1 + c(b-a))\ \forall \gamma_1, \gamma_2 \in [a,b], \\ \int_a^b f(x)\,\mathrm{d}x \leq 1 + c(b-a) \Big\}$$

and set

$$f_0^c := (1 + c(b-a)) f_0. \tag{15}$$

By construction, $\hat{f}_n^c$ and f_0^c belong to $\mathscr{F}_c$. Let us show that the functional J is maximized over $\mathscr{F}_c$ by f_0^c. Note that

$$J(f_0^c) = \ln(1 + c(b-a)) + \int_I \ln f_0(x) f_0(x)\,\mathrm{d}x > -\infty$$

since the product $y \ln y$ is bounded for bounded positive y.

Take any $f \in \mathscr{F}_c$ such that $J(f) > -\infty$. Since for any $y, y_0 > 0$

$$\ln y - \ln y_0 = \frac{y - y_0}{y_0} - \frac{(y-y_0)^2}{2\xi^2}$$

for some ξ between y and y_0 due to Taylor's Theorem, we have

$$\begin{aligned} J(f_0^c) - J(f) &= \int_I [\ln f_0^c(x) - \ln f(x)] f_0(x)\,\mathrm{d}x \\ &= \int_I \left[\frac{f_0^c(x) - f(x)}{f_0^c(x)} + \frac{(f(x) - f_0^c(x))^2}{2\xi^2(x)} \right] f_0(x)\,\mathrm{d}x \\ &= \int_I \frac{f_0^c(x) - f(x)}{1 + c(b-a)}\,\mathrm{d}x + \int_I \frac{(f(x) - f_0^c(x))^2}{2\xi^2(x)} f_0(x)\,\mathrm{d}x. \end{aligned} \tag{16}$$

It is obvious that the second integral on the right side is nonnegative; the first integral is also nonnegative because $\int_I f_0^c(x)\,\mathrm{d}x = 1 + c(b-a)$, while $\int_I f(x)\,\mathrm{d}x \le 1 + c(b-a)$. Moreover, the right side in (16) is equal to zero if and only if $f = f_0^c$ on I and, hence, on all of $[a,b]$ (otherwise, if $f = f_0^c$ on I, but not on $[a,b] \setminus I$, then $\int_a^b f(x)\,\mathrm{d}x > 1 + c(b-a)$). Therefore, $f = f_0^c$ is the unique global maximizer of J over $\mathscr{F}_c$.

Take $f \in \mathscr{F}_c$ such that $f(x) \ge c$ for all $x \in [a,b]$. We have

$$\begin{aligned} J(f_0^c) - J(f) &= [J(f_0^c) - \mathscr{L}_n(f_0^c)] + [\mathscr{L}_n(f_0^c) - \mathscr{L}_n(f)] + [\mathscr{L}_n(f) - J(f)] \\ &= [J(f_0) - \mathscr{L}_n(f_0)] + [\mathscr{L}_n(f_0^c) - \mathscr{L}_n(f)] + [\mathscr{L}_n(f) - J(f)]. \end{aligned} \tag{17}$$

(Here we used the fact that $J(f_0^c) = \ln[1 + c(b-a)] + J(f_0)$ and $\mathscr{L}_n(f_0^c) = \ln[1 + c(b-a)] + \mathscr{L}_n(f_0)$.) The first and the third differences on the right side are differences between the averages of independent identically distributed random variables $\mathscr{L}_n(f_0)$ and $\mathscr{L}_n(f)$, and their corresponding mathematical expectations. Due to the Central Limit Theorem, $[\mathscr{L}_n(f) - J(f)]$ is asymptotically normally distributed with zero mean and variance

$$\begin{aligned} \sigma^2 &= \frac{1}{n} \mathrm{Var}\,[\ln f(X)] = \frac{1}{n} \left[\int_I (\ln f(x))^2 f_0(x)\,\mathrm{d}x - \left(\int_I \ln f(x) f_0(x)\,\mathrm{d}x \right)^2 \right] \\ &\le \frac{1}{n} \sup_{x \in I} (\ln f(x))^2 \int_I f_0(x)\,\mathrm{d}x = \frac{1}{n} \sup_{x \in I} (\ln f(x))^2. \end{aligned}$$

Since f is Lipschitz with Lipschitz constant $l(1+c(b-a))$ and bounded below by $c>0$, we have an estimate

$$\sigma^2 \le \frac{1}{n}\sup_{x\in I}(\ln f(x))^2 \le \frac{1}{n}K_1((\ln c)^2+(\ln l)^2+1)$$

with some constant K_1 independent of f.

The value of c does not have to be independent of n, so we set $c=n^{-\alpha}$ with some $\alpha>0$ to be specified later. Thus we have asymptotic estimate of variance

$$\sigma^2 \le \frac{K_1(\ln n^{-\alpha})^2}{n} = \frac{\alpha^2 K_1(\ln n)^2}{n} \to 0 \quad \text{as } n\to\infty. \tag{18}$$

Therefore, for any $z>0$ and any $f\in\mathscr{F}_c$, $f\ge c$, we have, asymptotically,

$$P((\mathscr{L}_n(f)-J(f))>z) = 1-\Phi\left(\frac{z}{\sigma}\right) \le 1-\Phi\left(\frac{\sqrt{n}}{\alpha\sqrt{K_1}\ln n}z\right),$$

where

$$\Phi(y)=\frac{1}{\sqrt{2\pi}}\int_{-\infty}^{y} \mathrm{e}^{-\frac{x^2}{2}}\,\mathrm{d}x$$

is the cumulative standard normal distribution.

In what follows, we will use the following fact: If $Y:=Y_1+Y_2$ is a sum of two (not necessarily independent) random variables, then for any $\beta>0$

$$P(Y>\beta)\le P(Y_1>\beta/2)+P(Y_2>\beta/2). \tag{19}$$

Indeed, for $Y>\beta$ it is necessary that either $Y_1>\beta/2$ or $Y_2>\beta/2$. Further, if events A and B are such that A implies B, then $P(A)\le P(B)$. Therefore

$$P(Y>\beta)\le P(Y_1>\beta/2 \text{ or } Y_2>\beta/2)\le P(Y_1>\beta/2)+P(Y_2>\beta/2),$$

which proves (19).

Taking (19) into account, for the sum of the first and the third differences on the right side of (17) we have the estimate

$$\begin{aligned} P[(J(f_0)-\mathscr{L}_n(f_0))+(\mathscr{L}_n(f)-J(f))>z] &\le P\left[J(f_0)-\mathscr{L}_n(f_0)>\frac{z}{2}\right] \\ &\quad +P\left[J(f)-\mathscr{L}_n(f)>\frac{z}{2}\right] \\ &\le 2\left[1-\Phi\left(\frac{\sqrt{n}}{2\alpha\sqrt{K_1}\ln n}z\right)\right]. \end{aligned} \tag{20}$$

Set $f=\hat{f}_n$ in (17) and notice that, for the second difference,

$$\mathscr{L}_n(f_0^c)-\mathscr{L}_n(\hat{f}_n^c)\le \ln[1+c(b-a)]. \tag{21}$$

Indeed,

$$\begin{aligned}\mathscr{L}_n(f_0^c) - \mathscr{L}_n(\hat{f}_n^c) =& \{\ln[1+c(b-a)] + \mathscr{L}_n(f_0)\} - [\mathscr{L}_n(\hat{f}_n^c) - \mathscr{L}_n(\hat{f}_n)] \\ & - [\mathscr{L}_n(\hat{f}_n) - \mathscr{L}_n(f_0)] - \mathscr{L}_n(f_0) \\ =& \ln[1+c(b-a)] - [\mathscr{L}_n(\hat{f}_n^c) - \mathscr{L}_n(\hat{f}_n)] - [\mathscr{L}_n(\hat{f}_n) - \mathscr{L}_n(f_0)].\end{aligned}$$

The difference $[\mathscr{L}_n(\hat{f}_n^c) - \mathscr{L}_n(\hat{f}_n)]$ on the right side is nonnegative because $\hat{f}_n^c \geq \hat{f}_n$, the difference $[\mathscr{L}_n(\hat{f}_n) - \mathscr{L}_n(f_0)]$ is nonnegative because $\hat{f}_n$ maximizes the log-likelihood $\mathscr{L}_n$, whence (21) follows. Combining now (16), (17), and (21), we obtain

$$\int_I \frac{f_0^c(x) - \hat{f}_n^c(x)}{1+c(b-a)} \, \mathrm{d}x + \int_I \frac{(\hat{f}_n^c(x) - f_0^c(x))^2}{2\xi^2(x)} f_0(x) \, \mathrm{d}x \leq Y_n + \ln[1+c(b-a)], \quad (22)$$

where Y_n is a random variable (as a function of realization of $x_1, \ldots, x_n$) defined by

$$Y_n := [J(f_0) - \mathscr{L}_n(f_0)] + \left[\mathscr{L}_n(\hat{f}_n^c) - J(\hat{f}_n^c)\right] \to 0 \quad \text{as } n \to \infty$$

due to (20).

Taking into account that $\hat{f}_n^c(x) = \hat{f}_n(x) + O(c)$ and $[\hat{f}_n^c(x) - f_0^c(x)]^2 = [\hat{f}_n(x) - f_0(x)]^2 + O(c)$, we can write (22) as

$$\int_I [f_0(x) - \hat{f}_n(x)] \, \mathrm{d}x + \int_I \frac{[\hat{f}_n(x) - f_0(x)]^2}{2\xi^2(x)} f_0(x) \, \mathrm{d}x \leq Y_n + O(n^{-\alpha}). \quad (23)$$

The right side of (23) converges to zero in probability as $n \to \infty$; therefore, so does each term on the left side since both integrals are necessarily nonnegative. From the convergence

$$\int_I [f_0(x) - \hat{f}_n(x)] \, \mathrm{d}x \to 0 \quad \text{as } n \to \infty,$$

it follows that $\int_I \hat{f}_n(x) \, \mathrm{d}x \to 1$ as $n \to \infty$, hence, for any $x \in [a,b] \setminus I$, $\hat{f}_n(x) \to 0 = f_0(x)$.

To prove the convergence $\hat{f}_n(x) \to f_0(x)$ on the set I, consider the estimate

$$\int_I \frac{[\hat{f}_n(x) - f_0(x)]^2}{2\xi^2(x)} f_0(x) \, \mathrm{d}x \leq Y_n + O(n^{-\alpha}),$$

which follows from (23). Since ξ is a function between f_0^c and $\hat{f}_n^c$, we have

$$0 < \xi \leq \max_{y \in I} \left\{ f_0^c(y), \hat{f}_n^c(y) \right\} = K_2(l+1) + O(c),$$

with some uniform constant K_2. Therefore

$$\int_I [\hat{f}_n(x) - f_0(x)]^2 f_0(x) \, \mathrm{d}x \leq 2[K_2(l+1)]^2 [Y_n + O(n^{-\alpha})]. \quad (24)$$

The remainder of the proof of Theorem 1 is based on the following propositions.

Proposition 1. *Let $\Omega = \bigcup_i [a_i, b_i]$ be a union of finitely or countably many nonoverlapping intervals with $a \le a_i < b_i \le b$ for all i. Further, let $\varphi_n : \Omega \to R$ be a sequence of uniformly bounded nonnegative Lipschitz functions with a common Lipschitz constant $\ell > 0$, $\psi : \Omega \to R$ be continuous and positive on Ω except at the endpoints c_i and d_i (where $\psi(c_i) = \psi(d_i) = 0$), and let*

$$\int_\Omega \varphi_n(x)\psi(x)\,\mathrm{d}x \le a_n$$

for some sequence $a_n \to 0$ (deterministically). Then the sequence φ_n converges uniformly to zero on Ω, and, for any $x \in \Omega$ such that $\psi(x) > 0$, we have the estimate

$$\varphi_n(x) \le \sqrt{\frac{16\ell}{3\psi(x)} a_n}. \tag{25}$$

Proposition 2. *Let $\varphi_n : [c,d] \to R$ be a sequence of uniformly bounded nonnegative Lipschitz functions with a common Lipschitz constant $\ell > 0$ and let*

$$\int_c^d \varphi_n(x)\,\mathrm{d}x \le a_n$$

for some sequence $a_n \to 0$ (deterministically). Then for any $x \in [c,d]$, we have the estimate

$$\varphi_n(x) \le \sqrt{\frac{8\ell}{3} a_n}. \tag{26}$$

The proof of these propositions can be found in the Appendix.

Since the closure of I has the same structure as Ω in Proposition 1, the right side of (24) tends to zero in probability as $n \to \infty$ and the difference $(\hat{f}_n - f_0)^2$ is Lipschitz with Lipschitz constant $K_3 l(l+1)$, we conclude that $f_n \to f_0$ uniformly in probability on $[a,b]$. Estimate (25) implies that

$$[\hat{f}_n(x) - f_0(x)]^2 \le \left\{ \frac{16 \times 2K_3 l(l+1) K_2^2 (l+1)^2}{3 f_0(x)} [Y_n + O(n^{-\alpha})] \right\}^{1/2}$$

or, equivalently,

$$|\hat{f}_n(x) - f_0(x)| \le \left\{ \frac{32 K_3 K_2^2 l (l+1)^3}{3 f_0(x)} [Y_n + O(n^{-\alpha})] \right\}^{1/4}.$$

Therefore, for any $z > 0$

$$
\begin{aligned}
P(|\hat{f}_n(x) - f_0(x)| > z) &\le P\left[\frac{32K_3K_2^2 l(l+1)^3}{3f_0(x)}(Y_n + O(n^{-\alpha})) > z^4\right] \\
&= P\left[Y_n + O(n^{-\alpha}) > \frac{3f_0(x)}{32K_3K_2^2 l(l+1)^3} z^4\right] \\
&\le P\left[Y_n > \frac{1}{2}\frac{3f_0(x)}{32K_3K_2^2 l(l+1)^3} z^4\right] \\
&\quad + P\left[O(n^{-\alpha}) > \frac{1}{2}\frac{3f_0(x)}{32K_3K_2^2 l(l+1)^3} z^4\right],
\end{aligned}
$$

where the last inequality is due to (19).

Since α in $O(n^{-\alpha})$ can be set arbitrarily large, we can assume without loss of generality that the second probability on the right side of the formula above is zero. Due to (20), we arrive at the estimate

$$
P(|\hat{f}_n(x) - f_0(x)| > z) \le 2\left\{1 - \Phi\left[\frac{Kf_0(x)}{l(l+1)^3}\frac{\sqrt{n}}{\ln n} z^4\right]\right\},
$$

where $K = [(3/128)\alpha\sqrt{K_1}K_3K_2^2]^{-1}$. This proves (6).

The proof (7) similarly follows from Proposition 2 and the estimate

$$
\int_I [f_0(x) - \hat{f}_n(x)]\,\mathrm{d}x \le Y_n + O(n^{-\alpha}),
$$

which is equivalent to

$$
\int_{[a,b]\setminus I} \hat{f}_n(x)\,\mathrm{d}x \le Y_n + O(n^{-\alpha}).
$$

Appendix

In this section, we prove Propositions 1 and 2.

Proof of Proposition 1. Let us notice first that for any $x \in \Omega$ such that $\psi(x) > 0$ we have $\varphi_n(x) \to 0$. Indeed, take a sufficiently small interval I_x containing x such that $\psi \ge \gamma_1 > 0$ on I_x. If, arguing by the contrary, we assume $\varphi_n(x) \ge \gamma_2 > 0$ for all n, then, from Lipschitz continuity of φ_n, we can show that $\int_{I_x} \varphi_n(x)\psi(x)\,\mathrm{d}x$ does not converge to zero as $n \to \infty$, which is not possible. Convergence $\varphi_n(x) \to 0$ at the points where $\psi(x) = 0$ (the set of such points has measure zero by assumption) follows from Lipschitz continuity of φ_n. Hence, $\varphi_n(x) \to 0$ pointwise on Ω.

It is not difficult to establish that the convergence $\varphi_n \to 0$ on Ω is, actually, uniform. Indeed, if it was not the case, there would exist $\varepsilon > 0$ and a subsequence from

$n = 1,2,\dots$ (we do not relabel) such that $\sup_{x\in\Omega} |\varphi_n(x)| \geq \varepsilon$. However, since the sequence φ_n, $n = 1,2,\dots$ is uniformly bounded and equicontinuous, it must contain a uniformly convergent subsequence converging to the pointwise limit, which we excluded by the assumption $\sup_{x\in\Omega} |\varphi_n(x)| \geq \varepsilon$.

Let us prove estimate (25). Take $x^* \in \Omega$ such that $\psi(x^*) > 0$. Then on the interval $[x^*, x^* + \delta]$ for sufficiently small δ we have $\psi(x) \geq \frac{\psi(x^*)}{2}$ (if x^* happens to coincide with the right endpoint of the set Ω, consider the interval $[x^* - \delta, x^*]$ instead). For n sufficiently large we have $\delta \geq \varphi_n(x^*)/(2\ell)$ and, since φ_n is Lipschitz with Lipschitz constant ℓ, we have

$$\varphi_n(x) \geq \varphi_n(x^*) - \ell(x - x^*), \quad x \geq x^*$$

Therefore

$$a_n \geq \int_\Omega \varphi_n(x)\psi(x)\,\mathrm{d}x \geq \frac{\psi(x^*)}{2}\int_{x^*}^{x^*+\delta} \varphi_n(x)\,\mathrm{d}x \geq \frac{\psi(x^*)}{2}\int_{x^*}^{x^*+\varphi_n(x^*)/(2\ell)} \varphi_n(x)\,\mathrm{d}x$$
$$\geq \frac{\psi(x^*)}{2}\int_{x^*}^{x^*+\varphi_n(x^*)/(2\ell)} [\varphi_n(x^*) - \ell(x - x^*)]\,\mathrm{d}x = \frac{\psi(x^*)}{2}\frac{3\varphi_n^2(x^*)}{8\ell},$$

which implies the required estimate

$$\varphi_n(x^*) \leq \sqrt{\frac{16\ell}{3\psi(x^*)}a_n}$$

Proof of Proposition 2. Similarly to the proof of Proposition 1, we have

$$a_n \geq \int_{x^*}^{x^*+\delta} \varphi_n(x)\,\mathrm{d}x \geq \int_{x^*}^{x^*+\varphi_n(x^*)/(2\ell)} \varphi_n(x)\,\mathrm{d}x$$
$$\geq \int_{x^*}^{x^*+\varphi_n(x^*)/(2\ell)} [\varphi_n(x^*) - \ell(x - x^*)]\,\mathrm{d}x = \frac{3\varphi_n^2(x^*)}{8\ell},$$

which implies the required estimate (26).

References

1. Cox, D., O'Sullivan, F.: Asymptotic analysis of penalized likelihood and related estimators. The Annals of Statistics, **18**(4), 1676–1695 (1990)
2. Cramer, H: Mathematical Methods of Statistics. Princeton University Press, Princeton (1999)
3. Eggermont, P.P.B., LaRiccia, V.N.: Optimal convergence rates for good's nonparametric maximum likelihood density estimator. The Annals of Statistics, **27**(5), 1600–1615 (1999)
4. Fougéres, A.-L.: Estimation de densites unimodales. The Canadian Journal of Statistics/La Revue Canadienne de Statistique, **25(3)**, 375–387 (1997)
5. Good, I.J., Gaskins, R.A.: Nonparametric roughness penalties for probability densities. Biometrika, **58**, 255–277 (1971)
6. Härdle, W., Janssen, P., Serfling, R.: Strong uniform consistency rates for estimators of conditional functionals. The Annals of Statistics, **16**(4), 1428–1449 (1995)

7. Huang, J., Wellner, J.A.: Estimation of a monotone density or monotone hazard under random censoring. Scandinavian Journal of Statistics, **22**(1), 3–33 (1995)
8. Marshall, A., Proschan, F.: Maximum likelihood estimation for distributions with monotone failure rate. The Annals of Mathematical Statistics, **36**(1), 69–77 (1965)
9. de Montricher, G.F., Tapia, R.A., Thompson, J.R.: Nonparametric maximum likelihood estimation of probability densities by penalty function methods. The Annals of Statistics, **3**(6), 1329–1348 (1975)
10. Mordukhovich, B.S.: Methods of approximation in problems of optimization and control (in Russian). Nauka, Moscow (1988)
11. Mordukhovich, B.S.: Variational analysis and generalized differentiation. Springer, Berlin (2005)
12. Rosenblatt, M: Remarks on some non-parametric estimates of a density function. The Annals of Mathematical Statistics, **27**, 832–837 (1957)
13. Samiuddin, M, El-Sayyad, G.M.: On nonparametric kernel density estimates. Biometrika, **77**(4), 865–874 (1990)
14. Schucany, W.R., Sommers, J.P.: Improvement of kernel type density estimators. Journal of the American Statistical Association, **72**(358), 420–423 (1977)
15. Silverman, B.W.: On weak and strong uniform consistency of the kernel estimate of a density function and its derivatives. The Annals of Statistics, **6**, 177–184 (1978)
16. Silverman, B.W.: Density Estimation for Statistics and Data Analysis. Chapman and Hall, London (1992)
17. Vinter, R.B.: Optimal Control. Birkhäuser, Boston (2000)
18. Wegman, E.: Maximum likelihood estimation of a unimodal density function. The Annals of Mathematical Statistics, **41**(2), 457–471 (1970)
19. Wegman, E.: Maximum likelihood estimation of a probability density function. Sankhya: The Indian Journal of Statistics, Series A, **37**(2), 211–224 (1975)
20. Whittle, P.: On the smoothing of probability density functions. Journal of the Royal Statistical Society, Series B, **20**(2), 334–343 (1958)

Parametric Variational System with a Smooth-Boundary Constraint Set

J.-C. Yao and N.D. Yen

Abstract Solution stability of parametric variational systems with smooth-boundary constraint sets is investigated. Sufficient conditions for the lower semicontinuity, Lipschitz-like property, and local metric regularity in Robinson's sense of the solution map are obtained by using a calculus rule for the normal second-order subdifferential from B.S. Mordukhovich (*Variational Analysis and Generalized Differentiation, Vol. I: Basic Theory, Vol. II: Applications*, Springer, Berlin, 2006) and the implicit function theorems for multifunctions from G.M. Lee, N.N. Tam and N.D. Yen (J Math Anal Appl 338:11–22, 2008).

1 Introduction

Let X be a Banach space with the dual X^* and the second dual X^{**}. The canonical pairing between a Banach space and its dual is denoted by $\langle , \rangle$. Given a C^2-smooth function $\psi : X \to \mathbb{R}$, we put

$$C = \{x \in X \, : \, \psi(x) \le 0\}. \tag{1}$$

Let there be given also a single-valued map $f : X \times P \to X^*$, where P is a subset of a normed space and $f(\cdot, p)$ is a C^1-smooth function for each $p \in P$. Consider the variational system

$$0 \in f(x,p) + N(x;C) \tag{2}$$

J.-C. Yao
Department of Applied Mathematics, National Sun Yat-Sen University, Kaohsiung 80424, Taiwan
e-mail: yaojc@math.nsysu.edu.tw

N.D. Yen
Institute of Mathematics, Vietnamese Academy of Science and Technology,
18 Hoang Quoc Viet, Hanoi 10307, Vietnam
e-mail: ndyen@math.ac.vn

R.S. Burachik and J.-C. Yao (eds.), *Variational Analysis and Generalized Differentiation in Optimization and Control*, Springer Optimization and Its Applications 47,
DOI 10.1007/978-1-4419-0437-9_11,

with $p \in P$ being a parameter. Here $N(x;C)$ denotes the *normal cone* in the sense of Mordukhovich to a C at x (see [12, Definition 1.1] and a definition given below), which coincides with the Clarke normal cone to a C at x (see [4, Corollary 2, pp. 56–57] and Lemma 3.7 in Sect. 3). Since the boundary ∂C of C is given by

$$\partial C = \{x \in X : \psi(x) = 0\}$$

and ψ is a C^2-smooth function, we will call (2) a *parametric variational system with a smooth-boundary constraint set*. The solution set of (2) will be denoted by $S(p)$.

Since it is not assumed that the function ψ is convex, C can be convex or non-convex as well. If C is convex, then

$$N(x;C) = \{x^* \in X^* : \langle x^*, u - x\rangle \le 0 \text{ for all } u \in C\}$$

and (2) can be rewritten equivalently as a *variational inequality* in the classical form

$$x \in C, \quad \langle f(x,p), u - x\rangle \ge 0, \quad \forall u \in C. \tag{3}$$

Thus (2) is a generalization of (3).

Stability of the solution map of parametric variational systems in general, and of parametric variational inequalities in particular, has attracted attention of many researchers. We refer to Robinson [17] for a pioneering paper and to Mordukhovich [12, Chap. 4] (see also a recent paper by Aragón Artacho and Mordukhovich [1]) for significant results together with fresh and comprehensive information on the subject.

Our aim in this chapter is to investigate a new way of deriving verifiable sufficient conditions for the lower semicontinuity, Lipschitz-like property, and local metric regularity in Robinson's sense of the solution map of (2). We will combine a calculus rule for the normal second-order subdifferential from [12, Theorem 1.127] with the implicit function theorems for multifunctions from [9]. This approach allows us to avoid computing coderivative of $f(x,p)$ on both variables; thus it is somewhat different from the one used in [12, Chap. 4] (see also [21] and the references therein). Some technical difficulties related to the appearance of a metric projection operator in the implicit function theorems of [9] will be overcome by employing the specific structure of (2) and the generalized differentiation theory from [12].

It is well known that (1) lower semicontinuity of the solution map is an important stability sign of a parametric system, (2) the Lipschitz-like property of multifunctions (called also the pseudo-Lipschitz property, the Aubin property, the Aubin continuity property) was introduced by Aubin [2], and (3) the local metric regularity in Robinson's sense of implicit multifunction (called also the Robinson robust stability [3]) has the origin in the classical paper of Robinson [16]. The interested reader is referred to [3] for a recent study on the relationships between the Robinson robust stability and the Aubin continuity property of implicit multifunctions. (Note that the solution map $S(\cdot)$ of (2) is an implicit multifunction of a special type.)

Concerning the special constraint set in (1), we would like to make some remarks. If C is the solution set of a system of finitely many inequalities or, more generally, C is the solution set of a generalized inequality system of the form

$$C = \{x \in X : \Psi(x) \in K\}, \tag{4}$$

where $\Psi : X \to Z$ is a map between Banach spaces and $K \subset Z$ is a closed convex cone, then ∂C may have many "corners" (roughly speaking, a corner is a point where the contingent cone – see for example [12, Definition 1.8] – to ∂C is not a linear subspace of X). In this case, the stability analysis of (2) is rather complicated. Various index sets are needed to describe the normal and Fréchet coderivatives of the normal-cone mapping $N(\cdot;C) : X \rightrightarrows X^*$. The case where Ψ is an affine operator and K is a polyhedral convex set have been considered, e.g., by Dontchev and Rockafellar [5], Henrion and Römisch [8], Yao and Yen [18,19] (X is a finite dimensional Euclidean space), Henrion, Mordukhovich, and Nam [6] (X is a finite or infinite dimensional Banach space). The case where Ψ is a nonlinear mapping and X is finite dimensional has been considered by Mordukhovich and Outrata [13], Henrion, Outrata, and Surowiec [7]. In contrast to the case of the generalized inequality system (4), in (1) there is only one inequality given by a C^2-smooth function. If $\bar{x} \in \partial C$ and $\nabla \psi(\bar{x}) \neq 0$, then the contingent cone to ∂C is a linear subspace of codimension 1 in X. The differentiability analysis of the normal-cone operator $N(\cdot;C)$ is simpler than that in the just mentioned papers: no index set is needed. Meanwhile, many mechanical bodies can be represented in the form (1). Hence (2) can model certain equilibrium problems in mechanics. Since we have studied [18,19] the solution stability of variational inequalities with a polyhedral convex constraint set, here we will focus our attention on the behavior of the solution maps of parametric variational systems with smooth-boundary constraint sets. The obtained results help us to understand deeper the value of the central result of the second-order subdifferential calculus in general Banach spaces [12, Theorem 1.127]. The interested reader is referred to [12, pp. 167–170] for detailed comments on the development of the second-order generalized differentiation theory.

After some preliminaries, we derive formulae for computing the normal coderivative of the multifunction $f(\cdot,p) + N(\cdot;C)$ in Sect. 2. Stability of the solution map $S(\cdot)$ is studied in Sect. 3. An analysis of Theorem 3.1, the main result of this chapter, is given in Sect. 4.

We now present several notations and notions that will be needed in the sequel.

For a subset $\Omega \subset X$, the symbols $\overline{\Omega}$ and $\operatorname{int}\Omega$, respectively, denote the closure of Ω and the interior of Ω. The distance from $x \in X$ to Ω is

$$\operatorname{dist}(x,\Omega) := \inf\{\|x-u\| : u \in \Omega\},$$

where $\inf\emptyset := +\infty$. Denote by $B_\rho(x)$ and B_X, respectively, the closed ball centered at x with radius ρ and the closed unit ball in X. If $A : X \to Y$ is a bounded linear operator, then A^* denotes the adjoint of A. The symbol $\mathscr{L}(X,Y)$ stands for the set of bounded linear operators mapping X into Y. Let $\overline{\mathbb{R}} = [-\infty,\infty]$.

A multifunction $\Phi : X \rightrightarrows Y$ between Banach spaces is said to be *lower semicontinuous* at $x \in \operatorname{dom}\Phi := \{x \in X : \Phi(x) \neq \emptyset\}$ if for any open set $V \subset Y$ satisfying $V \cap \Phi(x) \neq \emptyset$ there exists a neighborhood U of x such that $V \cap \Phi(u) \neq \emptyset$ for all $u \in U$. One says that Φ is *lower semicompact* on its effective domain around $\bar{x} \in X$

if there exists a neighborhood U of $\bar{x}$ such that for any $x \in U$ and any sequence $x_k \xrightarrow{\mathrm{dom}\,\Phi} x$, there is a sequence $y_k \in \Phi(x_k)$, $k = 1,2,\ldots$, which contains a subsequence convergent in the norm topology of Y. Here the notation $x_k \xrightarrow{\mathrm{dom}\,\Phi} x$ means $x_k \to x$ and $x_k \in \mathrm{dom}\,\Phi$ for all $k \in \mathbb{N}$.

A Banach space X is called *Asplund* if every convex continuous function $\varphi : U \to \mathbb{R}$ defined on an open convex subset U of X is Fréchet differentiable on a dense subset of U; see [12, Definition 2.17]. The class of Asplund spaces is large. For instance, any reflexive Banach space is an Asplund space. The calculus of normal cones, coderivatives, and subdifferentials in Asplund spaces is simpler than that in general Banach spaces [12, Chap. 3].

For a multifunction $\Phi : X \rightrightarrows X^*$, the expression $\mathrm{Lim\,sup}_{x\to\bar{x}}\,\Phi(x)$ denotes the sequential Kuratowski–Painlevé upper limit of $\Phi(x)$ as $x \to \bar{x}$ with respect to the norm topology of X and the weak* topology of X^*, i.e.,

$$\mathop{\mathrm{Lim\,sup}}_{x\to\bar{x}} \Phi(x) = \Big\{x^* \in X^* : \exists \text{ sequences } x_k \to \bar{x},\ x_k^* \xrightarrow{w^*} x^*, \\ \text{with } x_k^* \in \Phi(x_k) \text{ for all } k = 1,2,\ldots\Big\}.$$

Let us recall some basic concepts of variational analysis from [12].

The set $\widehat{N}_\varepsilon(x;\Omega)$ of the Fréchet ε-normals to Ω at $x \in \overline{\Omega}$ is given by

$$\widehat{N}_\varepsilon(x;\Omega) = \left\{x^* \in X^* : \limsup_{u \xrightarrow{\Omega} x} \frac{\langle x^*, u-x\rangle}{\|u-x\|} \le \varepsilon\right\}. \tag{5}$$

One puts $\widehat{N}_\varepsilon(x;\Omega) = \emptyset$ for all $\varepsilon \ge 0$ whenever $x \notin \overline{\Omega}$. The set

$$N(\bar{x};\Omega) := \mathop{\mathrm{Lim\,sup}}_{x\to\bar{x},\ \varepsilon\downarrow 0} \widehat{N}_\varepsilon(x;\Omega)$$

is the *normal cone* in the sense of Mordukhovich to Ω at $\bar{x}$. If $\bar{x} \notin \overline{\Omega}$, then one puts $N(\bar{x};\Omega) = \emptyset$.

Let $\Phi : X \rightrightarrows Y$ be a set-valued map between Banach spaces. The multifunction $D_N^*\Phi(\bar{x},\bar{y}) : Y^* \rightrightarrows X^*$ defined by

$$D_N^*\Phi(\bar{x},\bar{y})(y^*) := \{x^* \in X^* : (x^*, -y^*) \in N((\bar{x},\bar{y}); \mathrm{gph}\,\Phi)\}, \quad y^* \in Y^*, \tag{6}$$

where $\mathrm{gph}\,\Phi := \{(x,y) \in X \times Y : y \in \Phi(x)\}$ denotes the graph of Φ, is said to be the *normal coderivative* (called also the *limiting coderivative* and the *coderivative in the sense of Mordukhovich*) of Φ at $(\bar{x},\bar{y})$. We put $D_N^*\Phi(\bar{x},\bar{y})(y^*) = \emptyset$ whenever $(\bar{x},\bar{y}) \notin \overline{\mathrm{gph}\,\Phi}$.

Suppose that $\varphi : X \to \overline{\mathbb{R}}$ is finite at $\bar{x} \in X$. The set

$$\partial\varphi(\bar{x}) := \{x^* \in X^* : (x^*, -1) \in N((\bar{x}, \varphi(\bar{x})); \mathrm{epi}\,\varphi)\},$$

where $\text{epi}\,\varphi := \{(x,\alpha) \in X \times \mathbb{R} : \alpha \geq \varphi(x)\}$ denotes the epigraph of φ, is said to be the *basic subdifferential* (or the *limiting subdifferential*) of φ at $\bar{x}$. Let $\bar{v} \in \partial\varphi(\bar{x})$, i.e., $(\bar{x},\bar{v})$ belongs to the graph of the subdifferential mapping $\partial\varphi : X \rightrightarrows X^*$, $x \mapsto \partial\varphi(x)$. The map $\partial_N^2\varphi(\bar{x},\bar{v}) : X^{**} \rightrightarrows X^*$ with the values

$$\partial_N^2\varphi(\bar{x},\bar{v})(u) := (D_N^*\partial\varphi)(\bar{x},\bar{v})(u), \quad u \in X^{**},$$

is called the *normal second-order subdifferential* of φ at $\bar{x}$ relative to $\bar{v}$; see [12, Definition 1.118].

As it will be seen in the next sections, normal second-order subdifferentials are very useful for studying solution stability/sensitivity of (2).

2 Formulae for Coderivative

Using some results from [12, Chap. 1], we will compute the coderivative of the multifunction

$$F(\cdot,p) : X \rightrightarrows X^*, \quad F(x,p) := f(x,p) + N(x;C), \tag{7}$$

where $f(x,p)$ and $N(x;C)$ are as in (2).

Consider the *normal-cone operator* $N(\cdot;C) : X \rightrightarrows X^*$, where C is given by (1). Formulae for computing the normal coderivative of $N(\cdot;C)$ and a point $(\bar{x},\bar{v}) \in \text{gph}N(\cdot;C)$ will be obtained by using the central result of the second-order subdifferential calculus in general Banach spaces which is stated as follows. (We mention only the formula for computing the normal second-order subdifferential of the composite function, omitting the formula for the mixed second-order subdifferential.)

Theorem 2.1. (see [12, Theorem 1.127]). *Let $\bar{v} \in \partial(\varphi \circ g)(\bar{x})$ with $g : X \to Y$ and $\varphi : Y \to \overline{\mathbb{R}}$. Assume that g is C^1-smooth around $\bar{x}$ with the surjective derivative $\nabla g(\bar{x}) : X \to Y$ and the derivative mapping $\nabla g : X \to \mathscr{L}(X,Y)$ is strictly differentiable at $\bar{x}$. Let $y^* \in Y^*$ be a unique element satisfying*

$$\bar{v} = \nabla g(\bar{x})^* y^* \quad \text{and} \quad y^* \in \partial\varphi(\bar{y}) \quad \text{with } \bar{y} := g(\bar{x}).$$

Then for all $u \in X^{**}$ one has

$$\partial_N^2(\varphi \circ g)(\bar{x},\bar{v})(u) \subset \nabla^2\langle y^*, g\rangle(\bar{x})^* u + \nabla g(\bar{x})^* \partial_N^2\varphi(\bar{y},y^*)(\nabla g(\bar{x})^{**} u).$$

Moreover, the latter becomes an equality if the range of $\nabla g(\bar{x})^*$ is w^*-extensible [12, Definition 1.122] *in X^*. This is true under one of the following conditions:*

(a) *The range of $\nabla g(\bar{x})^*$ is complemented in X^* (it occurs, in particular, when the kernel of $\nabla g(\bar{x})$ is complemented in X).*

(b) *The closed unit ball of X^{**} is weak* sequentially compact (it occurs, in particular, when either X is reflexive or X^* is separable).*

Let $\bar{x} \in C$ and $\bar{v} \in X^*$ be such that $\bar{v} \in N(\bar{x};C)$, that is $(\bar{x},\bar{v}) \in \mathrm{gph}N(\cdot;C)$. Setting $Y = \mathbb{R}$, $g = \psi$, and $\varphi(y) = 0$ if $y \le 0$, $\varphi(y) = +\infty$ if $y > 0$ (i.e., $\varphi : Y \to \overline{\mathbb{R}}$ is the indicator function of the half-line $\mathbb{R}_-$), we remark that $\varphi \circ \psi$ coincides with the indicator function of C. Hence

$$N(x;C) = \partial(\varphi \circ \psi)(x), \quad \forall x \in X.$$

It follows that

$$D_N^* N(\cdot;C)(\bar{x},\bar{v})(u) = \partial_N^2(\varphi \circ \psi)(\bar{x},\bar{v})(u), \quad \forall u \in X^{**}.$$

If $\psi(\bar{x}) < 0$, then $\bar{y} < 0$ and $\varphi(\bar{y}) = \{0\}$. This implies $\bar{v} = 0$. Besides, since

$$\partial(\varphi \circ \psi)(x) = N(x;C) = \{0\}, \quad \forall x \in \mathrm{int}C,$$

we have

$$\partial_N^2(\varphi \circ \psi)(\bar{x},\bar{v})(u) = \big(D_N^*\partial(\varphi \circ \psi)\big)(\bar{x},\bar{v})(u) = \{0\}, \quad \forall u \in X^{**}.$$

Suppose now that $\psi(\bar{x}) = 0$ and $\nabla\psi(\bar{x}) \neq 0$. By the chain rule in [12, Proposition 1.112],

$$\partial(\varphi \circ \psi)(\bar{x}) = \nabla\psi(\bar{x})^*\partial\varphi(\bar{y})$$

with $\bar{y} := \psi(\bar{x})$. Since $\bar{v} \in N(\bar{x};C)$, we infer that there is $y^* \in \partial\varphi(\bar{y})$ such that

$$\bar{v} = \nabla\psi(\bar{x})^* y^*. \tag{8}$$

As $\nabla\psi(\bar{x}) \neq 0$, $\nabla\psi(\bar{x}) : X \to \mathbb{R}$ is surjective, hence $\nabla\psi(\bar{x})^* : \mathbb{R} \to X^*$ is injective according to [12, Lemma 1.18]. In this case, for each $\bar{v} \in N(\bar{x};C)$ there is a unique $y^* \in \partial\varphi(\bar{y})$ satisfying (8). Since

$$\ker\nabla\psi(\bar{x}) := \{h \in X : \nabla\psi(\bar{x})(h) = 0\}$$

is a closed linear subspace of X with codimension 1, $\ker\nabla\psi(\bar{x})$ is complemented in X, i.e., there exists a closed linear subspace $L \subset X$ such that $X = \big(\ker\nabla\psi(\bar{x})\big) \oplus L$. By virtue of Theorem 2.1, for all $u \in X^{**}$ one has

$$\partial_N^2(\varphi \circ \psi)(\bar{x},\bar{v})(u) = \nabla^2\langle y^*,\psi\rangle(\bar{x})^*u + \nabla\psi(\bar{x})^*\partial_N^2\varphi(\bar{y},y^*)(\nabla\psi(\bar{x})^{**}u). \tag{9}$$

As

$$\partial\varphi(y) = \begin{cases} \{0\} & \text{if } y < 0, \\ [0,+\infty) & \text{if } y = 0, \\ \emptyset & \text{if } y > 0, \end{cases}$$

for every $\alpha \in \mathbb{R}$ and $\bar{y} = \psi(\bar{x}) = 0$, we find that

$$
\begin{aligned}
&\partial_N^2 \varphi(\bar{y}, y^*)(\alpha) \\
&= \Big\{\beta \in \mathbb{R} : (\beta, -\alpha) \in N\big((\bar{y}, y^*); \mathrm{gph}\,\partial\varphi\big)\Big\} \\
&= \begin{cases} \{\beta : (\beta, -\alpha) \in \mathbb{R} \times \{0\}\} & \text{if } y^* > 0 \\ \{\beta : (\beta, -\alpha) \in (\mathbb{R}_+ \times \mathbb{R}_-) \cup (\mathbb{R}_- \times \{0\}) \cup (\{0\} \times \mathbb{R}_+)\} & \text{if } y^* = 0 \\ \emptyset & \text{if } y^* < 0 \end{cases} \\
&= \begin{cases} \mathbb{R} & \text{if } y^* > 0, \alpha = 0 \\ \emptyset & \text{if } y^* > 0, \alpha \neq 0 \\ \mathbb{R}_+ & \text{if } y^* = 0, \alpha > 0 \\ \mathbb{R} & \text{if } y^* = 0, \alpha = 0 \\ \{0\} & \text{if } y^* = 0, \alpha < 0 \\ \emptyset & \text{if } y^* < 0. \end{cases}
\end{aligned}
$$

Consequently, for $\alpha := \nabla\psi(\bar{x})^{**}u = \langle u, \nabla\psi(\bar{x})\rangle$ (the value of $u \in X^{**}$ at $\nabla\psi(\bar{x}) \in X^*$) and $\bar{y} = 0$,

$$
\nabla\psi(\bar{x})^* \partial_N^2 \varphi(\bar{y}, y^*)(\nabla\psi(\bar{x})^{**}u) = \begin{cases} \mathbb{R}\nabla\psi(\bar{x}) & \text{if } y^* > 0, \alpha = 0 \\ \emptyset & \text{if } y^* > 0, \alpha \neq 0 \\ \mathbb{R}_+\nabla\psi(\bar{x}) & \text{if } y^* = 0, \alpha > 0 \\ \mathbb{R}\nabla\psi(\bar{x}) & \text{if } y^* = 0, \alpha = 0 \\ \{0\} & \text{if } y^* = 0, \alpha < 0 \\ \emptyset & \text{if } y^* < 0. \end{cases}
$$

Combining this with (9) gives

$$
\partial_N^2 (\varphi \circ \psi)(\bar{x}, \bar{v})(u) = \begin{cases} y^* \nabla^2 \psi(\bar{x})^* u + \mathbb{R}\nabla\psi(\bar{x}) & \text{if } y^* > 0, \langle u, \nabla\psi(\bar{x})\rangle = 0 \\ \emptyset & \text{if } y^* > 0, \langle u, \nabla\psi(\bar{x})\rangle \neq 0 \\ \mathbb{R}_+\nabla\psi(\bar{x}) & \text{if } y^* = 0, \langle u, \nabla\psi(\bar{x})\rangle > 0 \\ \mathbb{R}\nabla\psi(\bar{x}) & \text{if } y^* = 0, \langle u, \nabla\psi(\bar{x})\rangle = 0 \\ \{0\} & \text{if } y^* = 0, \langle u, \nabla\psi(\bar{x})\rangle < 0 \\ \emptyset & \text{if } y^* < 0. \end{cases}
$$

Setting

$$
\Delta := D_N^*\big(f(\cdot, p) + N(\cdot; C)\big)\big(\bar{x}, f(\bar{x}, p) + \bar{v}\big)(u) \tag{10}
$$

and invoking the sum rule in [12, Theorem 1.62(ii)] we have

$$
\begin{aligned}
\Delta &= \nabla_x f(\bar{x}, p)^* u + D_N^* N(\cdot; C)(\bar{x}, \bar{v})(u) \\
&= \nabla_x f(\bar{x}, p)^* u + \partial_N^2 (\varphi \circ \psi)(\bar{x}, \bar{v})(u),
\end{aligned}
$$

where $\nabla_x f(\bar{x}, p)$ is the Fréchet derivative of $f(\cdot, p)$ at $\bar{x}$. Therefore, if $\psi(\bar{x}) < 0$ then

$$\Delta = \begin{cases} \nabla_x f(\bar{x}, p)^* u & \text{for } \bar{v} = 0 \\ \emptyset & \text{otherwise} \end{cases} \tag{11}$$

and, in the case where $\psi(\bar{x}) = 0$,

$$\Delta = \begin{cases} \nabla_x f(\bar{x}, p)^* u + y^* \nabla^2 \psi(\bar{x})^* u + \mathbb{R} \nabla \psi(\bar{x}) & \text{if } y^* > 0, \langle u, \nabla \psi(\bar{x}) \rangle = 0 \\ \emptyset & \text{if } y^* > 0, \langle u, \nabla \psi(\bar{x}) \rangle \neq 0 \\ \nabla_x f(\bar{x}, p)^* u + \mathbb{R}_+ \nabla \psi(\bar{x}) & \text{if } y^* = 0, \langle u, \nabla \psi(\bar{x}) \rangle > 0 \\ \nabla_x f(\bar{x}, p)^* u + \mathbb{R} \nabla \psi(\bar{x}) & \text{if } y^* = 0, \langle u, \nabla \psi(\bar{x}) \rangle = 0 \\ \nabla_x f(\bar{x}, p)^* u & \text{if } y^* = 0, \langle u, \nabla \psi(\bar{x}) \rangle < 0 \\ \emptyset & \text{if } y^* < 0. \end{cases} \tag{12}$$

Theorem 2.2. *Let* $(\bar{x}, \bar{v}) \in \operatorname{gph} N(\cdot; C)$. *If* $\psi(\bar{x}) < 0$ *then* $\bar{v} = 0$, *and for all* $u \in X^{**}$ *one has* (11) *where* Δ *is given by* (10). *If* $\psi(\bar{x}) = 0$ *and* $\nabla \psi(\bar{x}) \neq 0$, *then for all* $u \in X^{**}$ *one has* (12) *where* Δ *is given by* (10) *and* y^* *is defined uniquely via* $\bar{x}$ *and* $\bar{v}$ *by* (8).

3 Stability of the Solution Map

The main result of this chapter can be stated as follows.

Theorem 3.1. *Assume that both* X *and* X^* *are Asplund spaces. Let* $\bar{x}$ *be a solution of the parametric variational system with a smooth-boundary constraint set* (2) *at a given parameter* $p = \bar{p} \in P$, *i.e.,* $\bar{x} \in S(\bar{p})$, *where* C *is defined by* (1) *with* ψ *being* C^2*-smooth. Suppose that* $f(\cdot, p)$ *is a* C^1*-smooth function for each* $p \in P$, *the derivative mapping* $\nabla_x f : X \times P \to X^*$ *is continuous in the norm topology of* X^*, *and the following conditions hold:*

(a1) *There is* $\bar{\alpha} > 0$ *such that* $\langle u, \nabla_x f(\bar{x}, \bar{p})^* u \rangle \geq \bar{\alpha} \|u\|^2$ *for every* $u \in X^{**}$.

(a2) *There exists a neighborhood* V_0 *of* $\bar{x}$ *such that for any* $x \in V_0 \cap \partial C$ *one has* $\nabla \psi(x) \neq 0$ *and* $\langle u, \nabla^2 \psi(x)^* u \rangle \geq 0$ *for all* $u \in X^{**}$ *satisfying* $\langle u, \nabla \psi(x) \rangle = 0$.

(a3) *There are neighborhoods* U *of* $\bar{p}$ *and* V *of* $\bar{x}$ *such that for every* $(x, p) \in V \times U$ *the map* $f(x, \cdot)$ *is lower semicontinuous at* p.

Then there exist a neighborhood $U_1 \subset U$ *of* $\bar{p}$, *an open convex neighborhood* $V_1 \subset V$ *of* $\bar{x}$ *such that:*

(i) $\widetilde{S}(p) := S(p) \cap V_1$ *is nonempty for every* $p \in U_1$.

(ii) *The multifunction* $\widetilde{S}$ *is lower semicontinuous on* U_1.

(iii) $S(\cdot)$ *is locally metrically regular in Robinson's sense around* $(\bar{x}, \bar{p}, 0)$, *i.e., there exist neighborhoods* U_2 *of* $\bar{p}$, V_2 *of* $\bar{x}$, *and constants* $\gamma > 0$, $\mu > 0$ *such that*

$$\operatorname{dist}(x, S(p)) \leq \gamma \operatorname{dist}(0, F(x, p))$$

for every $(x,p) \in V_2 \times U_2$ *with* $\mathrm{dist}(0,F(x,p)) < \mu$, *where* $F(x,p)$ *is defined by* (7).

In addition, if

(a4) $f(x,\cdot)$ *is locally Lipschitz at* $\bar{p}$ *uniformly with respect to* x *in a neighborhood of* $\bar{x}$, *i.e., there exist neighborhoods* U_3 *of* $\bar{p}$, V_3 *of* $\bar{x}$, *and a constant* $\bar{\ell} > 0$ *such that*

$$\|f(x,p') - f(x,p)\| \le \bar{\ell}\|p' - p\| \quad \text{for all } x \in V_3 \text{ and } p, p' \in U_3,$$

then the following property is valid:

(iv) $S(\cdot)$ *is Lipschitz-like around* $(\bar{p},\bar{x})$, *i.e., exist neighborhoods* U_4 *of* $\bar{p}$, V_4 *of* $\bar{x}$, *and a constant* $\ell > 0$ *such that*

$$S(p') \cap V_4 \subset S(p) + \ell\|p' - p\|B_X \quad \forall p, p' \in U_4.$$

Let X, P be as above, Y a Banach space, $F : X \times P \rightrightarrows Y$ a multifunction. Let $(\bar{x},\bar{p}) \in X \times P$ be such that $0 \in F(\bar{x},\bar{p})$. The set-valued map $G : P \rightrightarrows X$ given by

$$G(p) := \{x \in X \, : \, 0 \in F(x,p)\} \tag{13}$$

is called the *implicit multifunction* defined by the inclusion $0 \in F(x,p)$.

To obtain Theorem 3.1, we will rely on Theorem 2.2 and the implicit function theorems from [9], which are grouped in the forthcoming statement.

Theorem 3.2. *Let* X,Y *be Asplund spaces,* P *a subset of a normed space,* $F : X \times P \rightrightarrows Y$ *a multifunction,* $(\bar{x},\bar{p}) \in X \times P$ *a pair such that* $0 \in F(\bar{x},\bar{p})$. *Let* $F_p(\cdot) := F(\cdot,p)$. *Suppose that there exists* $\rho > 0$ *such that for each* $p \in P$ *the sets* $[\mathrm{gph}\, F_p(\cdot)] \cap [B_\delta(\bar{x}) \times Y]$ *and* $\mathrm{dom}\, F_p(\cdot) \cap B_\delta(\bar{x})$ *are closed. Besides, suppose that there exist open neighborhoods* U *of* $\bar{p}$, V *of* $\bar{x}$, W *of* $0 \in Y$ *such that*

(A1) *There is a constant* $c > 0$ *satisfying* $\|y^*\| \le c\|x^*\|$ *for all* $(x,y,p) \in V \times W \times U$, $y \in F_p(x)$, $y^* \in Y^*$, $x^* \in D^*F_p(x,y)(y^*)$.

(A2) *For any* $p \in U$ *and* $x \in V$, *the multifunction* $\Pi(0,F_p(\cdot))$ *defined by*

$$\Pi(0,F_p(x)) := \{v \in F(x,p) \, : \, \|v\| = \mathrm{dist}(0,F(x,p))\} \quad (x \in \mathrm{dom}\, F_p(\cdot))$$

has nonempty values and is lower semicompact on its effective domain around x.

(A3) *For every* $(x,p) \in V \times U$, *the map* $F(x,\cdot)$ *is lower semicontinuous at* p *whenever* $p \in \mathrm{dom}\, F(x,\cdot)$.

Then the implicit multifunction (13) *has the following properties:*

(i) *There exist a neighborhood* U_1 *of* $\bar{p}$ *and an open convex neighborhood* V_1 *of* $\bar{x}$ *such that* $\widetilde{G}(p) := G(p) \cap V_1$ *is nonempty for every* $p \in U_1$.

(ii) *The multifunction* $\widetilde{G}$ *is lower semicontinuous on* U_1.

(iii) G *is locally metrically regular in Robinson's sense around* $(\bar{x},\bar{p},0)$ *with the constant* $\gamma := c$, *i.e., there exist neighborhoods* U_2 *of* $\bar{p}$, V_2 *of* $\bar{x}$, *and a constant* $\mu > 0$ *such that*

$$\mathrm{dist}(x,G(p)) \le \gamma\, \mathrm{dist}(0,F(x,p))$$

for every $(x,p) \in V_2 \times U_2$ *with* $\mathrm{dist}(0,F(x,p)) < \mu$.

In addition, if

(A4) *$F(x,\cdot)$ is locally Lipschitz at $\bar{p}$ uniformly with respect to x in a neighborhood of $\bar{x}$, i.e., there exist neighborhoods U_3 of $\bar{p}$, V_3 of $\bar{x}$, and a constant $\ell_1 > 0$ such that*

$$F(x,p') \subset F(x,p) + \ell_1 \|p' - p\| B_Y \quad \forall x \in V_3,\ \forall p, p' \in U_3,$$

then

(iv) *G is Lipschitz-like around $(\bar{p},\bar{x})$ with the constant $\ell := 2\ell_1 c$, i.e., there exist neighborhoods U_4 of $\bar{p}$, V_4 of $\bar{x}$ such that*

$$G(p') \cap V_4 \subset G(p) + \ell \|p' - p\| B_X \quad \forall p, p' \in U_4.$$

Remark 3.3. In the formulation of the implicit function theorems in [9], there is an requirement that *F is nonempty-valued around* $(\bar{x},\bar{p})$, but the proofs remain valid if one requires that $\mathrm{dom} F(\cdot,p) \cap B_\delta(\bar{x})$ is closed for every $p \in P$ and applies the Ekeland principle for $v_p(\cdot) := \mathrm{dist}(0, F(\cdot,p))$ on the set $B_\rho(\bar{x}) \cap (\mathrm{dom} F(\cdot,p))$ with a sufficiently small $\rho \in (0,\delta]$, where δ is prescribed in the formulation of the theorem.

Remark 3.4. Under the additional condition that *F is lower semicontinuous at* $(\bar{x},\bar{p})$, Theorem 3.2 of [9] asserts a property stronger than (iii) in the above Theorem 3.2. Namely, in the notation of this chapter, that theorem infers that *G is metrically regular near $(\bar{x},\bar{p})$ with the constant $\gamma := c$, i.e., there exist neighborhoods U_2 of $\bar{p}$ and V_2 of $\bar{x}$ such that*

$$\mathrm{dist}(x, G(p)) \le \gamma\, \mathrm{dist}(0, F(x,p)) \quad \forall (p,x) \in U_2 \times V_2.$$

As observed in [21], for obtaining the local metric regularity (iii), it suffices to use (A3) and some assumptions on the family of multifunctions $F(\cdot,p)$, $p \in P$.

Remark 3.5. In the implicit function theorems of [9], (A2) was formulated in a stronger form: For any $p \in U$ and $x \in V$, *the multifunction $\Pi(0, F_p(\cdot))$ is lower semicompact around x*. The latter means that there exists a neighborhood U_x of x such that for any $u \in U_x$ and any sequence $u_k \to u$, there is a sequence $y_k \in \Pi(0, F_p(u_k))$, $k = 1,2,\ldots$, which contains a subsequence convergent in the norm topology of Y. Note that the proofs of Theorems 3.1–3.3 in [9] are valid under our (A2). This is because the proof of Theorem 6.1 in [14] (which was recalled in [9, Theorem 2.3]) works well under the assumption that the solution map in question is lower semicompact on its effective domain around the given point. The interested reader is referred to [14, 21] for more details.

The proof of Theorem 3.1 requires another auxiliary result that is an easy consequence of the following theorem.

Theorem 3.6. (see [12, Corollary 1.15 and Theorem 1.17]). *Let $f : X \to Y$ be a mapping between Banach spaces and $\Omega \subset Y$ be a subset with $\bar{y} = f(\bar{x}) \in \Omega$. If f is strictly differentiable at $\bar{x}$ with surjective derivative, then*

$$N\big(\bar{x}; f^{-1}(\Omega)\big) = \big(\nabla f(\bar{x})\big)^* N(\bar{y};\Omega).$$

Lemma 3.7. *If* $\psi(\bar{x}) < 0$*, then* $N(\bar{x};C) = \{0\}$*. If* $\psi(\bar{x}) = 0$ *and* $\nabla\psi(\bar{x}) \neq 0$*, then*

$$N(\bar{x};C) = \mathbb{R}_+\nabla\psi(\bar{x}) := \{\lambda\nabla\psi(\bar{x}) : \lambda \geq 0\}. \tag{14}$$

Proof. Setting $\Omega = (-\infty, 0]$, we see that $C = \psi^{-1}(\Omega)$. If $\psi(\bar{x}) < 0$, then $N(\bar{x};C) = \{0\}$ because $\bar{x} \in \operatorname{int} C$. Suppose that $\psi(\bar{x}) = 0$, i.e., $\bar{x} \in \partial C$. For $\bar{y} := \psi(\bar{x})$, we have $N(\bar{y};\Omega) = \mathbb{R}_+$. If $\nabla\psi(\bar{x}) \neq 0$ then, according to Theorem 3.6,

$$N(\bar{x};C) = \{\nabla\psi(\bar{x})^* y^* : y^* \in \mathbb{R}_+\}.$$

As

$$\begin{aligned}\langle\nabla\psi(\bar{x})^* y^*, x\rangle = \langle y^*, \nabla\psi(\bar{x})x\rangle &= y^*(\nabla\psi(\bar{x})x)\\ &= (y^*\nabla\psi(\bar{x}))x\end{aligned}$$

for every $x \in X$, we get

$$N(\bar{x};C) = \{y^*\nabla\psi(\bar{x}) : y^* \in \mathbb{R}_+\} = \mathbb{R}_+\nabla\psi(\bar{x}),$$

as claimed in (14). □

Proof of Theorem 3.1.

We set $Y = X^*$ and define $F(x,p)$ by formula (7). Due to the assumptions of the theorem, X and Y are Asplund spaces. It is clear that each claim of the theorem follows from the corresponding one in the set of assertions (i)–(iv) of Theorem 3.6. Thus, we only have to check the fulfillment of the conditions (A1)–(A4) and verify the existence of $\delta > 0$ such that, for every $p \in P$, the sets $[\operatorname{gph} F_p(\cdot)] \cap [B_\delta(\bar{x}) \times Y]$ and $\operatorname{dom} F_p(\cdot) \cap B_\delta(\bar{x})$ are closed. The latter is satisfied with any choice of $\delta > 0$ because $\operatorname{dom} F_p(\cdot) = C$ by Lemma 3.7.

If $\bar{x} \in \operatorname{int} C$, then we choose $\delta > 0$ as small as $B_\delta(\bar{x}) \subset \operatorname{int} C$. It is easy to see that

$$[\operatorname{gph} F_p(\cdot)] \cap [B_\delta(\bar{x}) \times Y] = [\operatorname{gph} f(\cdot,p)] \cap [B_\delta(\bar{x}) \times Y],$$

and the set on the right-hand side is closed. If $\bar{x} \in \partial C$ then $\nabla\psi(\bar{x}) \neq 0$ by (a2). In this case, we choose $\delta > 0$ as small as $\nabla\psi(x) \neq 0$ for every $x \in B_\delta(\bar{x})$. For any $p \in P$ and sequences $x_k \xrightarrow{B_\delta(\bar{x})} x$, $x_k^* \to x^*$ with $x_k^* \in f(x_k,p) + N(x_k;C)$ for all $k \in \mathbb{N}$, we have $(x,x^*) \in \operatorname{gph} F(\cdot,p)$, i.e., $x^* \in f(x,p) + N(x;C)$. Indeed, if there exists a subsequence $\{x_{k_j}\}$ of $\{x_k\}$ such that $x_{k_j} \in \operatorname{int} C$ for all j, then $x_{k_j}^* = f(x_{k_j},p)$ for all j. Hence, letting $j \to \infty$ yields

$$x^* = f(x,p) \in f(x,p) + N(x;C).$$

If there is a subsequence $\{x_{k_j}\}$ of $\{x_k\}$ such that $x_{k_j} \in \partial C$ for all j, then

$$x_{k_j}^* = f(x_{k_j},p) + \lambda_{k_j}\nabla\psi(x_{k_j}), \tag{15}$$

where, in agreement with Lemma 3.7, $\lambda_{k_j} \geq 0$ is uniquely defined by the last equality. Since $\nabla\psi(x) \neq 0$ by the choice of δ and the inclusion $x \in B_\delta(\bar{x})$, the sequence λ_{k_j} must be bounded; otherwise the estimates

$$\|x^*_{k_j}\| = \|f(x_{k_j},p) + \lambda_{k_j}\nabla\psi(x_{k_j})\| \geq \lambda_{k_j}\|\nabla\psi(x_{k_j})\| - \|f(x_{k_j},p)\|$$

would lead to a contradiction. Thus, we may suppose that $\lambda_{k_j} \to \lambda$. Passing (15) to the limit as $j \to \infty$, we get

$$x^* \in f(x,p) + \lambda\nabla\psi(x) \in f(x,p) + N(x;C),$$

which shows that $[\mathrm{gph}\, F_p(\cdot)] \cap [B_\delta(\bar{x}) \times Y]$ is closed.

To see that (A1) is fulfilled, we invoke (a1) and the assumed continuity of $\nabla_x f(\cdot)$ on $X \times P$ to find $\alpha \in (0,\bar{\alpha})$ and neighborhoods $\tilde{U}$ of $\bar{p}$ and $\tilde{V}$ of $\bar{x}$ such that

$$\langle u, \nabla_x f(x,p)^* u\rangle \geq \alpha\|u\|^2 \quad \forall (x,p) \in \tilde{V} \times \tilde{U},\ \forall u \in X^{**}. \tag{16}$$

There is no loss of generality in assuming that $\tilde{V} \subset V_0$, where V_0 is prescribed by (a2). Put $c = 1/\alpha$. Given any $(x,y,p) \in \tilde{V} \times Y \times \tilde{U}$, $y \in F_p(x)$, $u \in X^{**}$, $x^* \in D^*F_p(x,y)(u)$, let us show that

$$\|u\| \leq c\|x^*\|. \tag{17}$$

Since $D^*F_p(x,y)(u) \neq \emptyset$ we infer that $x \in \mathrm{dom}\, F_p = C$. If $x \in \mathrm{int}\, C$, then applying Theorem 2.2 for $(\bar{x},\bar{v}) := (x, y - f(x,p))$, from (11) we derive $\bar{v} = 0$ and $x^* = \nabla_x f(x,p)^* u$. Then, due to (16),

$$\alpha\|u\|^2 \leq \langle u, \nabla_x f(x,p)^* u\rangle \leq \|u\|\,\|x^*\|.$$

Hence (17) is valid. Suppose now that $x \in \partial C$, i.e., $\psi(x) = 0$. Using Theorem 2.2 once more for $(\bar{x},\bar{v}) := (x, y - f(x,p))$, we obtain (12). As $x^* \in \Delta$, only four cases can occur:

Case 1. One has $x^* \in \nabla_x f(x,p)^* u + y^*\nabla^2\psi(x)^* u + \mathbb{R}\nabla\psi(x)$ with $y^* > 0$ and $\langle u, \nabla\psi(x)\rangle = 0$.

Case 2. One has $x^* \in \nabla_x f(x,p)^* u + \mathbb{R}_+\nabla\psi(x)$ with $\langle u, \nabla\psi(x)\rangle > 0$.

Case 3. One has $x^* \in \nabla_x f(x,p)^* u + \mathbb{R}\nabla\psi(x)$ with $\langle u, \nabla\psi(x)\rangle = 0$.

Case 4. One has $x^* = \nabla_x f(x,p)^* u$ with $\langle u, \nabla\psi(x)\rangle < 0$.

In Case 1, we find an $\beta \in \mathbb{R}$ such that

$$x^* = \nabla_x f(x,p)^* u + y^*\nabla^2\psi(x)^* u + \beta\nabla\psi(x).$$

Then, taking account of (16) and (a2), we get

$$\begin{aligned}\|u\|\,\|x^*\| \geq \langle u, x^*\rangle &= \langle u, \nabla_x f(x,p)^* u + y^*\nabla^2\psi(x)^* u + \beta\nabla\psi(x)\rangle \\ &= \langle u, \nabla_x f(x,p)^* u\rangle + y^*\langle u, \nabla^2\psi(x)^* u\rangle \\ &\geq \alpha\|u\|^2,\end{aligned}$$

establishing (17). Cases 2–4 can be treated in the same manner. We have seen that assumption (A1) of Theorem 3.2 is satisfied with $c = 1/\alpha$, $V = \tilde{V}$, $W = Y$, and $U = \tilde{U}$.

We now check (A2) with V, W, and U being selected as above. Without loss of generality, we may assume that $V \subset B_\delta(\bar{x})$, where $\delta > 0$ was chosen at the beginning of the proof. First, let us show that $\operatorname{dom} \Pi(0, F_p(\cdot)) \cap B_\delta(\bar{x}) = C \cap B_\delta(\bar{x})$ for every $p \in U$. Of course, $\operatorname{dom} \Pi(0, F_p(\cdot)) \cap B_\delta(\bar{x}) \subset C \cap B_\delta(\bar{x})$. To obtain the reverse inclusion, fix any $x \in C \cap B_\delta(\bar{x})$. If $x \in \operatorname{int} C$, then $\Pi(0, F_p(x)) = \{f(x,p)\}$ because $N(x;C) = \{0\}$. If $x \in \partial C$ and $\nabla\psi(x) \neq 0$, then $\Pi(0, F_p(x)) \neq \emptyset$. Indeed, observe that $\operatorname{dist}(0, F(x,p)) \leq \|f(x,p)\|$ and let

$$x_k^* = f(x,p) + \lambda_k \nabla\psi(x) \in F(x,p) = f(x,p) + \mathbb{R}_+ \nabla\psi(x) \quad (k \in \mathbb{N},\ \lambda_k \geq 0)$$

be such that $\lim_{k\to\infty} \|x_k^*\| = \operatorname{dist}(0, F(x,p))$. Since the sequence $\{x_k^*\}$ is bounded and $\nabla\psi(x) \neq 0$, we may assume that $\lambda_k \to \lambda \in \mathbb{R}_+$. Then

$$x^* := \lim_{k\to\infty} x_k^* = f(x,p) + \lambda \nabla\psi(x) \in F(x,p).$$

From what which has already been said it follows that $x^* \in \Pi(0, F_p(x))$. We have thus shown that $\operatorname{dom} \Pi(0, F_p(\cdot)) \cap B_\delta(\bar{x}) = C \cap B_\delta(\bar{x})$. Next, to check the lower semicompactness of $\Pi(0, F_p(\cdot))$ on C around any $x \in V \cap C$, we fix any sequence $x_k \xrightarrow{C} x$. For each $k \in \mathbb{N}$, select a vector

$$x_k^* = f(x_k, p) + \lambda_k \nabla\psi(x_k) \in \Pi(0, F_p(x_k)) \tag{18}$$

with $\lambda_k \geq 0$. Since $\|x_k^*\| \leq \|f(x_k,p)\|$ and $\lim_{k\to\infty} \|f(x_{k_j},p)\| = \|f(x,p)\|$, $\{\|x_k^*\|\}$ is a bounded sequence. Hence the estimates

$$\|x_k^*\| = \|f(x_k,p) + \lambda_k \nabla\psi(x_k)\| \geq \lambda_k \|\nabla\psi(x_k)\| - \|f(x_k,p)\| \quad (k = 1,2,\ldots)$$

and the property $\lim_{k\to\infty} \nabla\psi(x_k) = \nabla\psi(x) \neq 0$ guarantee that the sequence $\{\lambda_k\}$ is bounded. Consequently, the latter has a subsequence λ_{k_j} converging to some $\lambda \geq 0$. From the equality in (18) it follows that

$$x_{k_j}^* \to x^* := f(x,p) + \lambda \nabla\psi(x) \in F_p(x)$$

as $j \to \infty$. If $x^* \notin \Pi(0, F_p(x))$, then we would find $\widetilde{\lambda} \geq 0$ such that

$$\widetilde{x}^* := f(x,p) + \widetilde{\lambda} \nabla\psi(x) \in F_p(x)$$

satisfies the inequality

$$\|\widetilde{x}^*\| < \|x^*\|. \tag{19}$$

Putting

$$\widetilde{x}_{k_j}^* = f(x_{k_j},p) + \widetilde{\lambda} \nabla\psi(x_{k_j}),$$

we note that $\widetilde{x}^*_{k_j} \in F_p(x_{k_j})$ and $\widetilde{x}^*_{k_j} \to \widetilde{x}^*$. By the choice of $x^*_{k_j}$, we have $\|x^*_{k_j}\| \leq \|\widetilde{x}^*_{k_j}\|$. Passing to the limit as $j \to \infty$, from the last inequality we get $\|x^*\| \leq \|\widetilde{x}^*\|$, contrary to (19). We have obtained the inclusion $x^* \in \Pi(0, F_p(x))$, which justifies the lower semicompactness of $\Pi(0, F_p(\cdot))$ on C around x and completes the proof, because (a3) certainly implies (A3) and (a4) yields (A4). □

4 An Analysis of Theorem 3.1

The assumption "X and X^* are Asplund spaces" in Theorem 3.1 is satisfied if X is a reflexive Banach space. Assumptions of this kind had appeared in the literature; see, e.g., [12, Theorems 4.54 and 4.65(ii)]. In general, X^* is not necessarily Asplund, when X is an Asplund space. Indeed, $X = c_0$ is an Asplund space because $X^* = l_1$ is a separable Banach space (see [15, Theorem 2.12]). However, $X^* = l_1$ is not an Asplund space [15, p. 13]. Thus, the assumption "X and X^* are Asplund spaces" specifies a class of Asplund spaces to which Theorem 3.1 can be applied to.

The next example shows that the "positive definiteness of $\nabla_x f(\bar{x}, \bar{p})$" in (a1) cannot be replaced by a weaker assumption on "positive semidefiniteness of $\nabla_x f(\bar{x}, \bar{p})$".

Example 4.1. Let $X = \mathbb{R}$, $P = \mathbb{R}_+$, $\psi(x) = x^2 - 1$, $f(x,p) = -px$, $\bar{x} = -1$, $\bar{p} = 0$. It is easy to verify that, except for (a1), all the other assumptions of Theorem 3.1 are satisfied. Note that we still have $\langle u, \nabla_x f(\bar{x}, \bar{p})^* u \rangle \geq 0$ for every $u \in X^{**}$. For any $p > 0$, it holds $S(p) \cap [-1, 0) = \emptyset$. Hence all the properties (i)–(iv) in Theorem 3.1 are not valid.

Interestingly, if the assumption on "positive semidefiniteness of $\nabla^2 \psi(x)$ on the tangent space to ∂C *at any* $x \in \partial C$ *sufficiently near to* $\bar{x}$" in (a2) is violated then, in general, the assertions (i)–(iv) are no longer valid. (As far as we understand, the just mentioned assumption does not imply that C is locally convex near $\bar{x}$.)

Example 4.2. Let $X = \mathbb{R}^2$, $P = (-1, +\infty)$, $\psi(x) = x_1^3 - x_2$ and $f(x,p) = ((p+1)x_1, x_2 - p)$ for any $x = (x_1, x_2) \in X$ and $p \in P$. Let $\bar{x} = 0$ and $\bar{p} = 0$. It is easy to verify that, except for (a2), all the other assumptions of Theorem 3.1 are satisfied. Note that we still have $\nabla \psi(\bar{x}) \neq 0$ and $\langle u, \nabla^2 \psi(\bar{x})^* u \rangle \geq 0$ for any $u \in X^{**}$ satisfying $\langle u, \nabla \psi(\bar{x}) \rangle = 0$. For any $p \geq 0$, $S(p) = \{(0, p)\}$ is a singleton. For $p \in (-1, 0)$, the solution set

$$S(p) = \left\{ (x_1, x_2) \, : \, x_1 = t^{1/3}, \; x_2 = t, \; p \leq t \leq 0 \right\}$$

is a curve. Note that the properties (i) and (ii) in Theorem 3.1 are valid, but (iv) is not. The reason is that the condition "$\nabla \psi(x) \neq 0$ and $\langle u, \nabla^2 \psi(x)^* u \rangle \geq 0$ for all $u \in X^{**}$ satisfying $\langle u, \nabla \psi(x) \rangle = 0$" in (a2) is violated at any $x = (x_1, x_1^3) \in \partial C$ with $x_1 < 0$.

It is not difficult to show by examples that (a3) cannot be omitted if one wants to have (iii), while (a4) is essential for the validity of (iv).

The problem of minimizing a linear-quadratic function under a convex quadratic constraint is known as the *trust-region subproblem* (see [11] and the references therein). Several stability properties of the problem have been obtained in [10]. Theorem 3.1 asserts some facts about the stability of strongly convex linear-quadratic minimization under a C^2-smooth convex constraint. The latter corresponds to the easiest case of the trust-region subproblem where the objective function is strongly convex.

Example 4.3. Let $\varphi(x) = \frac{1}{2}\langle x, Ax\rangle + \langle b, x\rangle$, where $x \in \mathbb{R}^n$, $A \in \mathbb{R}^{n\times n}$ is a given symmetric matrix, $b \in \mathbb{R}^n$ a given vector. Let $C = \{x \in \mathbb{R}^n : \psi(x) \le 0\}$, where $\psi : \mathbb{R}^n \to \mathbb{R}$ is a C^2-smooth convex function. It is well known that if $x \in C$ is a local minimizer of φ on C then the variational system (2), where $p = (A, b) \in \mathbb{R}^{n\times n} \times \mathbb{R}^n$ and $f(x, p) = Ax + b$, is satisfied. If A is positive semidefinite, then the converse is valid. In the case under consideration, assumptions (a2)–(a4) of Theorem 3.1 automatically hold. Under the assumption (a1), which means that A is positive matrix, the solution set of our minimization problem is a singleton, say, $S(p) = \{x(p)\}$. Moreover, according to Theorem 3.1, $x(\cdot)$ is a locally Lipschitz function. This result is not new. For instance, it is immediate from [20, Theorem 2.1].

The next two numerical examples are designed to how Theorem 3.1 can be applied to concrete problems.

Example 4.4. Consider the system (2) with $X = P = \mathbb{R}$, $\psi(x) = x^2 - x$, $f(x, p) = \alpha x + p$, and $\alpha \in \mathbb{R}$ is a constant. Note that $C = [0, 1] \subset \mathbb{R}$. We study the local behavior of the solution map $S(\cdot)$ of (2) around the following points in its graph:

(a) $(\bar{p}, \bar{x}) = (-\alpha\beta, \beta)$ with $\beta \in (0, 1)$.

(b) $(\bar{p}, \bar{x}) = (\beta, 0)$ with $\beta \ge 0$.

First, let $(\bar{p}, \bar{x}) = (-\alpha\beta, \beta)$ with $\beta \in (0, 1)$. Since $\nabla_x f(\bar{x}, \bar{p}) = [\alpha]$, assumption (a1) is satisfied if and only if $\alpha > 0$. Other assumptions of Theorem 3.1 are fulfilled. Hence, for $\alpha > 0$, the map $S(\cdot)$ is Lipschitz-like around $(\bar{p}, \bar{x})$ and it is locally metrically regular in Robinson's sense around the point $(\bar{x}, \bar{p}, 0)$.

Next, suppose that $(\bar{p}, \bar{x}) = (\beta, 0)$ with $\beta > 0$. Condition (a1) of Theorem 3.1 is satisfied if and only if $\alpha > 0$. As $\psi(\bar{x}) = 0$, we have to verify condition (a2). Note that $\nabla\psi(\bar{x}) \neq 0$. Since $\{u \in X^{**} : \langle u, \nabla\psi(\bar{x})\rangle = 0\} = \{0\}$, (a2) is satisfied. Thus, by Theorem 3.1, the map $S(\cdot)$ is Lipschitz-like around $(\bar{p}, \bar{x})$ and it is locally metrically regular in Robinson's sense around the point $(\bar{x}, \bar{p}, 0)$ for any $\alpha > 0$.

Example 4.5. Consider the system (2) with $X = P = \mathbb{R}^2$, $\psi(x) = x_2^2 - x_1$, $f(x, p) = \begin{bmatrix} x_1 - x_2 + p_1 \\ 2x_2 + p_2 \end{bmatrix}$ for all $x = (x_1, x_2) \in X$ and $p = (p_1, p_2) \in P$. Note that

$$C = \{x = (x_1, x_2) : x_1 \ge x_2^2\}, \quad \partial C = \{x = (x_1, x_2) : x_1 = x_2^2\}.$$

Let $\bar{x} = (0, 0)$, $\bar{p} = (1, 0)$. Since $\nabla_x f(\bar{x}, \bar{p}) = \begin{bmatrix} 1 & -1 \\ 0 & 2 \end{bmatrix}$, assumption (a1) of Theorem 3.1 is fulfilled. As $\psi(\bar{x}) = 0$, we have to check condition (a2). Since

$\nabla^2\psi(x) = \begin{bmatrix} 0 & 0 \\ 0 & 2 \end{bmatrix}$ is positive semidefinite for all x, (a2) is satisfied with $V_0 := \mathbb{R}^2$. By Theorem 3.1, the map $S(\cdot)$ is Lipschitz-like around $(\bar{p}, \bar{x})$ and it is locally metrically regular in Robinson's sense around the point $(\bar{x}, \bar{p}, 0)$.

Acknowledgements This work was supported by the National Sun Yat-Sen University, Kaohsiung, Taiwan and the National Foundation for Science & Technology Development, Vietnam. The authors are indebted to Dr. N.Q. Huy, Mr. N.H. Chieu and Mr. T.D. Chuong for an useful discussion on Asplund spaces. We thank the referee for helpful comments.

References

1. F. J. Aragón Artacho and B. S. Mordukhovich, *Metric regularity and Lipschitzian stability of parametric variational systems*, Nonlinear Anal. 72 (2010), 1149–1170.
2. J.-P. Aubin, *Lipschitz behavior of solutions to convex minimization problems*, Math. Oper. Res. 9 (1984), 87–111.
3. N. H. Chieu, J.-C. Yao and N. D. Yen, *Relationships between the Robinson robust stability and the Aubin continuity property of implicit multifunctions*, Nonlinear Anal. 72 (2010), 3594–3601.
4. F. H. Clarke, *Optimization and Nonsmooth Analysis,* Wiley, New York, 1983.
5. A. L. Dontchev and R. T. Rockafellar, *Characterizations of strong regularity for variational inequalities over polyhedral convex sets*, SIAM J. Optim. 6 (1996), 1087–1105.
6. R. Henrion, B. S. Mordukhovich and N. M. Nam, *Second-order analysis of polyhedral systems in finite and infinite dimensions with applications to robust stability of variational inequalities*, SIAM J. Optim. 20 (2010), 2199–2227.
7. R. Henrion, J. Outrata and T. Surowiec, *On the co-derivative of normal cone mappings to inequality systems*, Nonlinear Anal. 71 (2009), 1213–1226.
8. R. Henrion and W. Römisch, *On M-stationary points for a stochastic equilibrium problem under equilibrium constraints in electricity spot market modeling*, Appl. Math. 52 (2007), 473–494.
9. G. M. Lee, N. N. Tam and N. D. Yen, *Normal coderivative for multifunctions and implicit function theorems*, J. Math. Anal. Appl. 338 (2008), 11–22.
10. G. M. Lee, N. N. Tam and N. D. Yen, *Stability of linear-quadratic minimization over Euclidean balls*, Institute of Mathematics, Vietnamese Academy of Science and Technology, E-Preprint No. 2007/10/02. (Submitted).
11. S. Lucidi, L. Palagi and M. Roma, *On some properties of quadratic programs with a convex quadratic constraint*, SIAM J. Optim. 8 (1998), 105–122.
12. B. S. Mordukhovich, *Variational Analysis and Generalized Differentiation, Vol. I: Basic Theory, Vol. II: Applications*, Springer, Berlin, 2006.
13. B. S. Mordukhovich, J. V. Outrata, *On second-order subdifferentials and their applications*, SIAM J. Optim. 12 (2001), 139–169.
14. B. S. Mordukhovich and Y. Shao, *Nonsmooth sequential analysis in Asplund spaces*, Trans. Am. Math. Soc. 348 (1996), 1235–1280.
15. R. R. Phelps, *Convex Functions, Monotone Operators and Differentiablity*, 2nd Edition, Spinger, Berlin, 1993.
16. S. M. Robinson, *Stability theory for systems of inequalities, II. Differentiable nonlinear systems*, SIAM J. Numer. Anal. 13 (1976), 497–513.
17. S. M. Robinson, *Generalized equations and their solutions, Part I: Basic theory*, Math. Program. Study 10 (1979), 128–141.
18. J.-C. Yao and N. D. Yen, *Coderivative calculation related to a parametric affine variational inequality. Part 1: Basic calculations.* Acta Math. Vietnam. 34 (2009), 155–170.

19. J.-C. Yao and N. D. Yen, *Coderivative calculation related to a parametric affine variational inequality. Part 2: Applications.* Pacific J. Optim. 3 (2009), 493–506.
20. N. D. Yen, *Hölder continuity of solutions to a parametric variational inequality*, Appl. Math. Optim. 31 (1995), 245–255.
21. N. D. Yen and J.-C. Yao, *Pointbased sufficient conditions for metric regularity of implicit multifunctions*, Nonlinear Anal. 70 (2009), 2806–2815.

Exact Penalty in Constrained Optimization and the Mordukhovich Basic Subdifferential

Alexander J. Zaslavski

Abstract In this chapter, we use the penalty approach to study two constrained minimization problems in infinite-dimensional Asplund spaces. A penalty function is said to have the exact penalty property if there is a penalty coefficient for which a solution of an unconstrained penalized problem is a solution of the corresponding constrained problem. We use the notion of the Mordukhovich basic subdifferential and show that the exact penalty property is stable under perturbations of objective functions.

1 Introduction

Penalty methods are an important and useful tool in constrained optimization. See, for example, [3–7, 11, 14, 17, 18, 20, 21] and the references mentioned there. In this chapter, we use the penalty approach to study two constrained nonconvex minimization problems with Lipschitzian (on bounded sets) objective functions. The first problem is an equality-constrained problem in an Asplund space with a locally Lipschitzian constraint function and the second problem is an inequality-constrained problem in an Asplund space with a locally Lipschitzian constraint function. Note that a Banach space is an Asplund space if and only if every separable subspace has a separable dual [13].

A penalty function is said to have the exact penalty property if there is a penalty coefficient for which a solution of an unconstrained penalized problem is a solution of the corresponding constrained problem. The notion of exact penalization was introduced by Eremin [9] and Zangwill [18] for use in the development of algorithms for nonlinear constrained optimization. For a detailed historical review of the literature on exact penalization see [3, 5, 7].

Alexander J. Zaslavski
Department of Mathematics, The Technion-Israel Institute of Technology, 32000 Haifa, Israel
e-mail: ajzasl@tx.technion.ac.il

R.S. Burachik and J.-C. Yao (eds.), *Variational Analysis and Generalized Differentiation in Optimization and Control*, Springer Optimization and Its Applications 47,
DOI 10.1007/978-1-4419-0437-9_12,

In [20], it was established the existence of a penalty coefficient for which approximate solutions of the unconstrained penalized problem are close enough to approximate solutions of the corresponding constrained problem. This is a novel approach in the penalty type methods.

Consider a minimization problem $h(z) \to \min$, $z \in X$ where $h : X \to R$ is a lower semicontinuous bounded from below function on a Banach space X. If the space X is infinite-dimensional or if the function h does not satisfy a coercivity assumption, then the existence of solutions of the problem is not guaranteed and in this situation we consider δ-approximate solutions. Namely, $x \in X$ is a δ-approximate solution of the problem $h(z) \to \min$, $z \in X$, where $\delta > 0$, if $h(x) \le \inf\{h(z) :\ z \in X\} + \delta$.

In [20] and [21], we consider minimization problems in a general Banach space and in a general Asplund space, respectively. Therefore, we are interested in approximate solutions of the unconstrained penalized problem and in approximate solutions of the corresponding constrained problem. Under certain mild assumptions, we show the existence of a constant $\bar{\lambda} > 0$ such that the following property holds:

For each $\varepsilon > 0$, there exists $\delta(\varepsilon) > 0$ which depends only on ε such that if x is a $\delta(\varepsilon)$-approximate solution of the unconstrained penalized problem whose penalty coefficient is larger than $\bar{\lambda}$, then there exists an ε-approximate solution y of the corresponding constrained problem such that $||y - x|| \le \varepsilon$.

This property, which will be called here as the generalized exact penalty property, implies that any exact solution of the unconstrained penalized problem whose penalty coefficient is larger than $\bar{\lambda}$ is an exact solution of the corresponding constrained problem. Indeed, let x be a solution of the unconstrained penalized problem whose penalty coefficient is larger than $\bar{\lambda}$. Then for any $\varepsilon > 0$, the point x is also a $\delta(\varepsilon)$-approximate solution of the same unconstrained penalized problem and in view of the property above there is an ε-approximate solution y_ε of the corresponding constrained problem such that $||x - y_\varepsilon|| \le \varepsilon$. Since ε is an arbitrary positive number we can easily deduce that x is an exact solution of the corresponding constrained problem. Therefore, our results also include the classical penalty result as a special case.

In [20], the existence of the constant $\bar{\lambda}$ for the equality-constrained problem was established under the assumption that the set of admissible points does not contain critical points of the constraint function. The notion of critical points used in [20] is based on Clarke's generalized gradients [20]. It should be mentioned that there exists also the construction of Mordukhovich basic subdifferential introduced in [12] which is intensively used in the literature. See, for example, [13, 14] and the references mentioned there. In [21], we generalize the results of [20] for minimization problems on Asplund spaces using the (less restrictive) notion of critical points via Mordukhovich basic subdifferential. In this chapter, we use the Mordukhovich basic subdifferential and show the stability of the generalized exact penalty property under perturbations of objective functions. Note that the stability of the generalized exact penalty property is crucial in practice. One reason is that in practice we deal with a problem that is an approximation of the problem we wish to consider. Another reason is that the computations introduce numerical errors.

2 The Main Result

Let X be an Asplund space and X^* its dual equipped with the weak* topology w^*.

If $F : X \to 2^{X^*}$ is a set-valued mapping between the Banach space X and its dual X^*, then the notation

$$\limsup_{x \to \bar{x}} F(x) := \Big\{ x^* \in X^* : \text{ there exist sequences } x_k \to \bar{x} \text{ and } x_k^* \xrightarrow{w^*} x^* \\ \text{as } k \to \infty \text{ with } x_k^* \in F(x_k) \text{ for all natural numbers } k \Big\} \quad (1)$$

signifies the sequential Painleve–Kuratowski upper limit with respect to the norm topology of X and the weak* topology of X^*.

For each $x^* \in X^*$ and each $r > 0$, set

$$B_*(x^*, r) = \{ l \in X^* : \ ||l - x^*|| \le r \}$$

and for each $x \in X$ and each $r > 0$ set

$$B(x, r) = \{ y \in X : \ ||y - x|| \le r \}.$$

In this chapter, to obtain a sufficient condition for the existence of an exact penalty we use the notion of Mordukhovich basic subdifferential introduced in [12] (see also [13, p. 82]). To meet this goal, we first present the notion of an ε-subdifferential (see [13, p. 87]).

Let $\phi : X \to R$, $\varepsilon > 0$ and let $\bar{x} \in X$. Then the set

$$\widehat{\partial}_\varepsilon \phi(\bar{x}) := \{ x^* \in X^* : \ \liminf_{x \to \bar{x}} [(\phi(x) - \phi(\bar{x}) - < x^*, x - \bar{x} >) ||x - \bar{x}||^{-1}] \ge -\varepsilon \} \quad (2)$$

is the analytic ε-subdifferential of ϕ at $\bar{x}$.

By Theorem 1.8.9 of [13, p. 92], the set

$$\partial \phi(\bar{x}) = \limsup_{x \xrightarrow{\phi} \bar{x}, \ \varepsilon \to 0^+} \widehat{\partial}_\varepsilon \phi(x) \quad (3)$$

is Mordukhovich basic (limiting) subdifferential of the function ϕ at the point $\bar{x}$.

It should be mentioned that in view of Theorem 2.34 of [13, p. 218],

$$\partial \phi(\bar{x}) = \limsup_{x \xrightarrow{\phi} \bar{x}} \widehat{\partial}_0 \phi(x).$$

Here we use the notation that $x \xrightarrow{\phi} \bar{x}$ if and only if $x \to \bar{x}$ with $\phi(x) \to \phi(\bar{x})$, where $\phi(x) \to \phi(\bar{x})$ is superfluous if ϕ is continuous at $\bar{x}$.

Let $f : X \to R$ be a locally Lipschitzian function. For each $x \in X$, set

$$\Xi_f(x) = \inf\{ ||l|| : \ l \in \partial f(x) \}. \quad (4)$$

(We suppose that infimum of an empty set is ∞.) It should be mentioned that an analogous functional, defined using the Clarke subdifferentials, was introduced in [19] and then used in [20].

A point $x \in X$ is a critical point of f if $0 \in \partial f(x)$.

A real number $c \in R$ is called a critical value of f if there exists a critical point x of f such that $f(x) = c$.

For each function $h : X \to R$ and each nonempty set $A \subset X$, set

$$\inf(h) = \inf\{h(z) : z \in X\}, \quad \inf(h;A) = \inf\{h(z) : z \in A\}.$$

For each $x \in X$ and each $A \subset X$, put

$$d(x,A) = \inf\{||x-y|| : y \in A\}.$$

Denote by $\mathscr{M}$ the set of all continuous functions $f : X \to R$. The set $\mathscr{M}$ is equipped with the uniformity [10] determined by the following base:

$$\begin{aligned}\mathscr{U}(M,q,r) = &\Big\{(f,g) \in \mathscr{M} \times \mathscr{M} : |f(x) - g(x)| \leq r \text{ for all } x \in B(0,M)\Big\}\\ &\cap \Big\{(f,g) \in \mathscr{M} \times \mathscr{M} : |(f-g)(x) - (f-g)(y)|\\ &\leq q||x-y|| \text{ for all } x,y \in B(0,M)\Big\},\end{aligned} \tag{5}$$

where $M,q,r > 0$. It is not difficult to see that this uniformity is metrizable and complete.

Let $\phi : [0,\infty) \to [0,\infty)$ be an increasing function such that

$$\lim_{t\to\infty} \phi(t) = \infty \tag{6}$$

and let $\bar{a} > 0$. Denote by $\mathscr{M}_{\phi,\bar{a}}$ the set of all functions $f \in \mathscr{M}$ such that

$$f(x) \geq \phi(||x||) - \bar{a} \text{ for all } x \in X. \tag{7}$$

Let $f_0 : X \to R$ be a function that is Lipschtzian on all bounded subsets of X and such that

$$f_0 \in \mathscr{M}_{\phi,\bar{a}}. \tag{8}$$

Let $g : X \to R$ be a locally Lipschitzian function.

We say that the function g satisfies the Palais–Smale (P–S) condition on a set $Z \subset X$ if for each norm-bounded sequence $\{z_i\}_{i=1}^{\infty} \subset Z$ such that the sequence $\{g(z_i)\}_{i=1}^{\infty}$ is bounded and $\liminf_{i\to\infty} \Xi_g(z_i) = 0$ there exists a norm convergent subsequence of $\{z_i\}_{i=1}^{\infty}$ [1,2,15,19].

Let $c \in R$ be such that $g^{-1}(c) \neq \emptyset$.

We consider the following constrained minimization problems

$$f(x) \to \min \text{ subject to } x \in g^{-1}(c) \tag{P_e}$$

and

$$f(x) \to \min \text{ subject to } x \in g^{-1}((-\infty, c]), \tag{P_i}$$

where $f \in \mathscr{M}_{\phi,\bar{a}}$ belongs to a neighborhood of f_0 in $\mathscr{M}$.

We associate with these two problems the corresponding families of unconstrained minimization problems

$$f(x) + \lambda|g(x) - c| \to \min, \quad x \in X \tag{$P_{\lambda e}$}$$

and

$$f(x) + \lambda \max\{g(x) - c, 0\} \to \min, \quad x \in X, \tag{$P_{\lambda i}$}$$

where $\lambda > 0$ and establish the existence of the exact penalty.

Fix

$$\theta \in g^{-1}(c). \tag{9}$$

By (6), there exists a positive number M_0 such that

$$M_0 > ||\theta|| + 2 \quad \text{and} \quad \phi(M_0 - 2) > f_0(\theta) + \bar{a} + 2. \tag{10}$$

For each $f \in \mathscr{M}_{\phi,\bar{a}}$ and each $\lambda > 0$, set

$$\psi^{(e)}_{f,\lambda}(z) = f(z) + \lambda|g(z) - c|, \quad z \in X, \tag{11}$$

$$\psi^{(i)}_{f,\lambda}(z) = f(z) + \lambda \max\{g(z) - c, 0\}, \quad z \in X. \tag{12}$$

The following theorem is our main result.

Theorem 2.1. *Assertion 1. Assume that there exists $\gamma_* > 0$ such that the functions g and $-g$ satisfy the (P–S) condition on the set $g^{-1}([c - \gamma_*, c + \gamma_*])$ and the following property holds:*

If $x \in g^{-1}(c)$ is a critical point of the function g or a critical point of the function $-g$, then $f_0(x) > \inf(f_0; g^{-1}(c))$.

Let $A = g^{-1}(c)$, $\psi_{f,\lambda} = \psi^{(e)}_{f,\lambda}$ for each $f \in \mathscr{M}_{\phi,\bar{a}}$ and each $\lambda > 0$ and let $q > 0$. Then there exist positive numbers $\bar{\lambda}$, r such that the following property holds:

(Q) for each $\varepsilon > 0$ there exists $\delta \in (0, \varepsilon)$ such that if $f \in \mathscr{M}_{\phi,\bar{a}}$ satisfies $(f, f_0) \in \mathscr{U}(M_0, q, r)$, $\lambda > \bar{\lambda}$ and if $x \in X$ satisfies $\psi_{f,\lambda}(x) \le \inf(\psi_{f,\lambda}) + \delta$, then there is $y \in A$ such that $||y - x|| \le \varepsilon$ and $f(y) \le \inf(f; A) + \delta$.

Assertion 2. Assume that there exists $\gamma_* > 0$ such that the function g satisfies the (P–S) condition on the set $g^{-1}([c, c + \gamma_*])$ and the following property holds:

If $x \in g^{-1}(c)$ is a critical point of the function g, then

$$f_0(x) > \inf\left(f_0; g^{-1}(-\infty, c]\right).$$

Let $A = g^{-1}((-\infty, c])$, $\psi_{f,\lambda} = \psi^{(i)}_{f,\lambda}$ for each $f \in \mathscr{M}_{\phi,\bar{a}}$ and each $\lambda > 0$ and let $q > 0$. Then there exist positive numbers $\bar{\lambda}$, r such that property (Q) holds.

By Theorem 2.1, the existence of an exact penalty depends on a triplet (f,g,c). For given functions f,g, it is interesting to obtain an information about the set of all real numbers c for which an exact penalty exists. If the space X is finite-dimensional and the functions f,g are Frechet differentiable, then by Theorem 2.1 and the classical Sard theorem [16] this set has a complement with the Lebesgue measure zero.

3 Proof of Theorem 2.1

We prove Assertions 1 and 2 simultaneously. Set

$$A = g^{-1}(c) \quad \text{and} \quad \psi_{f,\lambda} = \psi^{(e)}_{f,\lambda} \quad \forall f \in \mathscr{M}_{\phi,\bar{a}}\, \forall \lambda > 0 \tag{13}$$

in the case of Assertion 1 and

$$A = g^{-1}((-\infty,c]) \quad \text{and} \quad \psi_{f,\lambda} = \psi^{(i)}_{f,\lambda} \quad \forall f \in \mathscr{M}_{\phi,\bar{a}}\, \forall \lambda > 0 \tag{14}$$

in the case of Assertion 2. Clearly, the function $\psi_{f,\lambda}$ is continuous for all $f \in \mathscr{M}_{\phi,\bar{a}}$ and all $\lambda > 0$.

We show that there exist $\bar{\lambda} > 0$ and $r > 0$ such that the following property holds:

(P1) For each $\varepsilon \in (0,1)$ there exists $\delta \in (0,\varepsilon)$ such that for each $f \in \mathscr{M}_{\phi,\bar{a}}$ satisfying $(f,f_0) \in \mathscr{U}(M_0,q,r)$, each $\lambda > \bar{\lambda}$ and each $x \in X$ which satisfies

$$\psi_{f,\lambda}(x) \le \inf(\psi_{f,\lambda}) + \delta \tag{15}$$

the set

$$\{z \in A :\ ||x - z|| \le \varepsilon \text{ and } \psi_{f,\lambda}(z) \le \psi_{f,\lambda}(x)\} \tag{16}$$

is nonempty.

It is not difficult to see that the existence of $\bar{\lambda}, r > 0$ for which the property (P1) holds implies the validity of Theorem 2.1.

Let us assume that there are no $\bar{\lambda}, r > 0$ for which (P1) holds. Then for each natural number k there exist

$$\varepsilon_k \in (0,1), \quad \lambda_k > k, \quad f_k \in \mathscr{M}_{\phi,\bar{a}}, \quad x_k \in X \tag{17}$$

such that

$$(f_k, f_0) \in \mathscr{U}(M_0, q, k^{-1}), \tag{18}$$

$$\psi_{f_k,\lambda_k}(x_k) \le \inf(\psi_{f_k,\lambda_k}) + 2^{-1}\varepsilon_k k^{-1} \tag{19}$$

and

$$\{z \in A :\ ||z - x_k|| \le \varepsilon_k \text{ and } \psi_{f_k,\lambda_k}(z) \le \psi_{f_k,\lambda_k}(x_k)\} = \emptyset. \tag{20}$$

Let k be a natural number. It follows from (19) and Ekeland's variational principle [8] that there exists $y_k \in X$ such that

$$\psi_{f_k,\lambda_k}(y_k) \leq \psi_{f_k,\lambda_k}(x_k), \tag{21}$$

$$||y_k - x_k|| \leq \varepsilon_k/2, \tag{22}$$

$$\psi_{f_k,\lambda_k}(y_k) \leq \psi_{f_k,\lambda_k}(z) + k^{-1}||z - y_k|| \text{ for all } z \in X. \tag{23}$$

By (20), (21), and (22),

$$y_k \notin A \qquad \text{for all natural numbers } k. \tag{24}$$

In the case of Assertion 2, we obtain that

$$g(y_k) > c \qquad \text{for all natural numbers } k. \tag{25}$$

In the case of Assertion 1, we obtain that for each natural number k either $g(y_k) > c$ or $g(y_k) < c$.

In the case of Assertion 1 by extracting a subsequence and re-indexing we may assume that either $g(y_k) > c$ for all natural numbers k or $g(y_k) < c$ for all natural numbers k. Replacing g with $-g$ and c with $-c$ if necessary we may assume without loss of generality that (25) holds in the case of Assertion 1 too. Now (25) is valid in both cases.

In view of (5), (7), (9)–(14), (17)–(19), and (21) for all natural numbers k,

$$\begin{aligned}\phi(||y_k||) - \bar{a} &\leq f_k(y_k) \leq \psi_{f_k,\lambda_k}(y_k) \leq \psi_{f_k,\lambda_k}(x_k) \leq \inf(\psi_{f_k,\lambda_k}) + 2^{-1}\varepsilon_k k^{-1} \\ &\leq \psi_{f_k,\lambda_k}(\theta) + 2^{-1} = f_k(\theta) + 2^{-1} \\ &= 2^{-1} + f_0(\theta) + f_k(\theta) - f_0(\theta) \leq f_0(\theta) + k^{-1} + 2^{-1}.\end{aligned} \tag{26}$$

Equations (10) and (26) imply that

$$||y_k|| \leq M_0 - 2 \qquad \text{for all natural numbers } k. \tag{27}$$

It follows from (5) and (18) that the restriction of f_k to $B(0, M_0)$ is Lipschitz for all natural number k.

Let k be a natural number. Then by (25) and (27) there is an open neighborhood V_k of y_k in X such that

$$V_k \subset B(0, M_0 - 1), \ g(z) > c \text{ for all } z \in V_k. \tag{28}$$

By (11)–(14), (23), and (28), for all $z \in V_k$,

$$\begin{aligned}f_k(y_k) + \lambda_k(g(y_k) - c) &= \psi_{f_k,\lambda_k}(y_k) \leq \psi_{f_k,\lambda_k}(z) + k^{-1}||z - y_k|| \\ &= f_k(z) + \lambda_k(g(z) - c) + k^{-1}||z - y_k||.\end{aligned} \tag{29}$$

Put

$$\phi_k(z) = f_k(z) + \lambda_k g(z) + k^{-1}||z - y_k||, \ z \in V_k. \tag{30}$$

In view of (28) and (30), the function $\phi_k : V_k \to R$ is locally Lipschitz.

By (2), (3), (29), and (30),

$$0 \in \partial(\phi_k)(y_k). \tag{31}$$

It follows from (28), (30), (31), and Theorem 3.36 of [13] that

$$0 \in \partial f_k(y_k) + \lambda_k \partial g(y_k) + k^{-1}\partial(||\cdot - y_k||)(y_k). \tag{32}$$

By (5), (18), (28), Theorem 3.36 of [13], and Corollary 1.8.1 of [13],

$$\partial f_k(y_k) \subset \partial f_0(y_k) + \partial(f_k - f_0)(y_k) \subset \partial f_0(y_k) + qB_*(0,1).$$

Together with (32) and Corollary 1.8.1 of [13] this implies that

$$0 \in \partial g(y_k) + \lambda_k^{-1}\partial f_0(y_k) + \lambda_k^{-1} q B_*(0,1) + \lambda_k^{-1}k^{-1}\partial(||\cdot - y_k||)(y_k) \tag{33}$$

$$\subset \partial g(y_k) + \lambda_k^{-1}\partial f_0(y_k) + \lambda_k^{-1} q B_*(0,1) + \lambda_k^{-1}k^{-1}B_*(0,1).$$

Since the function f_0 is Lipschitzian on bounded subsets of X it follows from (27) and Corollary 1.8.1 of [13] that there exists $L > 0$ such that

$$\partial f_0(y_k) \subset B_*(0,L) \quad \text{for all natural numbers } k. \tag{34}$$

By (33) and (34), for all natural numbers k,

$$0 \in \partial g(y_k) + \lambda_k^{-1}B_*(0,L) + \lambda_k^{-1} q B_*(0,1) + \lambda_k^{-1}k^{-1}B_*(0,1)$$

and in view of (4) and (17),

$$\lim_{k\to\infty} \Xi_g(y_k) = 0. \tag{35}$$

By (7), (11)–(14), (17), (25), and (26) for all integers $k \geq 1$,

$$\lambda_k(g(y_k) - c) - \bar{a} \leq \phi(||y_k||) + \lambda_k(g(y_k) - c) - \bar{a} \leq f_k(y_k) + \lambda_k(g(y_k) - c)$$

$$= \psi_{f_k,\lambda_k}(y_k) \leq f_0(\theta) + k^{-1} + 2^{-1}$$

and

$$0 < g(y_k) - c \leq \lambda_k^{-1}\left[f_0(\theta) + \bar{a} + k^{-1} + 2^{-1}\right] \to 0 \text{ as } k \to \infty. \tag{36}$$

Hence there is a natural number k_0 such that for all integers $k \geq k_0$

$$g(y_k) \in (c, c + \gamma_*]. \tag{37}$$

By (27), (35), (37), and the (P–S) condition, there exists a strictly increasing sequence of natural numbers $\{k_j\}_{j=1}^{\infty}$ such that $\{y_{k_j}\}_{j=1}^{\infty}$ converges in the norm topology to $\bar{y} \in X$. In view of (36),

$$g(\bar{y}) = c. \tag{38}$$

By (5), (7)–(14), (18), (19), (21), and (27)

$$f_0(\bar{y}) = \lim_{j\to\infty} f_0(y_{k_j}) = \lim_{j\to\infty} f_{k_j}(y_{k_j}) \leq \limsup_{j\to\infty} \psi_{f_{k_j},\lambda_{k_j}}(y_{k_j})$$

$$
\begin{aligned}
&\leq \limsup_{j\to\infty} \inf(\psi_{f_{k_j},\lambda_{k_j}}) \leq \limsup_{j\to\infty} \inf(\psi_{f_{k_j},\lambda_{k_j}};A) \\
&= \limsup_{j\to\infty} \inf(f_{k_j};A) \leq \limsup_{j\to\infty} \inf(f_{k_j};A\cap B(0,M_0)) \\
&= \inf(f_0;A\cap B(0,M_0)) = \inf(f_0;A).
\end{aligned}
$$

Together with (13), (14), and (38), this implies that

$$f_0(\bar{y}) = \inf(f_0;A). \tag{39}$$

Since the function f_0 is Lipschitz on bounded subsets of X there exists $L_0 \geq 1$ such that

$$|f_0(z_1) - f_0(z_2)| \leq L_0||z_1 - z_2|| \text{ for all } z_1, z_2 \in B(0,M_0). \tag{40}$$

Let k be a natural number. By (5), (17), (18), (27)–(29), and (40) for all $z \in V_k \setminus \{y_k\}$,

$$
\begin{aligned}
(g(z)-g(y_k))||z-y_k||^{-1} &\geq -||z-y_k||^{-1}|f_k(y_k)-f_k(z)|\lambda_k^{-1} - k^{-1}\lambda_k^{-1} \\
&\geq -\lambda_k^{-1}||z-y_k||^{-1}(|f_0(z)-f_0(y_k)| + |(f_k-f_0)(z) \\
&\qquad - (f_k-f_0)(y_k)|) - \lambda_k^{-1}k^{-1} \\
&\geq -\lambda_k^{-1}||z-y_k||^{-1}(L_0||z-y_k|| + q||z-y_k||) - \lambda_k^{-1}k^{-1} \\
&\geq -k^{-1}L_0 - k^{-1}q - k^{-1} \geq -k^{-1}(L_0+q+1).
\end{aligned}
$$

By the relation above and the definition (2),

$$0 \in \widehat{\partial}_{\gamma_k} g_k(y_k)$$

with

$$\gamma_k = k^{-1}(1+L_0+q).$$

Together with (3) and the equality $\bar{y} = \lim_{j\to\infty} y_{k_j}$ in the norm topology, this implies that $0 \in \partial g(\bar{y})$. This contradicts to the relations (38) and (39). The contradiction we have reached proves the existence of $\bar{\lambda} > 0, r > 0$ for which the property (P1) holds.

This completes the proof of Theorem 2.1.

References

1. Bao, T. Q. and Mordukhovich, B. S.: Variational principles for set-valued mappings with applications to multiobjective optimization, Control and Cybernetics **36**, 531–562 (2007).
2. Bao, T. Q. and Mordukhovich, B. S.: Relative Pareto minimizers for multiobjective problems: existence and optimality conditions, Mathematical Programming **122**(2), 301–347 (2010).
3. Boukari, D. and Fiacco, A. V.: Survey of penalty, exact-penalty and multiplier methods from 1968 to 1993, Optimization **32**, 301–334 (1995).
4. Burke, J. V.: Calmness and exact penalization, SIAM Journal on Control and Optimization **29**, 493–497 (1991).

5. Burke, J. V.: An exact penalization viewpoint of constrained optimization, SIAM Journal on Control and Optimization **29**, 968–998 (1991).
6. Clarke, F. H.: Optimization and Nonsmooth Analysis. Wiley, New York (1983).
7. Di Pillo, G. and Grippo, L.: Exact penalty functions in constrained optimization, SIAM Journal on Control and Optimization **27**, 1333–1360 (1989).
8. Ekeland, I.: On the variational principle, Journal of Mathematical Analysis and Applications **47**, 324–353 (1974).
9. Eremin, I. I.: The penalty method in convex programming, Soviet Mathematics Doklady **8**, 459–462 (1966).
10. Kelley, J. L.: General Topology, D. Van Nostrand, New York (1955).
11. Khaladi, M. and Penot, J.-P.: Estimates of the exact penalty coefficient threshold, Utilitas Mathematica **42**, 147–161 (1992).
12. Mordukhovich, B. S.: Maximum principle in problems of time optimal control with nonsmooth constraints, Journal of Applied Mathematics and Mechanics **40**, 960–969 (1976).
13. Mordukhovich, B. S.: Variational Analysis and Generalized Differentiation, I: Basic Theory. Springer, Berlin (2006).
14. Mordukhovich, B. S.: Variational Analysis and Generalized Differentiation, II: Applications. Springer, Berlin (2006).
15. Palais, R: Lusternik–Schnirelman theory of Banach manifolds, Topology **5**, 115–132 (1966).
16. Sard, A.: The measure of the critical values of differentiable maps, Bulletin of the American Mathematical Society **48**, 883–890 (1942).
17. Ward, D. E.: Exact penalties and sufficient conditions for optimality in nonsmooth optimization, Journal Optimization Theory & Applications **57**, 485–499 (1988).
18. Zangwill, W. I.: Nonlinear programming via penalty functions, Management Science **13**, 344–358 (1967).
19. Zaslavski, A. J.: On critical points of Lipschitz functions on smooth manifolds, Siberian Mathematical Journal **22**, 63–68 (1981).
20. Zaslavski, A. J.: A sufficient condition for exact penalty in constrained optimization, SIAM Journal on Optimization **16**, 250–262 (2005).
21. Zaslavski, A. J.: Exact penalty in constrained optimization and critical points of Lipschitz functions, Journal of Nonlinear and Convex Analysis **10**, 149–156 (2009).

Printed by Printforce, the Netherlands